DAVID CH

D0908633

COLD REGION
STRUCTURAL
ENGINEERING

COLD REGION STRUCTURAL ENGINEERING

E. ERANTI

G. C. LEE

MCGRAW-HILL BOOK COMPANY

New York St. Louis San Francisco Auckland Bogotá
Hamburg Johannesburg London Madrid Mexico
Montreal New Delhi Panama Paris São Paulo
Singapore Sydney Tokyo Toronto

Library of Congress Cataloging-in-Publication Data

Eranti, Esa M.
  Cold region structural engineering.

  Bibliography: p.
  Includes index.
  1. Structural engineering—Cold weather conditions.
I. Lee, George C.   II. Title.
TA636.E73   1986   624.1'0911   85-16797
ISBN 0-07-037034-6

1234567890 DOC/DOC 8932109876

ISBN 0-07-037034-6

*The editors for this book were Joan Zseleczky and Ingeborg M. Stochmal,
the designer was Naomi Auerbach, and the production supervisor
was Sally Fliess. It was set in Caledonia by Progressive Typographers.*

Printed and bound by R. R. Donnelley & Sons, Inc.

CONTENTS

PREFACE

Because of the many research interests among its faculty, the Center for Cold Regions Engineering, Science, and Technology (CREST) was formally established at the State University of New York at Buffalo in 1979. The areas of expertise of the participants at the Center cover a wide range of fields in engineering and in the natural sciences. Shortly after the formation of CREST, the first author joined the engineering school as a research associate, working with the second author on a cold region structural engineering research project. During 1979 we held extensive discussions and concluded that there is a lack of published material summarizing available information on structural design and construction in cold regions. This led to our collaborative effort to collect and review available information and to the publication of two summary reports, "Introduction to Ice Problems in Civil Engineering" and "Introduction to Cold Regions Structural Design and Construction," which formed the basis of this book.

The field of cold region engineering covers a wide range of topical areas; however, it is not possible to summarize all the information in a single volume. Therefore, we do not claim that this book is, in any way, all inclusive. We have concentrated our efforts on some of the cold region engineering areas that have relatively practical significance. Special attention has been given to alternative engineering solutions and practical approaches, including simple design formulas, graphs, and tables. The theoretical backgrounds of selected problem areas are also briefly discussed as introductions to the subjects.

Because cold region engineering is multidisciplinary and international in nature, we have attempted to include most of the relevant information from North America, Scandinavia, and, to the extent possible, the Soviet Union, in order to provide the reader with a relatively uniform view of some of the most feasible approaches and solutions to cold region engineering problems.

Although much emphasis has been given to practical engineering approaches, this book can also be used as a reference book in cold region engineering courses. The list of references contained in this book is fairly extensive, and it should be useful for those interested in furthering their understanding of the current state of the art.

In our effort to collect and digest the available information, we received invaluable assistance from the U.S. Army Cold Regions Research and Engineering Laboratory and the Technical Research Centre of Finland. It would not have been possible for us to complete the manuscript without the technical information provided to us by these two organizations. Further, we would like to acknowledge the support of the State University of New York at Buffalo and of Erkki Juva Consulting Engineers and Finn-Stroi Ltd. of Finland. We would also like to express our appreciation to a number of individuals who assisted us in various capacities during the preparation of the manuscript, including Jenn-Shin Hwang, Helen Liu, Liisa Viitanen, Leena-Marjut Rautio, and Pat Doeing.

Esa Eranti
George C. Lee

COLD REGION STRUCTURAL ENGINEERING

1

ENVIRONMENT AND DEVELOPMENT

1.1 Introduction

Winter darkness, subzero temperatures, snow, ice, and frost are not desirable conditions for human activities. Still, in establishing communities and industry, people have been able to adjust even to the harshest arctic environments. In areas with more temperate climates there is a continuous development of infrastructures. Considering the increased cost of construction and maintenance in cold regions, it is particularly worthwhile to study and develop appropriate engineering principles and practices.

For engineering purposes, cold regions can be defined by the 0°C (32°F) isotherm for the average temperature of the coldest month. This also is the limit for substantial frost penetration and ice formation. In the northern hemisphere the area extends roughly from the North Pole to the 40th parallel (Fig. 1.1). Typical cold region engineering problems associated with freezing temperatures, snow, ice, and frost also occur south of this limit, but they are temporary or locally confined.

The environment becomes increasingly difficult for development and construction as latitudes advance beyond the 40th parallel. Three major

Figure 1.1 Climatological boundaries of cold regions. (*Adapted from Bates and Bilello, 1966; Wilson, 1967; Central Intelligence Agency, 1978.*)

climatological zones can be defined within cold regions: temperate, subarctic, and arctic (Fig. 1.1). One definition for the subarctic zone is that the mean temperature of the coldest month is less than $-3\,°C$ ($27\,°F$), with no more than 4 months having mean temperatures above $+10\,°C$ ($50\,°F$). One of the preferred boundaries for the arctic region is the $+10\,°C$ ($50\,°F$) isotherm for July.

To consider cold region engineering problems, it is desirable to distinguish between the well developed southern parts and the undeveloped arctic and subarctic areas. For the southern parts, cold region engineering problems may be handled by relatively modest modifications to ordinary practice. This is because the regions already have solid infrastructures in place, which can support construction activities, and because of

Figure 1.2 Part of the Prudhoe Bay, Alaska, oil complex at the North Slope of Alaska (*Courtesy of ARCO Alaska, Inc.*)

less severe environmental conditions. However, the total economic value of all the problems associated with cold region engineering is quite substantial. This is why great emphasis is being placed on the continued development of design and construction methods.

In the Arctic and in the northern parts of the subarctic areas, cold region problems require a new engineering approach. Severe environmental conditions often dictate special solutions in design and construction. A lack of industrial infrastructures and transportation networks adds to the difficulty and cost of construction. Recently the exploitation of the northern oil, gas, and mineral resources has necessitated rapid development. Large-scale projects have been executed under difficult conditions, such as the Tjumen and Prudhoe Bay, Alaska, oil and gas

Figure 1.3 City of Norilsk with huge mining complex in northwestern Siberia.

exploitations (Fig. 1.2) and the establishment of the mining complex of Norilsk, U.S.S.R. (Fig. 1.3). However, generally speaking, the development is still at an early stage. Much more industry and supporting infrastructure will have to be built. New challenges related to the technology, living conditions, environmental protection, and project execution have to be addressed.

1.2   The Climate

Temperatures in the arctic coastal areas are relatively even and low throughout the year, typically -20 to $-30°C$ (0 to $-20°F$) in January and $+4$ to $+8°C$ (40 to 50°F) in July, because of the balancing effect of the ocean (Fig. 1.4). In the arctic and subarctic areas further inland, much colder winters and warmer summers can be expected (Fig. 1.5). For example, in Verkhoyansk, U.S.S.R., the recorded absolute winter low has been $-68°C$ ($-90°F$) and the summer high $+37°C$ (100°F), which is a temperature range of over $100°C$ (180°F). Temperatures as low as $-25°C$ ($-10°F$) have been experienced even in the temperate zone. It should be noted that high temperatures do not automatically mean more favorable conditions. High winds are common along the arctic coast, whereas the interior areas of arctic and subarctic North America and Siberia are much calmer (Fig. 1.6). Winds add considerably to the physiological feeling of cold as measured by the windchill factor and shown in Fig. 1.7. Experience has shown that this index offers a reasonable estimate for human discomfort when adjustments are made for such factors as sunshine and degree of activity.

The total precipitation in the Arctic is generally low because low temperatures limit the moisture-bearing capacity of the air. A significant part of the precipitation occurs during the summer as rain, and thus the average maximum snow cover is quite thin compared to that of the subarctic zone and the northern parts of the temperate zone. However, the thickness of the snow cover is quite irregular in the Arctic, because high winds sweep the snow from exposed surfaces, accumulating it in large quantities in depressions and protected areas.

Poor visibility is an important factor that often disturbs operations such as air service and survey, especially in the arctic coastal areas. Fog is most common during the summer and fall as warm damp air moves over cooler surfaces such as the water. During the winter fog may be caused, for example, by leads in the ice cover. Local winter fog may also be created by living communities at low temperatures, typically $-30°C$ ($-20°F$) or lower. Large amounts of water vapor are released to the air with combustion products and from possible open water surfaces. The vapor su-

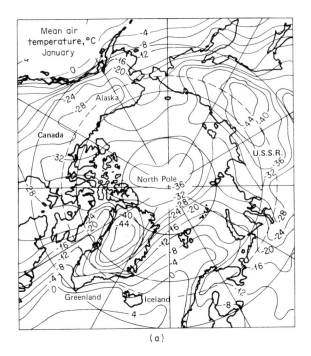

(a)

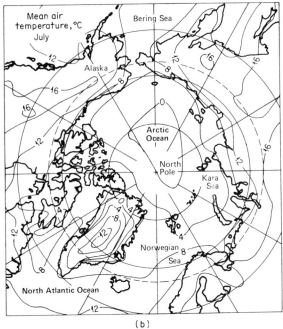

(b)

Figure 1.4 Mean air temperatures in January and July in the northern hemisphere. *(Central Intelligence Agency, 1978.)*

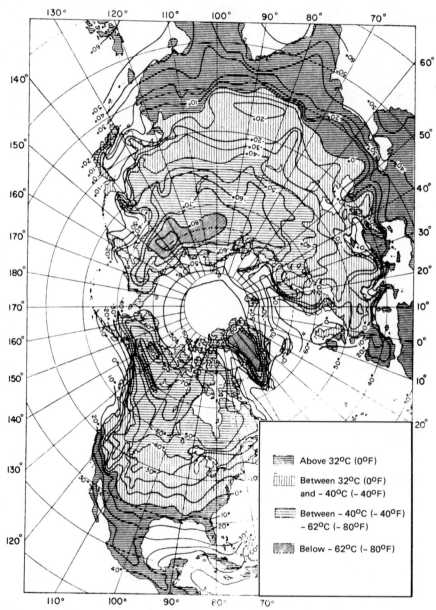

Figure 1.5 Lowest temperatures observed at major weather stations in the northern hemisphere. Lower temperatures than those indicated probably have occurred at other places, especially in mountains. (*U.S. Air Force, 1960.*)

Legend:

Above 32°C (0°F)

Between 32°C (0°F) and –40°C (–40°F)

Between –40°C (–40°F) –62°C (–80°F)

Below –62°C (–80°F)

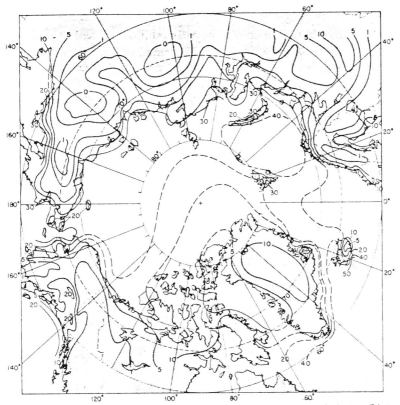

Figure 1.6 Mean percentage of surface winds exceeding 11 m/s (25 mi/h) in the Arctic, December through February. *(Wilson, 1969.)*

persaturates quickly in the cold air and forms the so-called ice fog. The phenomenon generally disappears with winds or clouds, which prevent the radiation loss to the outer atmosphere.

Blowing snow or heavy snowfall may also limit the visibility considerably. Whiteouts are a peculiar northern phenomenon, which causes loss of depth perception and horizon. Two factors must exist for a whiteout: a diffuse shadowless illumination (light and slowly falling snow) and a uniform white surface. Whiteouts create a considerable hazard to air service, as well as impairing ground operations.

Finally, the lack of daylight in polar regions restricts the visibility during the winter. On the other hand, there is plenty of daylight during the summer. The duration of daylight in the northern hemisphere can be estimated from Fig. 1.8.

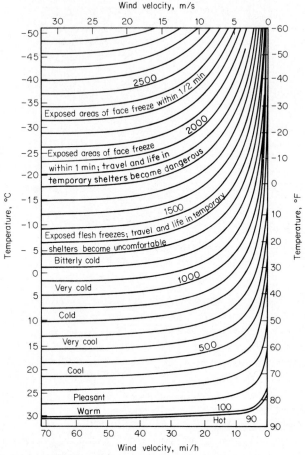

Figure 1.7 Windchill factor. Curves indicate the quantity of heat [kcal/(m²·h)] the atmosphere is capable of absorbing from an exposed surface. *(Siple, 1945.)*

1.3   The Environment

The environment of the north is characterized by snow, ice, and frost. The season of snow ranges from an irregular few months in the temperate zone up to 8 or even 10 months on the arctic coast. However, the average maximum depth of the snow cover in the arctic coastal areas is only 20 to 50 cm (8 to 20 in), whereas in the subarctic zone it typically ranges from 50 to 80 cm (20 to 32 in) and may exceed 1 m (40 in). Very heavy snowfalls occur also in some areas of the northern temperate zone. In glacier

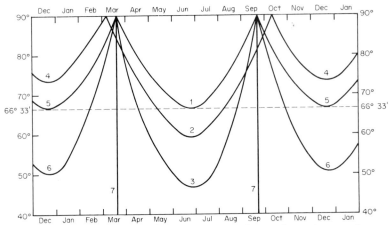

Figure 1.8 Duration of daylight in the northern hemisphere. 1 — continuous daylight; 2 — light night resulting from civil twilight; 3 — 16 hours daylight; 4 — continuous darkness; 5 — dark night with civil twilight at midday; 6 — 8 hours daylight; 7 — 12 hours daylight.

areas of Greenland and Antarctica the seasonal snow cover does not melt totally but accumulates and turns into ice under pressure.

The formation of an ice cover is quite similar to but somewhat slower than that of a snow cover. The maximum thickness of the ice cover in the subarctic zone is typically on the order of 1 m (3 ft). In the Arctic the ice growth may reach 2 m (7 ft) during the winter. At sea, ice may pile up and form pressure ridges that may extend as much as 40 m (130 ft) below sea level and up to 10 m (30 ft) above. Multiyear ice, typically 2 to 4 m (7 to 13 ft) thick, multiyear pressure ridges, icebergs, and ice islands are part of the environment in the arctic ocean and some of its neighboring seas.

The navigation window to the arctic coast and along the arctic rivers is generally open from 2 to 4 months because of ice. Even in the temperate zone, some major navigation routes are shut down annually for a period of time because of ice conditions. In rivers, severe flooding is often caused by ice jams that form in late fall during ice formation or in late spring during its breakup.

Flooding may take a very peculiar form in small streams that freeze to the bottom during the winter. Water penetrates to the surface and freezes in thin layers, called aufeis. At the end of the winter aufeis may cover considerable areas and gain several meters in thickness.

The existence of frost is governed by the thermal regime of the ground. The most important factors affecting the thermal regime are air temperatures, soil properties, absorption of solar radiation, insulation provided

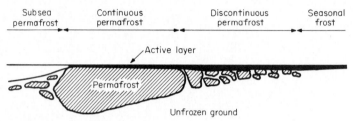

Figure 1.9 Typical frost profile (not to scale).

by snow, and heat flow from the inner parts of the earth. The depth of seasonal frost in the temperate zone is generally less than 3 m (10 ft) depending, among other things, on the thermal properties and the moisture content of the soil. In the subarctic zone the depth of seasonal frost increases, and sporadic and discontinuous permafrost [ground temperatures remain continuously below 0°C (32°F) for 2 or more years] appears as mean annual temperatures drop a few degrees below 0°C (32°F). As the mean annual temperatures approach −10°C (15°F), the permafrost layer becomes continuous, except for the thaw bulbs beneath larger lakes and rivers. The active layer subjected to annual freeze-thaw cycles begins to thin (Fig. 1.9). The maximum thickness of permafrost may be about 650 m (2100 ft) in northern Alaska. In Siberia subfreezing temperatures have been measured at a depth of 1500 m (5000 ft). Permafrost has also been found in significant magnitudes offshore along the coasts of the Arctic Ocean. In many cases this has been explained by the fact that the average annual temperature of the salty seawater may be significantly below 0°C (32°F).

One typical feature for cold regions is the softening of ground during the spring thaw. This is due to the moisture buildup blocked by the underlying frost. In the seasonal frost areas the situation is only temporary, but in the permafrost zone, where the downward drainage is blocked, a marshy surface layer with little bearing strength may be created for the entire summer.

Permafrost terrain often features polygon-shaped surface patterns resulting from a natural formation of ice wedges and small hills called pingos, which have been created when trapped water was frozen and expanded underground. Palsas, mounds of peat with a permafrost core, are commonly found in the areas of discontinuous permafrost. Solifluction, the low downward movement of thawed wet material, is also common in permafrost terrain. Finally there are areas where permafrost conditions are changing either naturally or as a consequence of human activities. Thawing of ice-rich soils creates an uneven topography with mounds, sinkholes, thaw lakes, and beaded streams.

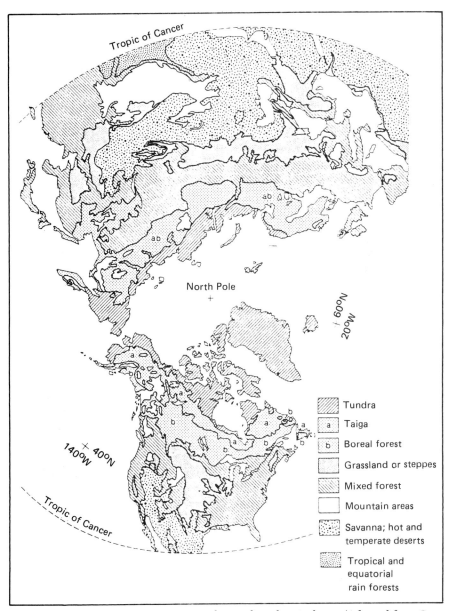

Figure 1.10 Major vegetation zones in the northern hemisphere. *(Adapted from Gray and Male, 1981.)*

The climate and the frost conditions naturally also have their impact on the vegetation (Fig. 1.10). Only few species can survive the short growing season, the nutrient-poor soil conditions, and the harsh climate of the Arctic. Tundra vegetation consists of scrubs, grasses, lichens, and mosses, which slowly vanish further north in the arctic deserts. South of the tundra is the taiga, a zone of small trees, which gradually changes into boreal forests and other vegetation zones of the temperate zone. The environmental characteristics of these regions have been discussed at a UNESCO symposium on the ecology of the subarctic regions (1970), for example.

1.4   Engineering Considerations

Because of the special environmental conditions in cold regions, engineering must, in many cases, adopt new principles and solutions ranging from material selection to general design and from construction methods to strategies in project execution.

Snow, ice, and frost are some of the most obvious special features that require consideration in cold region engineering. Snow loads and heavy icing have caused structural collapses. Blowing snow may cause significant maintenance problems, and in mountain areas avalanches may pose a hazard.

Figure 1.11   New problems have to be solved and innovative technological solutions created in arctic offshore oil exploration. The construction of the caisson-retained Tarsiut Island in the Canadian Beaufort Sea was a major step forward in this respect. *(Courtesy of Dome Petroleum Ltd.)*

Figure 1.12   Cold region road network is constantly being damaged because of frost action and requires extensive annual maintenance efforts. *(Courtesy of Finnish Roads and Waterways Administration.)*

Ice forces exerted on hydraulic structures and the bearing capacity of the ice cover are among the traditional ice problems. Problems of frazil ice and ice jams have also been tackled in order to prevent flooding and secure smooth water power generation. The exploration of arctic offshore hydrocarbon resources has recently added a new dimension to ice engineering (Fig. 1.11). Also, the increase in winter navigation has created new engineering problems connected with an accelerated growth of ice and the instability of the ice cover.

Frost problems are mostly encountered in foundation design and earthworks. Frozen soils are in general very hard and difficult to excavate and compact. In most cases soils are also frost-susceptible. Foundations have to extend below the active layer to avoid frost damage. Furthermore, in permafrost areas the thermal regime of the ground should be maintained in order to avoid thaw settlements (Figs. 1.12 and 1.13).

The cold climate restricts the applicability of several construction materials. Many materials such as ordinary steels or plastics become brittle in very low temperatures. Porous materials, such as concrete or some insulation boards, tend to deteriorate when they are subjected to consecutive freeze-thaw cycles in the presence of moisture. Finally, work phases such as welding, painting, roofing, and concrete pouring cannot be performed in all conditions without special protective measures.

Numerous other special aspects should be considered in cold region design, including thermal insulation, condensation, and fire safety in the winter. In cold environments the life span of wastes is long because of the

Figure 1.13 The Trans-Alaska Pipeline System connecting the Prudhoe Bay oil complex and the terminal of Valdez, Alaska, is among the largest and most sophisticated construction projects ever executed in cold regions. In the areas of seasonal frost the pipeline is buried, and in permafrost areas it is elevated above ground and supported by refrigerated piles. *(Courtesy of Glenn Johns.)*

Figure 1.14 New winter construction techniques were developed and tested in the city and mining complex of Kostomuksha, U.S.S.R. *(Courtesy of Finn-Stroi Oy.)*

slow biological and chemical decomposition processes. Long-lasting terrain disturbances may be accomplished by road construction or vehicular activity on the tundra if the vegetation is damaged or the drainage patterns are interrupted and the permafrost conditions are changed. The biological systems are quite simple and sensitive, and it is not impossible that links in the food chain are broken. It is thus important that the environmental impacts and risks of development be assessed and considered in the planning and design.

The actual construction in cold regions has its own peculiarities. The construction schedule is often adjusted to the seasonal timetable. Some construction systems and methods are favored in winter construction. Additional measures such as transportation, protection, and heating are often dictating factors (Fig. 1.14).

In a typical engineering project, construction materials, methods, and resources are considered simultaneously so that economy and productivity can be optimized together. In well-developed areas of even the southern subarctic zone this can be achieved with limited modifications and changes from the practice used in the more temperate zones. In the Arctic, however, engineering requires in many respects a new approach, innovative design, careful planning, special construction methods, and tight project control.

CHAPTER

2

SNOW AND ICING PROBLEMS

2.1 Characteristics of Snow Covers

The characteristics of snow covers vary greatly depending upon the location. In structural design and construction the factors that should be considered include snow precipitation, snow depth, snow properties, length of snow season, snow transport, and snow drifting. In mountainous terrain the risks of avalanches and snow creep should also be assessed.

Information about snow precipitation, snow thickness, snow loads, and the length of the snow season is often available in local codes and their commentaries. Figure 2.1 gives a rough idea of the existence of snow cover. The properties and characteristics of the snow covers in subarctic and arctic areas, however, differ. The key factor is snow transport, which depends, among other things, on wind speed, snow density, and how well the snow particles have bonded (Fig. 2.2). When the wind velocity is below 10 m/s (20 mi/h), the transport of even fresh falling snow is not very significant. However, blowing of loose surface snow is an important design consideration in flat treeless terrain when the wind speed exceeds 15 m/s (30 mi/h).

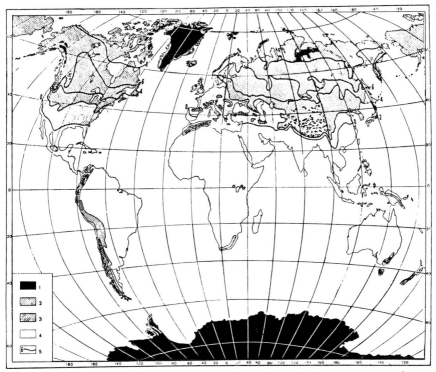

Figure 2.1 Extent of snow cover on the world surface. 1 — permanent cover of snow and ice; 2 — stable snow cover of varying duration forms every year; 3 — snow cover forms almost every year, but is not stable; 4 — no snow cover; 5 — duration of snow cover (months). *(Mellor, 1964.)*

Snow particles travel close to the snow surface in areas of laminar flow, and they are deposited especially in low-wind-speed regions in the vicinity of flow disturbances. Knowledge of airflow characteristics is thus important when snowdrifts are studied. When wind is associated with snowfall, snowdrift patterns are changed, especially at and near higher locations such as the roofs out of the reach of the normal snow transport path. The snow already deposited on the roof is only a limited source of drift accumulation, but during a snowstorm the supply is greatly increased.

In subarctic conditions the amount of snow transport is typically limited mostly because of vegetation. The density of freshly fallen snow in freezing temperatures is approximately 100 kg/m^3 (6 lb/ft^3) or less. When the snow cover reaches its maximum thickness, its average density will increase to about 200 to 300 kg/m^3 (12 to 20 lb/ft^3) due to the combined effect of compaction, moisture absorption, evaporation, and

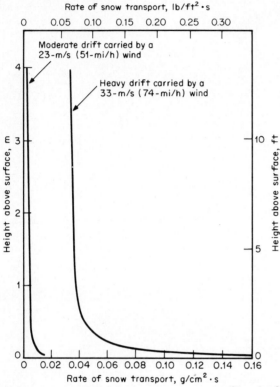

Figure 2.2 Typical arctic snow-transport profiles. (*Mellor, 1965.*)

melting. The density of compacted snow or old melting snow may exceed 500 kg/m³ (30 lb/ft³).

In arctic, treeless tundra, and high mountain regions, the snow particles are rounded because of the constant snow transport during the long winter. Even if the annual precipitation is small, snowdrifts may reach considerable dimensions. The average density of snowdrifts is high, on the order of 300 to 500 kg/m³ (20 to 30 lb/ft³). Due to the steep temperature gradients depth hoar, which is the result of vapor diffusion from a warm lower layer to colder layers above, is typical for arctic snow covers.

2.2   Snow Loads

The snow loads on individual structural elements, such as wires or elevated pipes, may be found based on the width and shape of the structures

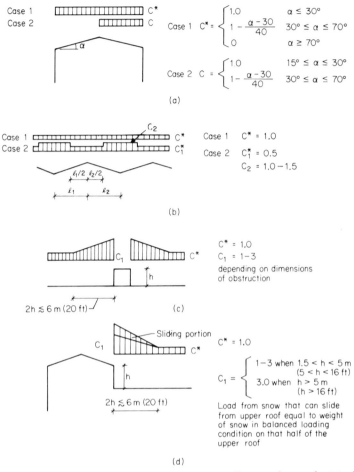

Figure 2.3 Typical snow accumulation coefficients for roofs. Wind reduction factor k may be used together with C° values. (a) Simple cable and hip roofs. (b) Multispan sloped roofs. (c) Major obstructions on flat roofs. (d) Two-level roofs.

rather than on local climatological conditions, but usually the snow load of a structure is given in the following form:

$$S = kCg \qquad (2.1)$$

where S = snow load
$\quad g$ = ground snow load based on snow thickness and average density
$\quad C$ = snow accumulation coefficient

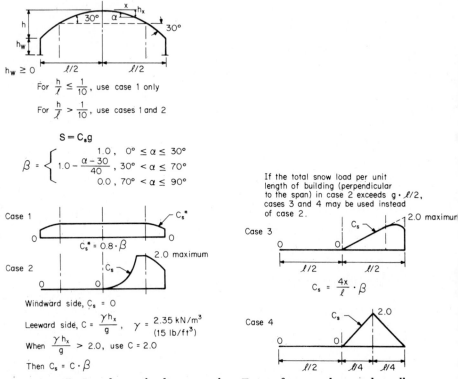

Figure 2.4 Proposed snow loading on arches. For roofs exposed to wind on all sides, all values of C_s° may be reduced by 25%. (*Adapted from Taylor, 1981.*)

k = wind reduction factor, accounting for snow transport off the elevated and exposed roofs

The structural importance factor and the thermal factor that account for the possible melting of snow can also be recognized in Eq. (2.1).

The ground snow load is given in local codes, and it usually has a value of between 0.5 and 4 kPa (10 to 80 lb/ft²). However, one should keep in mind that the average ground snow load normally increases with elevation because of longer winters and other meteorological effects. Thus in a mountain area the ground snow load given by a code may be exceeded.

The snow loads on roofs are generally smaller than the ground loads because of the effect of winds. The reduction factor k has a typical value

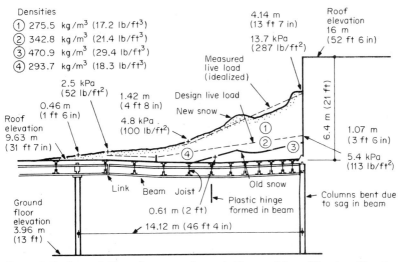

Densities
1. 275.5 kg/m³ (17.2 lb/ft³)
2. 342.8 kg/m³ (21.4 lb/ft³)
3. 470.9 kg/m³ (29.4 lb/ft³)
4. 293.7 kg/m³ (18.3 lb/ft³)

4.14 m (13 ft 7 in)
13.7 kPa (287 lb/ft²)

Roof elevation 16 m (52 ft 6 in)

Measured live load (idealized)

2.5 kPa (52 lb/ft²)
1.42 m (4 ft 8 in)   Design live load
0.46 m (1 ft 6 in)   New snow
Roof elevation 9.63 m (31 ft 7 in)
4.8 kPa (100 lb/ft²)

6.4 m (21 ft)

1.07 m (3 ft 6 in)

5.4 kPa (113 lb/ft²)

Link   Beam   Joist   Old snow
Plastic hinge formed in beam
0.61 m (2 ft)

Columns bent due to sag in beam

Ground floor elevation 3.96 m (13 ft)

14.12 m (46 ft 4 in)

Figure 2.5 Snow drifts on large warehouse in Boston, Mass. *(Courtesy Maurice A. Reidy Engineers, adapted from Templin and Schriever, 1982.)*

of 0.8 for not very well sheltered and 0.6 for well exposed roofs. However, because of the effects of snow accumulation, the average load may be exceeded. Some examples of typical accumulation coefficients are given in Figs. 2.3 and 2.4.

There are some situations where snow loads may catch the designer off guard, as shown in Figs. 2.5 and 2.6. In addition to ordinary code provisions, practical considerations based on local conditions must be allowed for. For example, in design one should pay full attention to the role of unbalanced snow loads as well as to the possibility of snow sliding from an upper roof to a lower one (Fig. 2.7). In the design of special structures, such as large domes, snow accumulation can be studied by small-scale model tests.

Water pressure combined with snow load is one important roof design consideration. The design snow load may be temporarily exceeded on flat roofs when heavy rain falls on snow and does not drain away rapidly. The problem may be greatly magnified in valleys and low areas of the roof due to snow meltwater and rain ponding if there is no adequate slope to the drain or if the drains are blocked with ice. These areas tend to deflect increasingly, allowing even deeper ponds to form. Local roof failures have been experienced due to such combined snow, meltwater, and rain loads. Leakage problems are even more common. Drains are often heat traced to prevent ice formation. Gutters and roof valleys may be provided with heating cables to secure proper drainage. An alternative solution is to cut the roof insulation locally and utilize thermal leaks.

Figure 2.6 Parking shelter collapsed in spring under the weight of accumulated wet snow. *(Courtesy of Lehtikuva Oy.)*

Figure 2.7 Example of heavy snow loads, Fort Wainwright, Alaska. Note how falling snow may increase load on patio roof. *(Courtesy of W. Tobiasson.)*

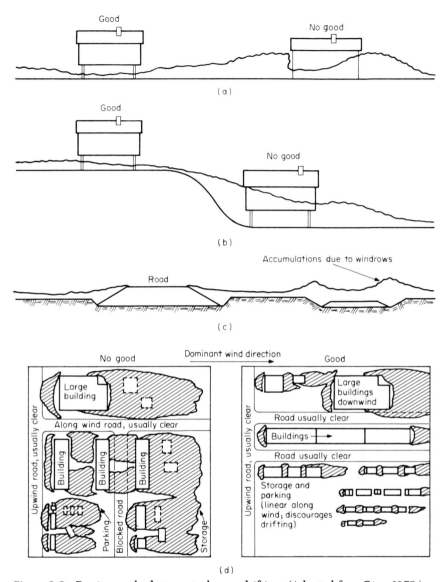

Figure 2.8 Passive methods to control snow drifting. *(Adapted from Rice, 1975.)*

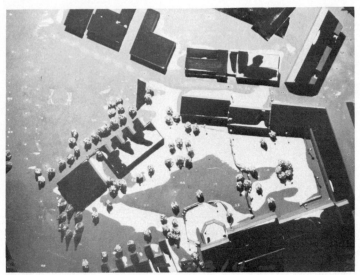

Figure 2.9 Snow accumulation and wind climate around a northern community, studied with small-scale model in wind tunnel. *(Courtesy of Technical Research Centre of Finland.)*

Lateral snow pressures have also caused some structural failures. The plowing of streets may, for example, damage traffic sign poles. In Alaska snow creep has caused failures of power transmission poles in steep slopes (Shira, 1978).

2.3   Snow Control

A variety of measures can prevent or minimize the structural, functional, or maintenance problems caused by snow deposits or drifting. These include site selection, special design considerations, erecting control structures, and snow melting. In subarctic regions snow drifting is only of local importance, but in treeless arctic tundra, where snow transport and drifting are almost continuous, control measures should be given great emphasis.

Some general design considerations in snow control are illustrated in Fig. 2.8. One should by all means avoid locating houses, roads, or other structures in depressions. Exposed plains or ridge tops should be considered in site selection in spite of the inconvenience caused by high winds. The common practice of using elevated structures in arctic permafrost areas is also helpful from the viewpoint of snow control.

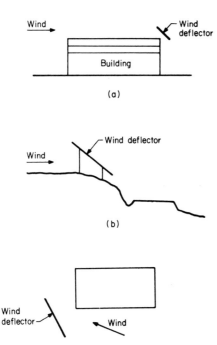

Wind

Wind deflector

Building

(a)

Wind deflector

Wind

(b)

Wind deflector

Wind

(c)

Figure 2.10 Use of wind deflectors to prevent snow accumulation where undesirable.

In road construction one should avoid sharp bends and use high subgrades. Thus the initial rate of snow deposition is minimized, and even if some snow has to be plowed from the road, the windrows do not reach such dimensions that the snow accumulation rate is significantly increased. In the design of communities the maintenance efforts are reduced if the structures are located parallel to the prevailing wind direction and no roads or structures are located immediately on the downwind side of large structures. The design of northern communities with respect to snow is further discussed in Velli et al. (1977). Small-scale model tests may prove useful in design (Fig. 2.9). Modeling techniques have been discussed in Anno (1984), Williams (1978), and Odar (1965), among others.

One can also use active methods in snow control. Snow drifting can be prevented in unfavorable locations, such as in front of doors, by using different kinds of wind deflectors (Fig. 2.10). Snow fences have been used to help in road maintenance (Fig. 2.11). These fences can also be used to gather snow for winter road construction. To be effective, snow fences should have sufficient height. Different kinds of strong cold-weather-resistant plastic strips are gradually replacing wood as the pri-

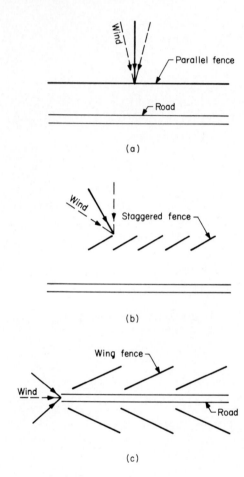

Figure 2.11 Basic arrangements of snow fences for protecting roads or railways. (*Pugh and Price, 1954.*)

mary fence material (Fig. 2.12). The fences do not have to be of the high-density type to be effective, as shown in Fig. 2.13.

Mechanical snow removal using different types of plowing equipment plays a dominant role in snow control measures aimed at maintaining the serviceability of roads, airfields, parking areas, and so on (Fig. 2.14). Salt is often used to melt the hard-packed snow and ice left on the road after plowing (Fig. 2.15). Some restricted areas such as bus terminals, bridges, or important crossroads can also be kept free of snow and ice by thermal methods using a heating cable network or hot fluid circulation in a pipe network (Fig. 2.16). A typical heat input requirement to overcome heat losses by convection, radiation, and evaporation and to melt the falling snow is on the order of 500 W/m² [150 Btu/(ft²·h)]. For further discus-

Figure 2.12 Traditional wooden snow fence in background is replaced by plastic strip fence at left. *(Courtesy of Finnish Roads and Waterways Administration.)*

sion of the different ice and snow control methods, the reader is referred to Gray and Male (1981), and for thermal control system design to the ASHRAE handbook (1980).

2.4   Construction on Snowfields

In the continental ice shelves of Greenland and Antarctica and in some other ice caps and glacierized areas the snow does not melt but slowly

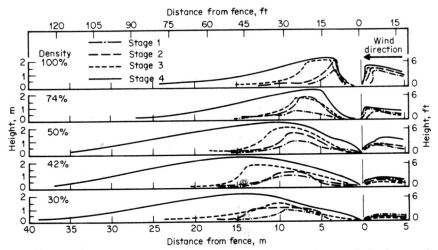

Figure 2.13 Snowdrifts generated by solid snow fences and vertical slat fences of various densities, all 1.87 m (6 ft 2 in) high with a gap of 20 ± 5 cm (8 ± 2 in) underneath. *(Price, 1961.)*

Figure 2.14 Snowplowing. *(Courtesy of D. Minsk.)*

turns into ice as the overburden pressure increases. The construction problems on these snowfields are unique. In addition to the severe climate conditions there are virtually no local resources available. Furthermore, the structures have to be founded on the thermally and mechanically unstable snow, and the typically extensive snowdrift phenomena have to be controlled. In foundation and tunnel design one must take into consideration the movements of snow layers in both the vertical and the horizontal direction. The movements are absolute as well as relative (with respect to the snow surface) as the snow undergoes viscoelastic deformation under overburden pressure. The natural process may be

Figure 2.15 Sand and salt spreading on road to increase abrasion and melt ice. *(Courtesy of Finnish Roads and Waterways Administration.)*

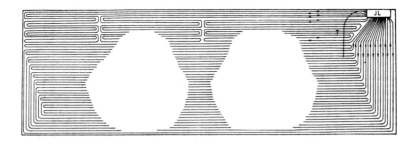

(a)

Pavement

Heating pipes

Insulation

Reinforced concrete

(b)

Figure 2.16 (a) Typical heating pipe network in slab construction. (b) Cross section of slab. *(Matilainen and Suontausta, 1970.)*

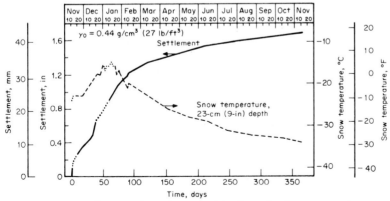

Figure 2.17 Settlement of 0.9- by 0.9-m² (3- by 3-ft²) footing with bearing pressure of 48 kPa (7 lb/in²) at Antarctica. *(Mellor, 1969.)*

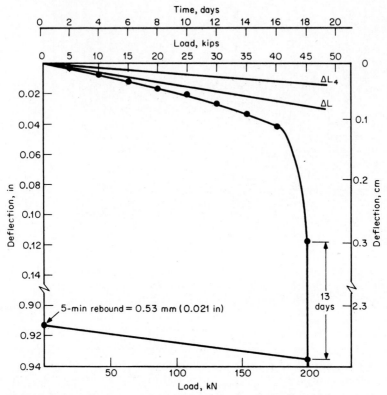

Figure 2.18 Deflection of closed-end steel pile installed into polar snow in Greenland. ΔL— elastic shortening of pile assuming pure end bearing; ΔL_4 — elastic shortening of pile assuming uniform skin friction. Pile diameter 0.25 m (10 in); embedded length 5.5 m (18 ft); bearing capacity was reached at 180 kN (40 kips). *(Kovacs, 1976.)*

disturbed as a result of construction activities. Snow undergoes acceler-ated creep under concentrated loads, and complex stress and strain fields form around any kind of mechanical or thermal disturbance (Mellor and Reed, 1967).

Separate or strip footings and friction piles are typical foundation structures used on snowfields. Bearing pressure values for moderate relative settlement rates are usually less than 50 kPa (7 lb/in²), and footings may be tied together to avoid separation (Fig. 2.17). For piles the allowable long-term skin friction values may be on the order of 10 kPa (1.5 lb/in²) (Fig. 2.18). It is interesting to note that negative skin friction may result at the upper portions of the pile when the new snow

Figure 2.19 DEW line station DYE-3 in Greenland. *(Courtesy of U.S Army Cold Regions Research and Engineering Laboratory.)*

settles faster than the pile. Snow density and temperature are the most important parameters in foundation design, but snow properties and foundation size should also be considered (Mellor, 1969; Kovacs, 1976). Because experience with foundation behavior on snowfields is limited, long-term bearing tests may be necessary for large-scale construction.

Some basic design alternatives for construction on snowfields are shown in Figs. 2.19 to 2.22. If snowdrifts are to be controlled, the structures have to be lifted well above the ground level. Small or temporary buildings can be removed when the surface of the snow cover reaches levels that are not acceptable. Larger buildings should be provided with a lifting mechanism that can also be used to balance the structure in case of uneven foundation settlements (Fig. 2.20). Airflow disturbances due to buildings, parked cars, and equipment can cause large snowdrifts in the surrounding areas, which may have to be leveled off occasionally.

In the alternative shown in Fig. 2.21, snow is allowed to drift around and eventually bury the structure. The thermal stability of the snow is maintained by ventilating the airspace between the protective steel arch and the heated facilities. The effective lifespan of any undersnow structure is limited, especially if the protective arches or possible steel tube linings are not designed to resist the total overburden pressure as the surrounding snow deforms.

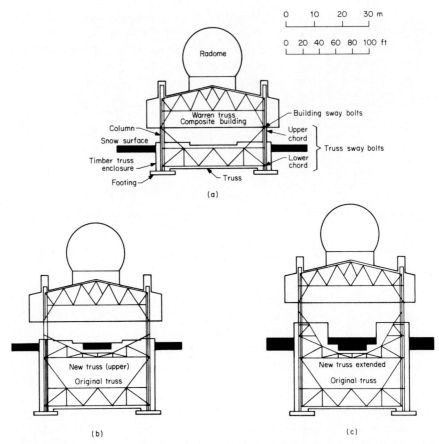

Figure 2.20 Cross sections of structural frames of **DEW** line ice-cap stations. (*a*) DYE-2 and DYE-3 as built in 1959–1960. (*b*) DYE-2 since fall 1970. (*c*) DYE-3 since fall 1972. (*Tobiasson et al., 1974.*)

2.5 Avalanches

An avalanche may be a small trickle of loose snow, a huge devastating slide of snow, ice, and rock, or anything in between (Fig. 2.23). An avalanche is initiated when the shear strength of the snow cover is exceeded over a sufficiently large area. This may occur when the creep of snow reaches the tertiary stage, but often the reason is an outside factor, such as a significant change in temperature, strong winds, heavy snowfall, an earthquake, or falling snow, rock, or ice. Avalanches have also been triggered by human activities such as skiing, blasting, or even by sound waves.

Figure 2.21   Arch-shaped structures at Byrd Station, Antarctica. *(Mellor, 1965.)*

A typical avalanche terrain has a deep snow cover and steep slopes. The characteristics of the slopes also have an important effect on avalanche occurrence. Uneven slopes and trees tend to anchor the snow cover in its place, whereas smooth slopes favor avalanches. Large snowslides occur usually when the slope angle is between 25 and 50°. There are usually no significant snow accumulations on steeper slopes. On gentle slopes only smooth wet slush runs occur sometimes. Most of the avalanches occur in well-defined areas. Central Europe is maybe the best known avalanche area, but there are also large avalanche areas in the western part of North America and in Asia.

The most important characteristic of an avalanche is its magnitude. The reaches of the largest observed avalanches have been measured in kilometers, and the volumes have exceeded 10^6 m^3 (3.5×10^7 ft^3). Such avalanches destroy everything in their path and may create impact pressures approaching 1 MPa (150 lb/in^2). The typical volume range of

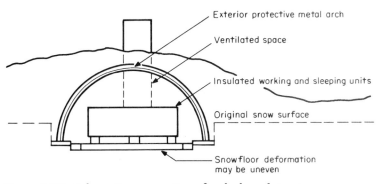

Figure 2.22   Schematic cross section of arch-shaped structure.

Figure 2.23   Dry slab avalanche. *(Courtesy of U.S. Forest Service.)*

an avalanche is between 10^3 and 10^5 m^3 (3.5×10^4 and 3.5×10^6 ft^3), and the corresponding impact pressures are on the order of 20 to 200 kPa (3 to 30 lb/in^2), although higher peak pressures may be experienced (Fig. 2.24). In loose snow the avalanche may begin in a small area, but the rupture of hard snow usually occurs over an extended area (sliding slabs), either along a layer boundary in the snow or along the ground surface. Different types of avalanches can be described by the U.S. Forest Service avalanche classification shown in Fig. 2.25. For a more detailed discussion of the characteristics and dynamics of avalanches the reader is referred to Mellor (1968) and Colbeck (1980).

Avalanches represent a serious potential hazard to human life and property. When necessary, they will have to be controlled. It is possible to trigger avalanches in advance by using explosives. For more permanent protection, structural solutions are available.

Some structural solutions for avalanche control are shown in Fig. 2.26. An extensive snow accumulation at the initial failure areas can be prevented using wind deflectors. Cut and fill terraces and supporting structures can be used as avalanche barriers. Supporting structures are typically arranged 15 to 30 m (50 to 100 ft) apart, depending on slope angle, slope smoothness, and snow depth. The snow force component parallel to the slope is approximated in Swiss practice (Mellor, 1968) by

$$P = \tfrac{1}{2} K N \gamma_s \left(\frac{h}{\cos \alpha}\right)^2 \tag{2.2}$$

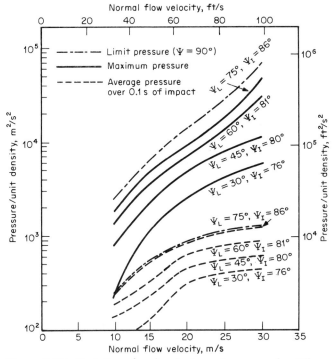

Figure 2.24 Variation of avalanche impact pressure for different leading-edge angles as a function of flow velocity and snow density. Wall is normal to slope. Ψ_L—initial angle of leading edge of avalanche; Ψ_I—angle of leading edge at impact against wall. *(Lang and Brown, 1980.)*

where N = glide factor (1.2 to 3.2, depending on slope smoothness and direction)

K = creep factor (typically 0.7 to 1.0, depending on slope angle and snow density)

γ_s = snow density

h = snow thickness

α = slope angle

Instead of preventing release of the slide, the control can also be accomplished by using deflective or protective structures. The most important loads on the avalanche gallery shown in Fig. 2.26 are the weight of the avalanche, the possible dynamic effects in case of a changing slope, the weight of the snow cover and previous avalanche debris in case they are not swept away, and the shear or friction force along the roof. Typical design values for normal roof loads may be up to 50 kPa

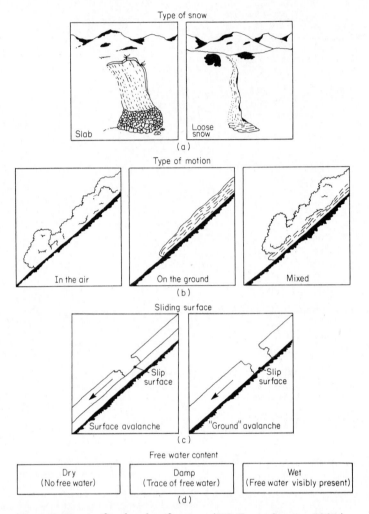

Figure 2.25   Avalanche classification. *(U.S. Forest Service, 1961.)*

(1000 lb/ft^2). The friction force along the roof is often considered to be 50% of the normal force caused by the sliding avalanche. The design principles of avalanche control measures are discussed thoroughly in Mellor (1968).

2.6   Icing on Structures

Icing on structures is caused by freezing of water droplets in subsequent layers on the surface of a structure exposed to the atmosphere. The

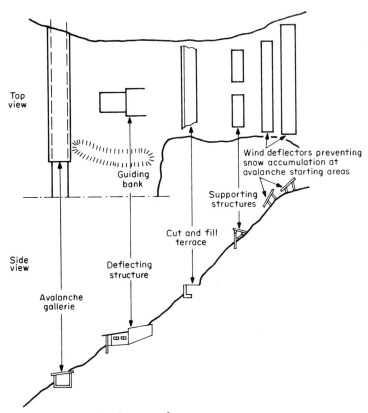

Top
view

Guiding
bank

Wind deflectors preventing
snow accumulation at
avalanche starting areas

Supporting
structures

Cut and fill
terrace

Side
view

Deflecting
structure

Avalanche
gallerie

Figure 2.26 Avalanche control measures.

source of this ice buildup may be fog or clouds containing tiny super-cooled water droplets, freezing rain, wet snow, or spray from breaking waves and wave crests. Icing can occur anywhere in cold regions, but the frequency and severity vary greatly with the location.

Icing on pavements is maybe the most familiar type of icing in cold regions. This occasional event may cause extremely hazardous driving conditions. Ice may form on the pavement under several conditions, but in a typical case the temperature of the pavement is significantly below the freezing point and the water droplets are supercooled or near the freezing point. A comprehensive analysis of the event is presented in Jumikis (1966).

Heavy atmospheric icing combined with strong winds may be the dominant factor in the design of slender structures such as radio antennas or power transmission lines (Figs. 2.27 and 2.28). Numerous collapses of such structures have been experienced. In the case of ships and ocean

(a)

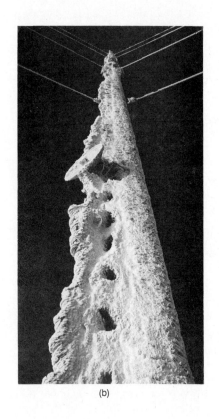

(b)

structures the source of icing is generally sea ice spray. Icing causes mainly operational difficulties, but in extreme cases also safety hazards (Fig. 2.29).

Ice can deposit on the surface of a structure in different forms and densities. Well supercooled tiny water droplets in fog or a cloud freeze very rapidly when they get in contact with a structure. As a result, very-low-density ice, known as soft rime, is formed. Hard rime forms when the freezing of droplets occurs slowly so that some flow of water has time to occur before crystallization. It has a density of 100 to 600 kg/m^3 (6 to 40 lb/ft^3). When water droplets have sufficient time to wet the surface of the structure before freezing, a glaze with densities running from 700 to 900 kg/m^3 (45 to 55 lb/ft^3) is formed. Icing deposits may also be combinations of these different forms and snow. When icing is formed from seawater, it contains brine, pockets of unfrozen saltwater.

The occurrence, severity, and type of atmospheric icing depend very much on temperature, wind speed, total water content of the air, and water droplet dimensions. The general principles of different forms of atmospheric icing are illustrated in Fig. 2.30. The formation of glaze is

Figure 2.27 Heavy icing on a radio tower (*a*), (*b*) and structure after collapse due to combined effects of ice and wind (*c*). (*Courtesy of Finnish Broadcasting Company.*)

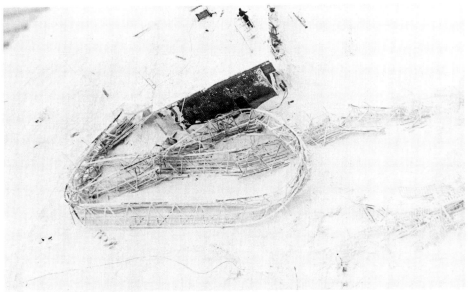

(c)

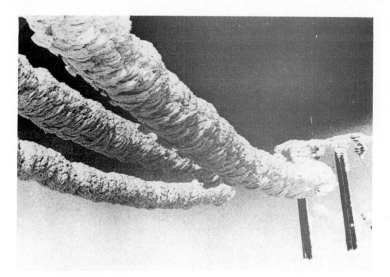

Figure 2.28 Exceptionally heavy ice buildup on conductors in a power transmission line. *(Courtesy of Horn Tron Ab.)*

most probable at temperatures between 0 and $-3°C$ (32 and 27°F). Lower temperatures and low wind speeds favor the formation of soft rime.

Similar approximate relationships can also be established for the severity of icing occurring at sea, as shown in Fig. 2.31. Naturally the peculiarities of the local wave climate and the structural considerations have some effect on the icing intensity. Sea icing is usually caused by spray

Figure 2.29 Heavy sea icing on a ship. *(Courtesy of Oy Wärtsilä Ab.)*

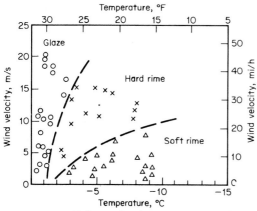

Figure 2.30 Relationship between meteorological conditions and type of icing. *(Kuroiwa, 1965.)*

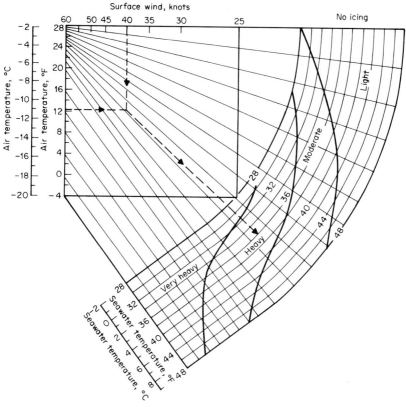

Category	Rate of icing per 3 h, mm (in)
Light	1–5 (0.05–0.2)
Moderate	5–8 (0.2–0.3)
Heavy	8–19 (0.3–0.75)
Very heavy	19+ (0.75+)

Figure 2.31 Nomogram giving the severity of sea icing as a function of wind speed, air temperature, and seawater temperature. *(Wise and Comiskey, 1980.)*

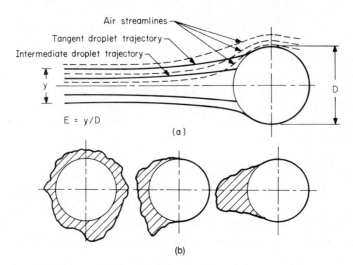

Figure 2.32   (*a*) Collection efficiency *E* and (*b*) typical ice accumulation shapes on cylindrical structures.

from waves breaking against ship hulls or other solid objects. Spray from wave crests is a significant source of icing at very high wind speeds. Contrary to atmospheric icing, the bulk of sea icing occurs at low elevations, generally less than 15 m (50 ft) above the peak water level, although waves breaking against structures may raise significant amounts of droplets above this elevation.

Theoretically speaking, the amount of icing on a structure, or the so-called icing efficiency, can be estimated by

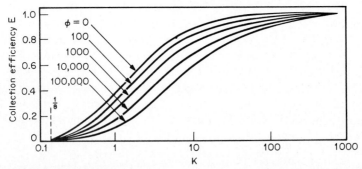

Figure 2.33   Collection efficiency *E* as determined by inertia parameter *K* and parameter ϕ. *(Langmuir and Blodgett, 1946.)*

$$\frac{dM}{dt} = ECDVW \qquad (2.3)$$

where M = mass of ice deposited in time t
$\quad E$ = collection efficiency coefficient
$\quad C$ = capture coefficient
$\quad V$ = wind velocity
$\quad W$ = liquid water content in air
$\quad D$ = width of structure

The collection efficiency coefficient corresponds to the fact that water droplets are deflected from their linear trajectories as they approach the structure (Fig. 2.32). It is the ratio of the mass of droplets striking the structure to the total mass of droplets that would have hit the structure if they had not been deflected. In Fig. 2.33 the collection efficiency E is given for a cylinder with radius R as a function of two variables, the inertia parameter K and parameter ϕ,

$$K = 1290 \frac{Vr^2}{R} \quad \left[\frac{cm^2}{s}\right] \qquad (2.4)$$

$$\phi = 0.175\,VR \quad \left[\frac{cm^2}{s}\right] \qquad (2.5)$$

where r is the radius of the droplets. It can be seen that theoretically no icing occurs when $K < 0.125$.

Not all the water droplets that strike the structure will necessarily freeze. Especially at high wind speeds, the heat of fusion of all incoming water droplets may not be dissipated before they flow to the outer surface and are carried away by the airflow. Data in Glukhov (1971) illustrated that while the capture coefficient is close to 1 at wind speeds ranging from 0 to 5 m/s (0 to 10 mi/h), it may decrease to about 0.2 at wind speeds exceeding 10 m/s (20 mi/h).

In practice there is not enough information available for theoretical icing computations to be used. Attempts have been made to describe areal risk and severity of atmospheric or spray icing on a large scale (for example, Bennett, 1959; Tattelman and Gringorten, 1973). In any case, icing design loads should be based on local codes and experience [for example, in Finland the ice load on a conductor is at least 12.5 to 25 N/m (1 to 2 lb/ft)] along with rational considerations given to site location, elevation, and exposure; the influence of the structural shape on the magnitude of icing; and the asymmetrical and uneven character of ice formations.

There are no easy solutions to control icing. Heat and freezing point depressants have been applied with some success, but generally this kind of protection requires too great an effort except in power conductors, where losses are adequate to melt the ice. In some cases, structures can be protected by flexible sheetings. Space truss structures can be enclosed by corrugated plastic sheets in order to control icing. Site selection is in many cases very important. For example, elevated power-line routing above treeline should be avoided, while valleys shielded from dominant moist wind directions are preferred. In coastal areas, protected structures should be located out of the reach of droplets from major wave-breaking elements. Sometimes this distance may be over 100 m (300 ft). If ice has to be removed mechanically, it is desirable to use simple structural shapes and act quickly before a strong bond can form between ice and structure. Coating with low-adhesion materials, such as some plastics, may also be used.

For further discussions on icing phenomena and their regional occurrence the reader is referred to the extensive summary reports by Minsk (1977, 1980) and to McLeod (1977).

3

ICE PROBLEMS

3.1 Ice and Engineering

Engineers have long been dealing with the problems associated with the bearing capacity of the ice cover and with ice forces on traditional hydraulic structures. In recent years the development of arctic offshore structures and increased winter navigation have added new dimensions to ice engineering. This chapter discusses the basics of ice mechanics and the whole range of ice-related civil engineering problems.

The occurrence and the thickness of ice are quite closely associated with the freezing index of rivers and lakes. In sea areas the situation is more complex because of the large heat capacity of the water mass and the effects of wind and currents. The minimum and maximum extents of sea ice are depicted in Fig. 3.1.

The occurrence of ice may take many different forms besides the ordinary land-fast level ice. Different types of ice jams may develop in rivers, especially in the spring and fall. On the open sea pack ice is in constant movement under the influence of wind and currents. It may raft and form ridges and rubble fields under pressure. Icebergs, ice islands, and other multiyear ice features may represent the ultimate hazard to offshore structures in some sea areas.

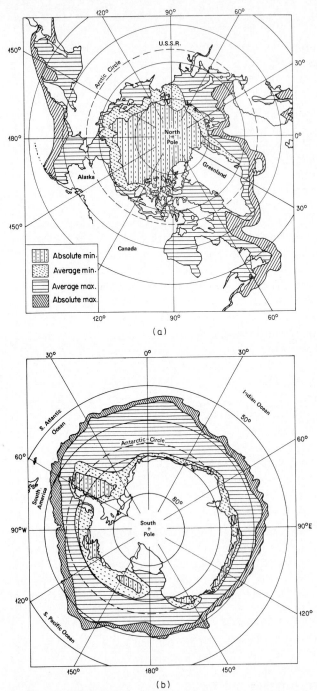

Figure 3.1 Minimum and maximum extents of sea ice of one-eighth or greater concentration. *(Central Intelligence Agency, 1978.)*

Ice is quite a complicated material. Its mechanical properties depend among other things on its crystal structure, the temperature, and the duration of loading. In short-term loading ice behaves elastically and fails in a brittle manner. If the loading rate is slow or the duration of loading is long, creep and plastic failure predominate. This is clearly experienced in the classical bearing capacity problem, where the deflection increases and the bearing capacity decreases with the duration of loading.

Ice forces on structures are the result of ice movements (Figs. 3.2 and 3.3). Vertical ice movements occur as a result of water-level fluctuations. Thermal expansion and contraction, wind and current friction, hydrodynamic forces, and gravity are factors that may contribute to the lateral movements of ice fields or floes. If the movements are slow or if the forces are exerted on the structure via an ice rubble, the loads are static in nature. Dynamic loads are created when ice impacts or crushes against the structure.

Ice control includes a wide variety of engineering approaches which aim at improving existing ice conditions, manipulating the behavior of ice, and reducing or eliminating ice forces on structures. For example, in rivers large amounts of frazil ice may form in open rapidly flowing sections during the fall and winter. Frazil ice then accumulates under the ice cover, partly blocking the flow and causing flooding. The formation of an ice cover can be facilitated and hence the extensive production of frazil ice prevented by changing the flowing conditions or by using an ice boom. Other examples of ice control include methods that prevent or reduce ice formation locally, barriers that stop moving ice floes, and different structural arrangements.

There are a number of ways to study ice engineering problems, including laboratory experiments, theoretical computations, model tests, in situ experiments, and field measurements. The problems are complex, however, and each approach has its own drawbacks.

Laboratory studies are often made to measure the deformation and strength properties of ice. These results are not usually directly applicable to engineering problems. There are several reasons for this. Test results seem to be quite sensitive to the testing methods, ice structure, and scale effects (smaller test specimens give higher strength values than larger ones). Furthermore the natural ice cover is generally nonuniform as it contains different ice layers, cracks, and impurities. The effects of these phenomena will not show up in small-scale laboratory tests. But laboratory experiments provide nevertheless valuable information on the general trends concerning ice behavior. The results also provide limits for strength and deformation parameters.

Theories from basic mechanics are often applied in the analysis of different ice engineering problems. Reasonable results are obtained with

Figure 3.2 Dock damaged by vertical ice forces. *(Courtesy of L. J. Zabilansky.)*

Figure 3.3 Ice crushing against lighthouse. *(Courtesy of Finnish Board of Navigation.)*

Figure 3.4 Model test in an ice tank simulating ice failure against a caisson structure and rubble-formation process. *(Courtesy of Wärtsilä Arctic Research Centre, WARC.)*

simple formulas when semiempirical correction factors or strength parameters are used. However, the limitations of these computations should be recognized. Basic theories of elasticity, plasticity, and fracture mechanics do not describe the behavior of ice very well. The effects of nonuniformity, viscoelasticity, and cracking are difficult to take into account, even in a sophisticated finite element analysis.

Model tests provide one possibility to study ice forces and ice processes (Fig. 3.4). However, model ice does not always behave like real

Figure 3.5 In situ measurement of the bending strength of ice with a cantilever beam test. *(Courtesy of Oy Wärtsilä Ab Helsinki Shipyard.)*

ice. It is not possible to create complete similitude (geometric, kinematic, and dynamic), even in the simple case of level ice crushing against a structure. For example, an attempt to simulate the properties of a pressure ridge is an even more difficult task. Model tests can nevertheless simulate natural ice processes in many cases with reasonable accuracy. They can also be used to verify engineering theories. As more field verification data become available, the reliability of model test analysis will improve, and test results can be used in design with greater confidence.

Different kinds of field measurements, full-scale experiments, and programs monitoring the performance of structures remain the most reliable sources of information in ice engineering (Fig. 3.5). However, fieldwork is expensive and often complicated. There are many important phenomena that have not yet been studied in a reliable manner in the field. In some cases it may take a long time before statistically reliable information is obtained because extreme ice events occur rarely.

In spite of these drawbacks, ice engineering already provides means to solve most problems in a reasonable manner. Good knowledge of local ice conditions and available experience form the basis of design. Laboratory tests, theoretical computations, model tests, fieldwork, or some combination of these approaches can be used to provide additional information for sound and economical engineering.

3.2   Description of Ice Covers

3.2.1   Ice Classification

Ice is a granular material. The typical size and shape of a grain, or actually an individual ice crystal, vary greatly. The orientation of grains is defined by the hexagonal molecule structure of ice. The axis of molecular symmetry is called the c axis and is perpendicular to the basal plane. The basal plane is of great importance for the strength and deformation properties of ice. The different types of ice are classified and described in Table 3.1 and discussed more thoroughly in Michel and Ramseier (1971).

The freezing process of lakes can begin after the temperature of the water body has dropped below 4°C (39°F). The density of water reaches its maximum at this temperature. In the absence of mixing currents the surface can cool down further and freeze while the main body of water remains at temperatures above freezing.

When the temperature of the surface water drops below 0°C (32 °F), tiny separate ice crystals form at the surface layer. In calm conditions these crystals grow laterally until they join adjacent crystals. Winds and

TABLE 3.1   Classification of Ice

Designation		Characteristics[*]
Primary ice (forms first)	P1	Calm surface and small temperature gradient. Grains usually large to extra large with irregular boundaries and preferred vertical c-axis orientation.
	P2	Calm surface and large temperature gradient. Grain size medium to extra large and crystal shape tabular to needlelike, with random or preferred vertical to random c-axis orientation.
	P3	Agitated surface nucleated from frazil. Grain size fine to medium and grain shape tabular, with random c-axis orientation.
	P4	Nucleation by snow. Grain size fine to medium and grain shape equiaxed, with random c-axis orientation.
Secondary ice (forms parallel to heat flow from primary ice)	S1	Columnar grained ice. Grain size usually large to extra large, increasing with depth, and grain shape irregular, with preferred vertical c-axis orientation.
	S2	Columnar grained ice. Grain size fine to extra large, increasing with depth. c-axis orientation may be initially random, but it becomes preferred horizontal with increasing depth.
	S3	Columnar grained ice. c-axis orientation preferred aligned horizontal.
	S4	Congealed frazil slush. Grain size fine to medium and crystal boundaries irregular. c-axis orientation is random.
	S5	Drained congealed frazil slush. Grain size fine to medium and grain shape angular. This ice has formed as water has drained through the ice cover, leaving the slush to be refrozen; it has low density.
Superimposed ice (forms at top of ice cover)	T1	Snow ice. Grain size fine to medium and grain shape round to angular, with random c-axis orientation. The density varies from 0.83 to 0.90 g/cm^3 (52 to 56 lb/ft^3).
	T2	Drained snow ice. Grain size fine to medium and grain shape rounded, with random c-axis orientation. The density is about 0.6 g/cm^3 (37 lb/ft^3)
	T3	Surface ice. Layers of columnar ice, which have formed at the top of the original primary ice.
Agglomerate ice	R	Agglomeration of individual ice pieces which have refrozen. May be associated with rafting or ridging.

[*] Grain size ranges: (a) fine, diameter less than 1 mm (0.04 in); (b) medium, diameter 1 to 5 mm (0.04 to 0.2 in); (c) large, diameter 5 to 20 mm (0.2 to 0.8 in); (d) extra large, diameter greater than 20 mm (0.8 in); (e) giant, dimensions in meters.

SOURCE: Adapted from Michel and Ramseier (1971).

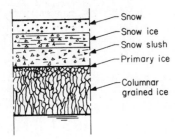

Figure 3.6 Typical cross section of ice cover in a lake.

waves or snowfall tend to produce a layer of ice crystals called frazil slush. These may further develop into slush ice floes before freezing into a solid ice cover.

The crystal orientation at the thin top layer of the ice cover is usually more or less random. However, ice tends to grow fastest in the direction perpendicular to the c axis. That is why grains with a horizontal c axis become dominant. The grains tend to become longitudinal and gain in size with increasing distance from the surface. This type of ice is called columnar grained ice (S2 or S3 ice).

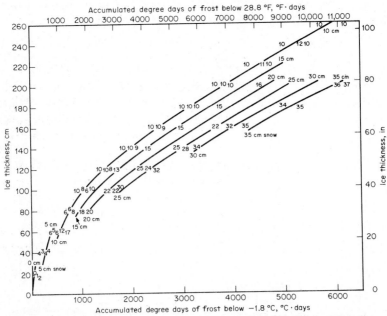

Figure 3.7 Ice growth curves measured at the Canadian arctic archipelago. Notice effect of snow. *(Bilello, 1960.)*

If the freezing conditions are calm and the freezing rate is sufficiently slow, the orientation of the first ice crystals at the surface of the ice cover tend to have vertical c-axis orientation. If the freezing process continues smoothly, the ice crystals retain their orientation through the ice cover. This type of ice is also columnar grained, but the basal planes of the crystals are parallel to the surface of the ice cover (S1 ice).

In rivers, frazil ice forms in turbulent supercooled conditions. It may develop into frazil slush or floes and form the initial ice cover. It may also deposit under the existing ice cover (S4 ice).

Yet another important ice type forms when water penetrates to the top of the ice cover due to the weight of snow (Fig. 3.6). When the flooded snow slush freezes, a new type of ice, called snow ice (T1), is formed. The grain size of snow ice is relatively small, usually less than 5 mm (0.2 in), and the orientation of the grains is random. Granular ice with somewhat similar grain structure can also be found in multiyear ice features and in dirty sea ice.

3.2.2   Some Features of Lake and Sea Ice Covers

The thickness growth of the ice cover can usually be predicted accurately on one-dimensional thermal analysis. However, the analysis would require detailed information on, among other things, meteorological factors and the thermal conductivity of the snow cover. That is why the thickness growth and the maximum thickness of the ice cover are usually estimated by using Stefan's equation in a simplified form:

$$h = \alpha\sqrt{\Sigma F} \qquad\qquad (3.1)$$

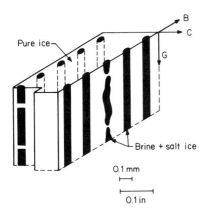

Pure ice

Brine + salt ice

0.1 mm

0.1 in

Figure 3.8   Idealized model of the structure of sea ice. *(Assur, 1958.)*

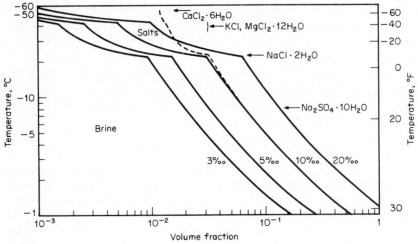

Figure 3.9 Brine volume as a function of ice salinity and temperature. (*Anderson, 1960.*)

where h = thickness of ice, cm (in)

ΣF = number of degree-days of freezing temperatures, °C·days (°F·days)

α = coefficient based on local experience

A typical value for the coefficient α for relatively windy shallow lakes with practically no snow cover is, according to Assur (1956), about 3.2 if metric units are used (0.9 for U.S. customary units), and for moderate snow cover it is on the order of 2.5 (0.7). At sea, in large lakes, and in rivers the value of the coefficient depends more on local conditions. For arctic sea ice the first approximation for α might be 2.5 (0.7) and for rivers with moderate flow, 2.0 (0.6). However, these values must be verified with field measurements (Fig. 3.7).

The most important factors affecting the thawing process of the ice cover are temperature, absorption of solar radiation, winds, currents, and meltwater. The thawing of lake ice covers usually begins at the shoreline. When meltwater flows to the shoreline, it weakens the ice and darkens the snow cover so that the solar radiation can be absorbed more effectively. When ice along the shores has already thawed and the ice cover can move slightly under the action of winds and currents, it may still carry considerable loads. Meltwater runs either directly to open water or to cracks and holes in the ice cover. The surface of the ice cover becomes white and reflects solar radiation. In the final phase of the thawing process the water content of the ice cover increases sharply and the ice becomes dark. Therefore the ice absorbs solar radiation more effectively. Winds or currents can break the ice cover easily at this stage.

Land-fast sea ice is quite similar to lake ice. It is usually columnar grained, although the thickness of the randomly oriented surface layer is often larger than that of lake ice. The growth rate also appears to be similar to that of lake ice. However, there are also differences, as discussed in Mellor (1964). When the salinity of water exceeds 24.7‰, water reaches its maximum density at the freezing point. Arctic seas commonly have salinities in the range of 30 to 35‰ and freezing points at −1.8 or −1.9°C (28.8 or 28.6°F). This means that circulation continues until the temperature of the entire water body has reached approximately the freezing point. The beginning of the freezing process may thus be delayed compared to that of lakes.

The dominant impurity in lake ice is air. It is trapped into ice in the form of air bubbles. Sea ice contains also brine, vertical pockets of saltwater, and, in very low temperatures, solid salts (Fig. 3.8). The volume of brine in a unit volume of ice is referred to as brine volume. It can be estimated from Fig. 3.9 based on the salinity and temperature of the ice.

The salinity of ice depends on the freezing rates and the age of the ice. New, rapidly frozen sea ice may have up to 20‰ salinity, while the salinity of slowly frozen sea ice may be only 4‰ (Fig. 3.10). When the temperature decreases, some brine freezes at the edge of the pocket and

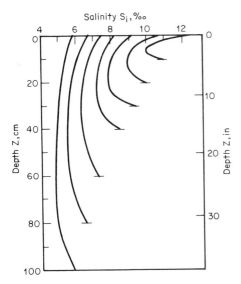

Figure 3.10 Salinity profiles in an ice sheet. *(Weeks and Assur, 1967.)*

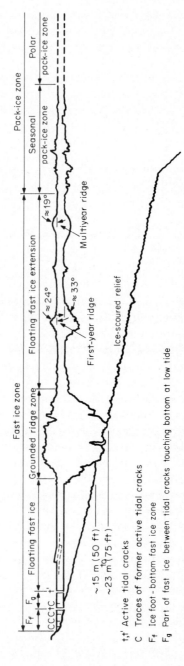

Figure 3.11 Late-winter ice zoning as found along the Alaska Beaufort Sea coast. (*Kovacs and Sodhi, 1979.*)

t,t′ Active tidal cracks
C Traces of former active tidal cracks
F_f Ice foot - bottom fast ice zone
F_g Part of fast ice between tidal cracks touching bottom at low tide

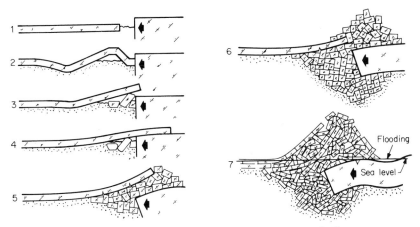

Figure 3.12 Sequence of events observed during formation of a floating pressure ridge. *(Kovacs and Sodhi, 1979.)*

the remaining brine increases in salinity until equilibrium is reached. In the winter the temperature at the surface is lower than that at the bottom of the ice cover. Therefore the brine pocket is not totally in equilibrium and brine tends to migrate downward. The salinity of ice slowly decreases with time. The reduction of ice salinity occurs faster through drainage in the spring when the ice warms up and brine mixes with meltwater and seawater traveling in the brine channels. The salinity of the top layer of the ice cover rapidly drops close to zero.

In large lakes and seas a land-fast level ice cover is typical only for areas near the shoreline (Fig. 3.11). Farther away from the shore a new ice cover can easily break into floes under the action of wind and waves, forming a field of pack ice. The individual floes may float freely or be tightly jammed together. The pack-ice field may refreeze and become rafted or heavily hummocked under pressure. Pressure ridges, long strips of piled or hummocked ice, represent another type of irregularity in ice fields. The ridges usually form at contact areas of two ice fields or along cracks in an ice field when they close under pressure (Fig. 3.12). Heavy ridging typically occurs at the edge areas of the land-fast ice zone.

The spring breakup of sea ice begins with local weakening and deterioration of the ice cover. Winds and currents may then easily break the weakened ice cover. Massive ice movements are typical for many sea areas and large lakes before the total decay of the ice cover.

In some sea areas the existence of multiyear ice floes, icebergs, or ice islands has to be taken into account in engineering considerations. Multiyear ice floes are those that survive the summer melting period. They

(a)

(b)

58

(c)

(d)

Figure 3.13   Different forms of sea ice. (*a*) Open arctic pack ice. (*b*) Hummocked ice. (*c*) Large pressure ridge. (*d*) Iceberg. (*Courtesy of Oy Wärtsilä Ab Helsinki Shipyard.*)

are typically 2 to 4 m (7 to 13 ft) thick, but also much thicker formations, such as consolidated multiyear ice ridges and hummock fields, are commonly encountered in the Arctic Ocean. Icebergs and ice islands are large pieces of glacier ice formed in snow metamorphosis and broken off from land-based ice or from an ice shelf. The sizes of icebergs and ice islands range from a few cubic meters to cubic kilometers in extreme cases. Different forms of sea ice are illustrated in Fig. 3.13.

Recently considerable work has been done to develop a general understanding of the laws governing sea ice growth, drift, and decay on a large scale (Colbeck, 1980). The external and internal forces of the ice cover, ice rheology, and ice strength are used as basic elements in the analysis. Separate attempts have also been made to use model techniques for predicting the behavior of ice covers in large lakes (Rumer et al., 1979).

3.2.3 Ice Covers in Rivers

Ice formation in rivers usually begins from the shores. In slowly flowing reaches of small rivers this border ice growth is dominant and ice forms rapidly over a large part or the entire surface of the river. A different type of ice-formation process occurs in more rapidly flowing sections of rivers involving frazil ice, drifting ice floes, and ice accumulations (Fig. 3.14). This kind of ice-formation process sometimes causes severe flooding and needs a more comprehensive discussion.

Figure 3.14   Ice-formation process in progress in Niagara River. *(Courtesy of New York Power Authority.)*

In rapidly flowing river sections water has to supercool before the ice-formation process can begin. First tiny frazil ice crystals appear in the water. Frazil ice then turns into slush and floes that drift downstream. This mass may jam in a river bend or accumulate in front of the solid ice cover. The ice-formation process continues rapidly until equilibrium in front of the accumulation is reached. The limiting condition used in Pariset et al. (1966) corresponds to "no spill," that is, the stagnation water level does not exceed the top corner of the accumulation. It is given by

$$v_c = \sqrt{2gh\frac{\rho_w - \rho_i}{\rho_w}\left(1 - \frac{h}{H}\right)} \tag{3.2}$$

where v_c = critical velocity of water in front of ice cover
ρ_w = density of water
ρ_i = density of ice
h = thickness of accumulated ice cover
H = water depth in front of ice cover
g = acceleration of gravity

If the equilibrium condition is not satisfied, the incoming frazil slush and floes go under the ice cover until an equilibrium ice thickness is reached. According to Eq. (3.2), the maximum value for the ratio h/H at the frontal edge of the ice cover is about ⅓. When this maximum value is reached, the ice cover can progress upstream only after the water level has increased and the water velocity decreased in front of the ice cover. This may occur, for example, due to head losses when incoming ice floes and frazil slush deposit under the ice cover in areas with sufficiently slow flow velocities (Fig. 3.15).

The internal pressures within an ice accumulation increase with the distance from the frontal edge of the ice cover but not indefinitely because the acting forces (water and wind friction, the gravitational component due to the slope of the river, and hydrodynamic forces at the frontal edge) are transmitted to banks by arching effects. The maximum pressure is proportional to the river width and depends, among other things, on the type of accumulation and the friction factors. In narrow rivers the ice accumulation may be able to resist the external forces and Eq. (3.2) offers not only an approximation for the thickness of the frontal edge, but also an estimate for the overall thickness of the unconsolidated ice accumulation. In wider rivers the initial internal resistance of the accumulation may not be adequate to resist the thrust, and the accumulation will thicken by shoving and piling until the equilibrium has been reached.

The problem of the stability of ice accumulations has been studied in Pariset et al. (1966), Michel (1971), and more recently in Calkins (1983).

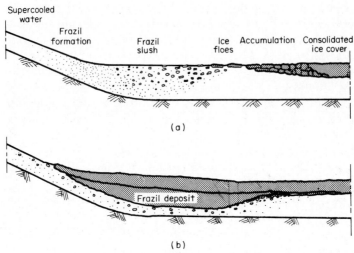

Figure 3.15 (*a*) Schematic representation of ice-formation process. (*b*) Development of a hanging dam.

The result of the analysis of Pariset et al. is shown in Fig. 3.16. The stability area corresponding to a certain accumulation thickness *h* is defined by the factor *X*, which is related to the discharge *Q*, the channel width *B*, the hydraulic water depth *H*, the mean water velocity *V*, and the Chézy coefficient *C*. In the analysis, the channel is assumed to be straight with uniform slope, roughness, cross section, and flow, and the ice accumulation is treated as cohesionless and fluidized. It has also been assumed that the Chézy coefficient has approximately the same value for the ice accumulation and the riverbed.

The Chézy coefficient is proportional to the hydraulic radius *R* according to the equation

$$C = KR^{1/6} \tag{3.3}$$

where *K* is the constant of proportionality. The hydraulic radius can usually be assumed to be one-half of the water depth. With $17 < K < 20$ m$^{1/3}$/s ($25 < K < 30$ ft$^{1/3}$/s) suggested for the initial ice formation period and $24 < K < 30$ m$^{1/3}$/s ($35 < K < 40$ ft$^{1/3}$/s) later under the solid ice cover, the latter assumption seems reasonable in many cases.

The analysis in Pariset et al. (1966) has been used successfully to predict the different stages of the formation process in nature. Two of the examples are considered here. Point A_0 represents an initially unstable situation in Fig. 3.16. The ice accumulation thickens by shoving and piling until it reaches point A_1 on the stability curve. If the flow velocity at the frontal edge of the cover is sufficiently large, incoming floes are carried under the ice cover, where they can deposit (see Uzuner, 1975). The ice cover may thicken in this manner to the limit of stability A_2, and even beyond this limit to A_3. Hence it will again be subjected to movement until a new equilibrium is reached through an increase in water depth H.

As a second example it is assumed that the water depth does not increase with discharge as in power channels upstream of a dam. Starting from the stable point B_0, the discharge can be increased to a point B_1 without change in the accumulation thickness h. A further increase in discharge to B_2 will then force the ice cover to thicken accordingly.

Finally it is noted that there is an absolute limit in discharge in Fig. 3.16 given by $Q^2 \approx 2.8 \times 10^{-3} \, C^2 B H^4$. If the discharge is greater than this limit, thickening and shoving will continue until a stable point is reached through the increase in depth H following the head losses that result from the piling up of ice downstream.

Generally the water velocities are sufficiently slow in rivers so that the initial freezing occurs smoothly as a result of border ice formation, formation of a thin ice accumulation, or both. The thickness growth of the ice cover occurs in a similar manner as in lakes, although the rate may be lower and more uneven because of the heat content of the flow. On the other hand, if there is a layer of frazil ice deposited beneath the columnar grained ice cover, the growth rate of a solid ice cover may be significantly

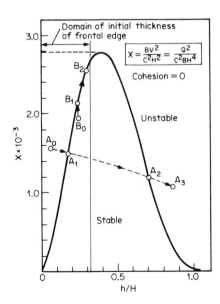

Figure 3.16 Stability diagram for ice accumulation. *(Pariset et al., 1966.)*

increased because the latent heat of solidification of frazil slush is considerably lower than that of water.

The importance of frazil ice in the winter regime of a river depends on hydraulic and meteorological conditions. In cold climates, large amounts of frazil form at rapidly flowing open river stretches. This frazil deposits on the bottom of the ice cover and it may also adhere on the river bottom. Sometimes the frazil formations may be several meters thick (Fig. 3.17), clogging the channel almost totally except for a small section that carries the flow. Flooding is a natural result of such formations. If the temperature of the incoming water is at the freezing point, the frazil production in open river stretches can be estimated from the equation

$$V = \frac{h_a A(\Sigma F)}{L(1 - e)} \tag{3.4}$$

where V = volume of frazil ice produced
 h_a = average heat transfer coefficient
 A = area of open water
 ΣF = number of degree-days of freezing temperature
 L = volumetric latent heat of fusion of ice, $= 307$ MJ/m^3 (8230 Btu/ft^3)
 e = porosity of frazil ice

The heat transfer coefficient h_a from the water surface is a complicated function of meteorological conditions (Michel, 1971). One simple approximation is given in Jobson (1973) as

Figure 3.17 Massive frazil ice formations. *(Courtesy of S. DenHartog.)*

$$h_a = 3.4 + 4.4v_a \tag{3.5}$$

where v_a is the wind velocity in m/s and h_a is given in $W/(m^2 \cdot {}^\circ C)$.

Assume that the air temperature is $-20\,^\circ C$ $(-4\,^\circ F)$ with an average wind velocity of 5 m/s (11 mi/h) and an open water area of 10,000 m^2 (2.5 acres). If the flowing water is at the freezing point, almost 3000 m^3 (100,000 ft^3) of frazil ice with a porosity of 0.5 is produced daily. This mass is transported downstream until the flow velocity is sufficiently low so that it can deposit underneath the ice cover.

Methods are being developed to simulate the fall and winter regimes of entire river systems. Hopper et al. (1978), for example, evaluated the change in the winter regime of the Burntwood River waterway, Manitoba, as a consequence of an artificial increase of winter flow. The flow conditions at freeze up were divided into different categories from 0.15 m/s (0.5 ft/s) basically corresponding to thermal ice growth to critical velocities where only border ice growth is possible. The results included time-based ranges of possible stages and flows, the size of ice dams, and the volume of water detained in lakes along the waterway.

The first sign of ice breakup in a river is the increase in daily discharge. This causes cracking of the ice cover and flooding along the shores. In rapidly flowing reaches of the river, ice floes come loose and drift downstream until they accumulate or jam in front of the solid ice cover. The characteristics of actual spring ice breakup in a river depend heavily on the local hydraulic and meteorological conditions. In some rivers, ice practically melts in place or at least deteriorates considerably before ice transport begins. The ice breakup occurs smoothly. However, especially in certain large north-flowing rivers, the ice breakup is known to be an abrupt process with large and strong ice floes, severe ice jams, and excessive flooding (Figs. 3.18 and 3.19).

3.3 Deformation and Strength Properties of Ice

3.3.1 Microstructure and Mechanical Properties

The behavior of ice under load depends among other things on its crystal structure, temperature, impurities, and the type of loading. The micromechanisms that govern the fracture or deformation of ice are not very well known and are reviewed here only briefly. For a more sophisticated discussion the reader is referred to Michel (1978) and Bogorodsky and Gavrilo (1980).

Ice crystals reject impurities such as air and salts during the freezing process. However, large amounts of impurities get trapped within the ice, especially when freezing is fast. The impurities concentrate along

Figure 3.18 Ice breakup in Yukon River. Track of a bridge pier can be recognized in large floe. *(Courtesy of F. D. Haynes.)*

Figure 3.19 Scene after severe ice jamming and flooding. *(Courtesy of C. L. Wuebben.)*

crystal boundaries together with a thin water film. Both the individual crystals and the crystal boundaries contribute to the mechanical properties of polycrystalline ice. Ice crystals deform and crack easily along basal planes when subjected to shear stresses, but the resistance against deformation and cracking along other planes is much higher. Boundaries with other crystals represent restraints to deformation, especially when the orientation of the crystals differs. On the other hand, stress concentration may form at crystal boundaries, causing nucleation of a crack in the crystal. There may also be some slip and cracking along the crystal boundaries.

The thawing of ice begins along crystal boundaries at temperatures below 0°C (32°F), weakening the connections between crystals. That is why the crystal boundaries have a more profound effect on the mechanical properties of ice at temperatures close to the thawing point.

It is difficult to define deformation and strength properties of ice, because the test results show considerable scatter depending on the ice type, test scale, and test arrangements. Hence the test results given in this context should be considered only as examples. However, they do illustrate in a general manner what kinds of effects different factors such as temperature, loading rate, and loading direction have on the mechanical properties of ice.

3.3.2 Elastic Behavior

Ice behaves elastically only when the duration of loading is very short. The elastic properties of ice are best measured dynamically using sonic or seismic methods, for example. The results show little dependence on temperature, crystal structure of the ice, or loading direction. However,

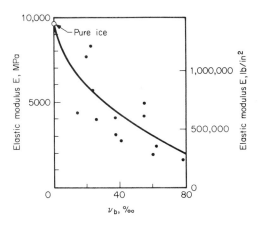

Figure 3.20 Elastic modulus E versus brine volume as determined by seismic measurements. (*Anderson, 1958.*)

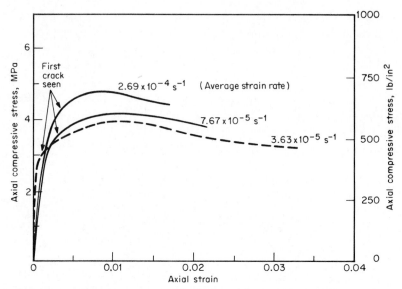

Figure 3.21 Stress-strain curves for low-speed compression tests on cylindrical specimens at $-7 \pm 1\,°C$ ($19.4 \pm 1.8\,°F$). *(Hawkes and Mellor, 1972.)*

brine volume and porosity do have some significance (Fig. 3.20). The measured values for Young's modulus E have been on the order of 10,000 MPa (1450 kips/in^2) for a Poisson ratio v of about 0.33. The shear modulus G is usually computed from the simple relationship of linear elasticity for isotropic material:

$$G = \frac{E}{2(1 + v)} \qquad (3.6)$$

In practice, the loading is seldom very fast, and ice will exhibit some viscoelastic behavior. Such properties can be studied by using constant stress-rate or strain-rate tests (Fig. 3.21). Young's modulus can be approximated from the initial tangent of the curve. Different values for the secant modulus can also be determined from the experimental curves. These values may be useful in practical applications as approximate nominal values for Young's modulus.

The measured values of the initial Young's modulus tend to be smaller than the values measured in dynamic experiments. They also depend on the stress or strain rate, the temperature, and the crystal structure of the ice. The effect of the loading direction is important because of the anisotropic nature of most types of ice. The secant modulus is even more sensitive to these factors. It is also dependent on the stress level and its sign (tension or compression).

3.3.3 Viscoelastic Behavior

If the period of loading to fracture is short, on the order of a few seconds, or if the stresses are low and the loading rate is moderate, it is usually reasonable to assume that ice behaves elastically. When the duration of loading increases, the inelastic behavior of ice must be considered.

The characteristics of the viscoelastic behavior of ice in tension or compression can be conveniently found using usually uniaxial creep (constant stress) or constant-strain-rate tests. These tests can also be used to analyze the long-term behavior of ice under shear stresses. The results depend very strongly on temperature but also on the structure, loading history, porosity, and brine volume of ice and on the loading direction.

Some typical shapes of creep curves for compression are shown in Fig. 3.22. The classical creep phases, primary, secondary, and tertiary creep, are best seen in curve 2. In the primary phase the strain rate decreases until it reaches a constant value in the secondary phase. Finally the strain rate begins to increase and ice fails in the tertiary phase. If the stress level is low, a long time is needed before even the secondary phase is reached (curve 3). On the other hand, at high stress levels secondary creep may disappear and primary creep is followed by tertiary creep (curve 1). The uniaxial creep of ice does not necessarily obey the classical curves. For example, the strain rate of columnar grained S2 ice may vary instead of constantly decreasing in the primary phase (curve 4).

Numerous creep tests have to be pursued at different temperatures and stress levels in order to develop a creep model for a specific ice type (Fig. 3.23). For primary creep ϵ_p it may be a simplified version of the equation

$$\epsilon_p = A(T, \sigma) + B(T, \sigma)t^{n(T,\sigma)} \tag{3.7}$$

For the secondary creep rate, a relation of the Arrhenius type has often been used in the form

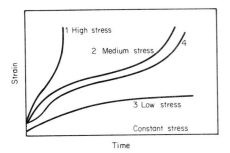

Figure 3.22 Typical creep curves for ice.

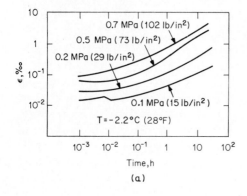

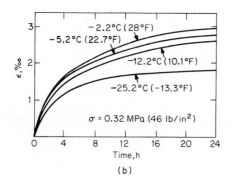

Figure 3.23 Example of the effects of (a) stress and (b) temperature on creep in compression. (*Leppävuori, 1977.*)

$$\dot{\epsilon}_s = A\sigma^n \exp\left(-\frac{Q_c}{RT}\right) \qquad (3.8)$$

where $\dot{\epsilon}_s$ = secondary creep rate
$\quad \sigma$ = applied stress
$\quad A, n$ = constants
$\quad Q_c$ = activation energy
$\quad R$ = Boltzmann's constant
$\quad T$ = temperature, K

When constant-strain-rate tests are used, the stress-strain curves are obtained for different strain rates and temperatures. Some typical curves for uniaxial compression are shown in Fig. 3.24. At relatively slow strain rates the stress increases until it reaches a maximum and then decreases to a more or less constant value. If the strain rate is very slow, there will be no obvious maximum point at the curve, and if the strain rate is fast, ice begins to fail at maximum stress.

The effect of the crystal structure on the behavior of ice is illustrated in Fig. 3.25. If the crystal structure and orientation are very regular and the

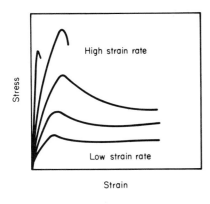

High strain rate

Stress

Low strain rate

Strain

Figure 3.24 Typical compressive stress-strain curves.

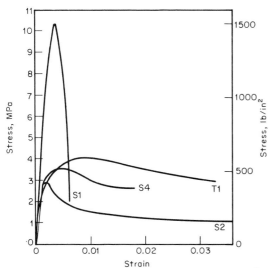

Figure 3.25 Compressive stress-strain curves for various ice types with $T = -9.5°C$ (14.9°F) and $\dot{\epsilon} = 1.67 \times 10^{-5}$ s^{-1}. T1 — snow ice; S1 — columnar grained ice with vertical c axis; S2 — columnar grained ice with horizontal c axis; S4 — congealed frazil slush. (*Gold, 1973.*)

loading direction is perpendicular or parallel to the basal planes of ice, there are no significant shear stresses along the basal planes. Ice may behave in a brittle manner even for slow strain rates, as shown for S1 ice.

3.3.4 Strength of Ice

Uniaxial constant-strain-rate tests can also be used to measure the strength of ice. Depending on the strain rate, three different failure zones can be distinguished. When the strain rate is low, plastic flow governs and no specific failure point can be observed. In this ductile zone strength is defined as the maximum stress achieved during the deformation of the sample. When the strain rate is high, ice fails in a brittle manner. In the transition zone between brittle and ductile zones, a failure of ice contains both brittle and plastic features. The ice strength usually reaches its maximum value in this zone.

An example of the effects of temperature and strain rate on the uniaxial compressive strength of columnar grained S2 ice loaded perpendicular to the growth direction of the ice cover is given in Fig. 3.26. Although there are difficulties in measuring the strength of ice at high loading rates, it is believed that the strength does not decrease considerably at strain rates higher than 10^{-2}. Crystal structure, brine volume, and porosity of the ice naturally have a significant influence on the compressive strength. One example of the effect of the loading direction on the ice strength is given in Fig. 3.27.

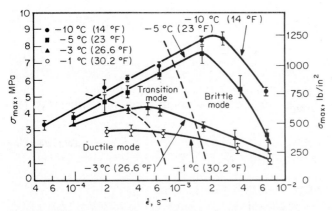

Figure 3.26 Compressive strength of columnar grained S2 ice versus strain rate at various temperatures. *(Adapted from Wu et al., 1976.)*

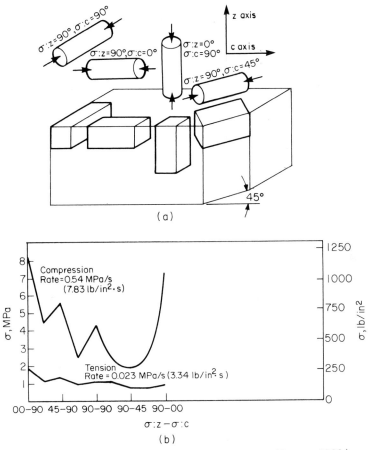

Figure 3.27 Sea ice strength at various orientations. *(Peyton, 1966.)*

At very low strain rates the behavior of ice in tension is very similar to that in compression (Fig. 3.28). However, the formation of the first crack leads to fracture, and that is why ice in tension behaves in a brittle manner at much lower strain rates than in compression. The tensile strength decreases with increasing brine volume. Increasing grain size also appears to have a negative effect on tensile strength. On the other hand, the strength seems to be relatively independent of loading rate and temperature.

The flexural strength of ice is measured with different types of beam experiments. The strength value σ_f is usually defined simply by

$$\sigma_f = \frac{6M}{bh^2} \tag{3.9}$$

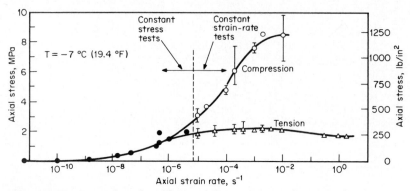

Figure 3.28 Strength and yield stress of snow ice as a function of strain rate. *(Hawkes and Mellor, 1972.)*

where h = thickness
$\quad\;\; b$ = width
$\quad\; M$ = maximum moment of ice beam

In fast laboratory experiments the assumption of linear stress distribution may be quite realistic. However, in nature the ice sheet is usually not in a uniform temperature. If the loading rate is not rapid, the viscoelastic properties of the ice become important. The stress distribution of a cantilever beam in a typical test might look like the one shown in Fig. 3.29. The failure occurs after a crack has formed on the tension side. Some results illustrating the effect of brine volume on the nominal bending strength are given in Fig. 3.30.

Shear stresses are important factors governing the strength of ice under any kind of loading. However, the behavior of ice under direct shear has not been studied extensively, partly because of the difficulties involved with a proper testing method. The shear strength of ice shows a significant dependence on temperature and loading rate. For columnar

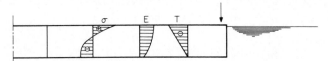

Figure 3.29 Idealized ice temperatures and stress distributions in cantilever beam test.

grained ice the dependence on the loading direction becomes important, as can be seen in Fig. 3.31.

It is hard to measure the adfreeze bond between ice and steel, concrete, or even wood because failure tends to occur through the ice in-

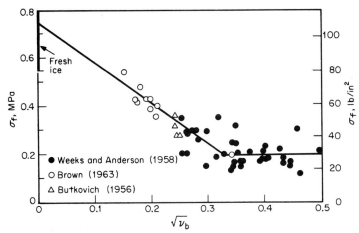

Figure 3.30 Relation between nominal bending strength and square root of ice brine volume as measured by in situ cantilever beam tests. *(Adapted from Weeks and Assur, 1967.)*

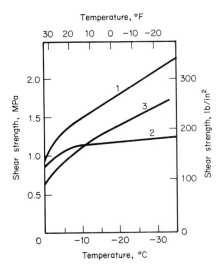

Figure 3.31 Shear strength of ice as a function of temperature. 1—columnar grained ice, loading perpendicular to long direction of grains; 2— columnar grained ice, loading parallel to long direction of grains; 3—snow ice. *(Butkovich, 1954.)*

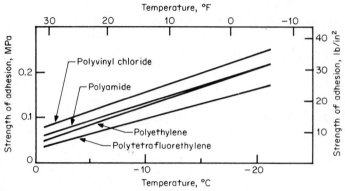

Figure 3.32 Adhesion strength of ice on some plastics as a function of temperature. *(Oksanen, 1980.)*

stead of the contact surface. However, the adfreeze strength of ice on some plastics seems to be considerably lower than the shear strength of ice. Results from laboratory measurements at different temperatures are shown in Fig. 3.32.

3.4 Bearing Capacity of Ice Cover

3.4.1 General Considerations

The ice cover has a considerable bearing capacity, which can be utilized for many kinds of engineering activities. Winter roads and aircraft runways are common examples. The ice cover supports workers and materials and provides a natural platform for various construction purposes, such as excavating, piling, and pipe laying. It can also be used as temporary support in construction or as a storage area. If the bearing capacity of the natural ice cover is not adequate, the ice cover can be thickened and strengthened artificially by flooding techniques and reinforcement. These techniques have been used for heavy transportation, storage, and oil-drilling purposes.

A good knowledge of the characteristics of the ice cover is needed in order to give a proper estimate of the bearing capacity. It is important to select the site of activities in areas where the thickness development of the ice cover is relatively uniform. Areas such as straits and inlets, outlets, and rapidly flowing parts of rivers should be avoided. Because the effective thickness of the ice cover may still vary considerably, careful and regular measurements are needed. At the same time, observations should be made on the ice structure and on a possible formation of open cracks.

In principle, predicting the bearing capacity of the ice cover requires very good information on the temperature- and time-related viscoelastic and strength properties of the ice in a complex numerical analysis. However, the theory of linear elasticity based on nominal values for Young's modulus and flexural strength seems to work reasonably well in most practical cases. Observations on deflections and cracking of a loaded ice cover will naturally make this kind of analysis more reliable. Because there are many unknown factors, conservative assumptions should be used, especially when working safety is to be considered.

3.4.2 Analytical Solutions

If the ice cover is assumed to behave elastically under load, and if the shear deformations are neglected, the governing differential equation for the ice cover is given by

$$\nabla^4 w = \frac{q - kw}{D} \tag{3.10}$$

where D = plate rigidity, $= Eh^3/12(1 - v^2)$
   q = applied distributed load
   k = subgrade reaction, $= 9.81$ kN/m^3 (62.4 lb/ft^3)
   E = Young's modulus
   v = Poisson ratio
   h = thickness of ice cover
   w = deflection

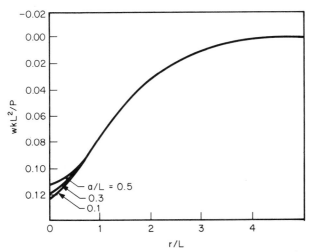

Figure 3.33 Deflection of ice cover under load. $P = q\pi a^2$.

Wyman (1950) has solved the problem for uniform load q distributed over a circular area of radius a (Fig. 3.33). The deflection is given as follows. For $r \leq a$,

$$w(r) = \frac{q}{k} + \frac{qa}{kL}\left[\text{ker}'\left(\frac{a}{L}\right)\text{ber}\left(\frac{r}{L}\right) - \text{kei}'\left(\frac{a}{L}\right)\text{bei}\left(\frac{r}{L}\right)\right]$$

and for $r \geq a$,

$$w(r) = \frac{qa}{kL}\left[\text{ber}'\left(\frac{a}{L}\right)\text{ker}\left(\frac{r}{L}\right) - \text{bei}'\left(\frac{a}{L}\right)\text{kei}\left(\frac{r}{L}\right)\right]$$

$$w_{\text{max}} = \frac{q}{k}\left[1 + \frac{a}{L}\text{ker}'\left(\frac{a}{L}\right)\right] \tag{3.11}$$

where

$$L = \left[\frac{Eh^3}{12k(1-v^2)}\right]^{1/4} \tag{3.12}$$

and ker$'$, kei$'$, ber$'$, and bei$'$ are the first derivatives of the modified Bessel functions ker, kei, ber, and bei. When the ratio a/L is small, the deflection under load can be approximated by

$$w_{\text{max}} = \frac{q\pi a^2}{8kL^2} \tag{3.13}$$

Radial and tangential moments and stresses can be derived from Eq. (3.11). Maximum stress under the load is given by

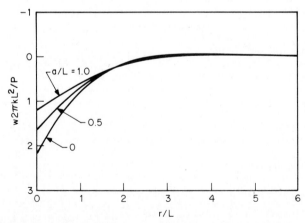

Figure 3.34 Deflection of radially cracked ice cover. (*Nevel, 1961.*)

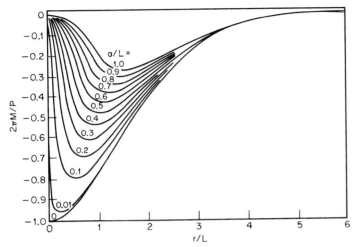

Figure 3.35 Moment distribution in radially cracked ice sheet. (*Nevel, 1961.*)

$$\sigma_{max} = \frac{3qaD \text{ kei}' (a/L)(1 + v)}{kh^2 L^3} \tag{3.14}$$

When the tensile strength of ice has been exceeded, radial cracks form at the bottom of the ice cover. Ice can still carry higher loads. The bearing capacity of the ice cover is considered to have reached its ultimate value when a circumferential crack forms at the top of the radially cracked ice cover. However, due to the interaction of wedges, breakthrough does not necessarily follow immediately after the formation of the first circumferential crack. Actually the breakthrough often resembles more a shearing than a bending type of failure. Nevel (1961) solved the problem of ultimate bearing capacity by studying the behavior of narrow infinite wedges on an elastic foundation. This solution is shown in Figs. 3.34 and 3.35.

The bearing capacity of a semi-infinite ice sheet can also be estimated using Nevel's solution. Panfilov (1960) has given a simple approximate solution for a semi-infinite ice plate loaded by a narrow strip load along the edge. The cracking load is given by

$$P_{cr} = 0.16 \left(1 + 2.30 \frac{b}{L}\right) \sigma_f h^2 \tag{3.15}$$

where b is the length of the load. The ultimate capacity is approximated by

$$P_{\text{max}} = 0.45 \left(1 + 0.84 \, \frac{b}{L} \right) \sigma_f h^2 \qquad (3.16)$$

The ice sheet was found to reach its full bearing capacity when the distance of the load is more than $3.5L$ from the edge. The results for infinite and semi-infinite ice sheets are compared in Fig. 3.36.

If the ice cover is subjected to a very long and narrow strip load p, the maximum bending stress under the load becomes

$$\sigma_{\text{max}} = \frac{3Lp}{\sqrt{2} \, h^2} \qquad (3.17)$$

The corresponding deflection is

$$w = \frac{p}{2 \sqrt{2} \, kL} \qquad (3.18)$$

The bearing capacity of the ice cover is not reached when a crack forms at the bottom of the ice cover. The load can be further increased until

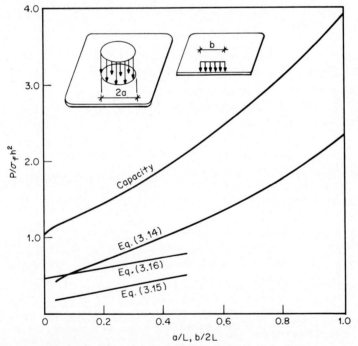

Figure 3.36 Comparison of cracking loads and bearing capacities for infinite and semi-infinite ice sheets.

cracks form at the top of the ice cover at a distance of about $1.1L$. The bearing capacity is given according to Meyerhof (1962) by

$$p_{max} = \frac{1.03\,\sigma_f h^2}{\sqrt{2}\,L - 0.75a} \tag{3.19}$$

where the width of the strip load is $2a$ and $a/L \ll 1$. The deflection can be approximated by

$$w = \frac{p}{\sqrt{2}\,kL} \tag{3.20}$$

In practice the loading condition is often more complicated than those described here. The loading may consist of several distributed loads with various forms, and the vicinity of structures or shores may have some effect on the maximum stress in the ice cover. General-purpose computer programs can be used to solve the stress distributions in this kind of situation.

Because the theory of thin plates is used, the maximum stress in uncracked ice sheet is overestimated when the radius a of the loaded area is small. On the other hand, the possibility of punching through failure becomes evident for highly concentrated loads, especially if the ice temperature is high and the loading static.

3.4.3 Moving Loads

The formation of cracks provides a convenient criterion for the allowable load when slowly moving vehicles strain the ice cover. If the probability of crack formation at the bottom of the ice cover is small, repetition of loads does not extend the cracks, divide the ice cover into small fractions, and thus diminish the bearing capacity. The factor of safety will be more than 3 for a solid ice cover and more than 1 in the vicinity of open cracks that do not transfer shear.

According to various observations in the field, the nominal short-term flexural strength σ_f in Fig. 3.36 for good-quality freshwater ice is on the order of 1.5 to 2.5 MPa (200 to 350 lb/in^2). The nominal shear strength for rapid loading conditions may be about 0.5 MPa (70 lb/in^2) or more, and Young's modulus is about 6000 MPa (870,000 lb/in^2). The load restrictions for Finnish public ice roads are shown in Fig. 3.37. The curves correspond to ice stresses of about 0.7 MPa (100 lb/in^2) or less for an uncracked ice sheet, and they are in relatively good agreement with the commonly used North American criterion $P = 0.35h^2$ (Gold, 1971), where h is in meters and P in meganewtons. (If pounds and inches are used, the multiplier is 50.) Finnish ice roads are monitored for ice qual-

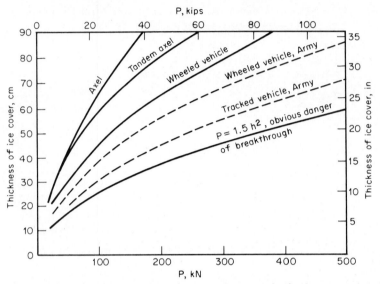

Figure 3.37 Weight restrictions for Finnish ice roads (freshwater ice).

ity, ice thickness, and cracking. Vehicle speeds are restricted, heavy vehicles are sparse, and further load restrictions are set in the fall and during prolonged periods of thawing temperatures. In order to prevent interaction between stress fields or the formation of a failure mechanism corresponding to strip loading, the distance between vehicles should be adequate, on the order of $5L$ or more.

In certain cases, considerably higher loads are allowed to act on the ice cover. Small factors of safety against breakthrough are used for example in military standards for emergency operations. In such cases remote control of vehicles is recommended. Ice strength and stresses can be determined more accurately and the factor of safety can be selected according to the seriousness of failure for special transportation purposes. The procedure may include numerical analysis for the correct loading condition (Nevel, 1978); observations of the ice thickness, quality, and cracking; and in situ loading experiments.

There are several factors that should be understood in order to use the ice cover effectively and safely for transportation purposes. First, a proper value for effective ice thickness has to be determined. The effective thickness is usually taken as the thickness of a uniform good-quality columnar grained ice layer (black ice). Parts, but not more than half, of the total thickness of other layers may contribute to the effective thickness if they are dense good-quality ice and well attached to the primary layer. It is possible to make exceptions to this rule for strongly layered ice covers using the theory of composite plates. However, the results should be verified with careful field experiments.

The bearing capacity of the ice cover depends also in many cases on its temperature. This is quite obvious in the case of sea ice since its brine volume increases strongly with temperature (Fig. 3.9). The flexural and shear strengths of sea ice seem to drop linearly about 2.5% for an increase of 1% in the square root of the brine volume (Fig. 3.30; Paige and Lee, 1966).

At temperatures close to and above 0°C (32°F), ice begins to lose its strength rapidly. Experience, examination of local conditions, and good engineering judgment are the best tools in determining the bearing capacity in this situation. A 10% reduction of the bearing capacity for each degree-day of temperatures above −1°C (30°F) can be used as a rule of thumb. Even more severe restrictions are required for saline ice as well as for static loads. One should be especially careful with layered ice covers. The boundary between the layers very easily loses its capacity to transfer shear and the layers begin to act separately.

Finally, although the strength of the ice cover increases with decreasing temperature, ice also becomes brittle. Cracks and internal stresses may be induced into the ice cover as a result of steep temperature drops. In this sense low temperatures may also have negative effects on the bearing capacity.

In the spring solar radiation weakens the surface layer of the ice cover. The effective thickness decreases and rotten ice becomes bumpy under traffic. It is possible to expand the operational time of ice roads slightly

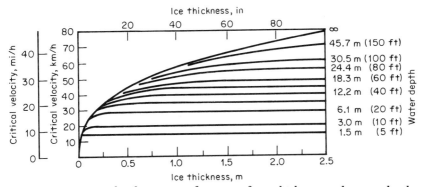

Figure 3.38 Critical velocity as a function of ice thickness and water depth. (*Nevel, 1970.*)

by protecting the road surface from solar radiation. Insulation or reflective materials such as sawdust and snow can be used.

The ice cover can be assumed to behave elastically under loads moving with velocities on the order of 1 km/h (0.6 mi/h) or more. The strength of the ice does not depend very much on the loading rate. However, speed limits are needed for roads in order to avoid the formation of a hydrodynamic wave that is in resonance with the load-induced deflection of the ice cover. Nevel's (1970) result for the critical velocity as a function of water depth and ice thickness is shown in Fig. 3.38.

Observations have shown that the deflection caused by a load moving at constant speed equal to the critical velocity is about two to three times the value at slow velocity (Fig. 3.39), and the amplification of maximum stress is even higher. In order to avoid the hydrodynamic magnification of stresses in the ice cover, the speed limit should be 40% or less of the critical velocity. If the velocity of the vehicle is at a critical range only momentarily, the situation is not as severe as in the case of constant critical velocity. When approaching the shore, especially low velocities

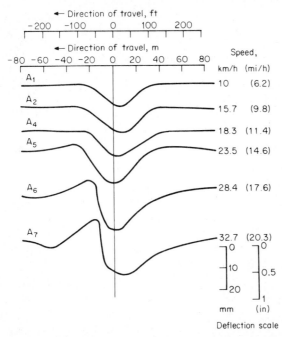

Figure 3.39 Deflection of 45- to 50-cm (18- to 20-in)-thick ice cover under a 7.7-ton vehicle. Water depth 5 to 10 m (15 to 30 ft); measurement point 6 m (20 ft) from loading path. *(Sundberg-Falkemark, 1963.)*

are needed in order to prevent the breaking of shore ice due to hydro-dynamic waves.

The effect of cracks on the bearing capacity of ice depends on their ability to transfer bending and shear. Single hairline cracks do not seem to have great significance, but open cracks more than a few millimeters wide may considerably reduce the bearing capacity, especially if water has been able to penetrate into the crack. This kind of wet crack may not be able to transfer any shear, and the allowable load should be reduced to that of a semi-infinite ice sheet. At the crossings of wet cracks the bearing capacity is further reduced. It is important to examine carefully the refreezing of the cracks before bearing capacity values for uncracked ice sheet are used.

The ice cover is often subjected to constant cracking along shores due to water-level fluctuations. Bridging the route to the solid ice cover improves the efficiency and safety of transportation. It also helps the situation in the spring because softening of the ice cover begins near the shoreline. Careful consideration and analysis are needed if the distance between shores is less than $8L$ (Nevel, 1978). Shores may induce additional stresses on the ice cover due to water-level fluctuations. The width of the uncracked ice cover may be less than $5L$. These effects may reduce the bearing capacity of ice significantly.

The ice cover may lose part of its bearing capacity if loading is repeated frequently. These fatigue effects become especially important if the intensity of loading is close to the cracking limit. It is good engineering practice to change occasionally the course of an ice road subjected to heavy traffic and give the ice cover time to recover from possible fatigue damage.

3.4.4 Behavior of the Ice Cover under Static Loads

The behavior of the ice cover under static loads is governed by the viscoelastic properties of ice, which depend strongly on the ice type and temperature. After the initial deflection, ice continues to deform, reaching a more or less constant rate, as shown in Fig. 3.40. The bearing capacity of ice decreases as the loading time increases. One example of this temperature-dependent relationship is shown in Fig. 3.41. For concentrated loads, the probability of shear failure increases with time, especially when the ice temperature is close to or at the thawing point. The necessity of parking restrictions on ice roads is thus obvious.

It is possible to give a design value for static load by estimating the failure load for given conditions and time, and dividing the critical load by a factor of safety. However, in many cases a more convenient criterion is to limit the maximum deflection of the ice sheet to the value of the

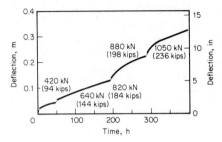

Figure 3.40 Deflection of ice cover under gradually filling pool. Ice thickness 1.33 m (52 in); loading radius 3.5 m (11 ft 6 in). *(Kingery, 1962.)*

freeboard (that is, the distance between the water table and the surface of the free-floating ice cover). Several reasons favor this criterion, although the deflection and time may be only fractions of those causing breakthrough. If the freeboard is exceeded, water may flood the ice cover through cracks and openings. This warms and weakens the ice and eliminates part of the buoyancy forces, making ice behavior more unpredictable. Flood water and its freezing may also be harmful to operations on ice. Finally, when using maximum deflection criteria, it is easy to compare the behavior of the ice to that predicted and to take the necessary steps to avoid breakthrough well in advance.

The behavior of ice under static loads could be analyzed using general-purpose finite element programs (Masterson et al., 1979). Unfortunately the viscoelastic properties of ice covers in nature are not yet known well enough for a reliable analysis in a general case. Frederking and Gold (1976) developed a semiempirical relationship between the deflection rate w and the initial maximum stress σ_f under load, as shown in Fig. 3.42.

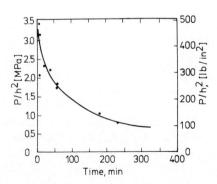

Figure 3.41 Bearing capacity of ice as a function of loading time. *(Assur, 1961.)*

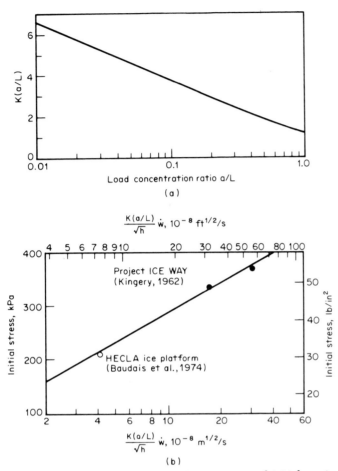

Figure 3.42 Relation between deflection rate and initial maximum bending stress of ice cover. (a) $K(a/L)$ as a function of load concentration ratio. (b) Design curve with reference points. (*Frederking and Gold, 1976.*)

The total deflection w is given by

$$w = w_0 + \dot{w}t \tag{3.21}$$

where w_0 is the sum of initial deflection and delayed elasticity.

Figures 3.41 and 3.42 should give reasonable estimates of the behavior of cold good-quality ice. However, the long-term bearing capacity decreases very rapidly when the air temperature approaches or exceeds the thawing limit (Fig. 3.43). The behavior of the ice cover should be monitored closely under such conditions.

(a)

(b)

Figure 3.43 Ice has failed under parked vehicle and lifting operation is in progress. (*a*) Supporting structure. (*b*) Lifting of machine. (*Courtesy of U. V. Laakso & Co.*)

3.4.5   Methods to Thicken and Strengthen the Ice Cover Artificially

The natural growth of the ice cover does not always satisfy engineering needs. However, the growth rate can be greatly increased by artificial flooding techniques. Layers of water of a thickness of up to about 10 cm (4 in) are pumped or sprayed on the ice cover. Each layer is allowed to freeze, cool, gain strength, and attach well onto the ice cover before the next flooding is applied. Ice platforms up to 6 m (20 ft) thick have been constructed using this technique (Fig. 3.44).

Ice core samples and small-scale strength measurements are necessary to confirm that the built-up ice layer will behave in a favorable way. Flooding with seawater causes problems because the salinity of built-up ice will be much higher than that of naturally formed ice. In warm temperatures, the brine content of the built-up layer will be high while the strength will be lowered. The ice buildup rates can be easily computed or estimated using the data of Adams et al. (1963) shown in Fig. 3.45. For better results an additional cooling period is required after each flooding.

Ice covers are commonly thickened in order to build up the capacity needed for special transportation purposes and for storage and parking areas. Conventional methods to estimate the bearing capacity should be used only if the ice cover is thickened over a large area around the load. In case of a concentrated load the necessary distance from the loading to the edge of the thickened ice cover may be on the order of $3L$. An example of the effect of the width of an ice bridge on the capacity is

Figure 3.44 Artificial thickening of ice cover by surface flooding. (*Courtesy of Fenco Consultants Ltd.*)

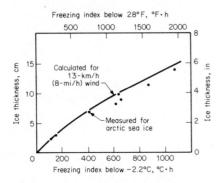

Figure 3.45 Rate of ice formation by surface flooding with seawater. *(Adams et al., 1963.)*

shown in Fig. 3.46. When the bridge is very narrow, the theory of a beam on an elastic foundation can be applied. If the bridge becomes wider, the solution approaches that for an infinite plate on an elastic foundation.

Artificially thickened floating ice pads have also been used as drilling platforms in the Canadian arctic islands (Baudais et al., 1974; Masterson et al., 1979). The principle is shown in Fig. 3.47. This kind of drilling platform can be used only in stable ice conditions because the allowed lateral movement in drilling operations is only 5% of the total water depth. The latest development is to use urethane foam blocks within the ice pads (Maclean et al., 1981). Considering the weight of urethane,

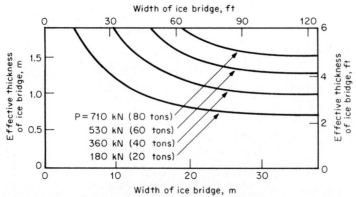

Figure 3.46 Effect of the width of an artificially thickened ice bridge on the bearing capacity. Curves are for freshwater ice, 1.4 MPa (200 lb/in²) bending strength. For crawler tractor-trailer combination, total weight P, the safety factor is 3.0; for cracked condition it is 1.0. *(Kivisild et al., 1975.)*

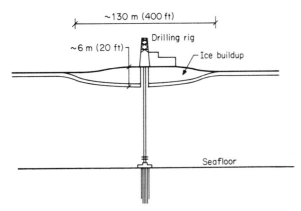

Figure 3.47 Principle of floating drilling platform constructed using artificial thickening of ice cover.

about 30 kg/m³ (1.87 lb/ft³), the load-bearing capacity can be increased significantly, and the pad construction time can also be reduced.

Reinforcement of steel, wood, or fiberglass is sometimes used to strengthen built-up ice covers. Tests have shown (Karri, 1979; Ohstrom and DenHartog, 1976) that it is possible to increase the short-term flexural strength of reinforced ice beams by at least three times compared to the strength of nonreinforced ice beams of the same dimensions. Furthermore, moderate reinforcement seems to tie the ice pieces together, thus preventing a sudden brittle breakthrough. The failure usually occurs slowly as a result of yielding of ice in compression. However, because the shear strength and the fatigue behavior of a cracked reinforced ice sheet are not well known, the design load is usually based on the behavior of an uncracked ice sheet, although a small factor of safety against major cracking is used. Capacity analysis in the cracked stage should be used as a check.

3.5   Ice Forces on Structures

3.5.1   Vertical Ice Forces Due to Water-Level Fluctuations

When the water level changes, vertical loads are exerted on structures that are frozen fast to the ice cover. These loads can be critical to many types of hydraulic structures such as piles, piers, caissons, and sheet pilings. Local damage may also be caused to structures such as water intakes, lockports, and rivetments.

The analysis for estimating vertical ice loads on structures is quite similar to determining the bearing capacity of the ice cover. In this case the loading is a line load along the edge of the ice sheet. Furthermore, the slope of deformation is zero along the contact surface between ice and the structure before the formation of cracks.

The vertical ice force on a cylindrical structure corresponding to a water-level change w is according to Kerr (1975), given by

$$P = \frac{2\pi a D w}{L^3} \left\{ \frac{[\text{kei}'\,(a/L)]^2 + [\text{ker}'\,(a/L)]^2}{\text{kei}\,(a/L)\,\text{ker}'\,(a/L) - \text{kei}'\,(a/L)\,\text{ker}\,(a/L)} \right\} \quad (3.22)$$

where a = radius of structure
$\quad\quad L$ = characteristic length, Eq. (3.12)
$\quad\quad D$ = plate rigidity, Eq. (3.10)

The maximum force corresponding to the bending strength of ice σ_f is, after Wortley (1978), given by

$$P = \frac{1}{3}\,\pi\sigma_f h^2 \left(\frac{a}{L}\right) \frac{[\text{kei}'\,(a/L)]^2 + [\text{ker}'\,(a/L)]^2}{\text{kei}\,(a/L)\,\text{kei}'\,(a/L) + \text{ker}\,(a/L)\,\text{ker}'\,(a/L)} \quad (3.23)$$

Some numerical examples for the uncracked condition are shown in Figs. 3.48 and 3.49.

It must be emphasized that the formation of the circumferential crack in the vicinity of the structure does not necessarily correspond to the maximum vertical load on the structure. If the crack can transfer shear, the load may increase to the value corresponding to the formation of a circumferential crack at some distance from the structure. In that case, Nevel's (1961) solution can be used to estimate the ultimate load and the corresponding deflection (Figs. 3.34 and 3.35).

The nominal values of Young's modulus and the bending strength of ice depend on the rate and amplitude of water-level fluctuations and on the ice temperature. Only relatively small changes in water level can cause some cracking in the vicinity of the structure. These water-level changes may occur very fast in larger lakes and near the seashore. Values on the order of 1.4 MPa (200 lb/in²) for maximum flexural strength and 5000 MPa (700,000 lb/in²) for Young's modulus are commonly used. Large changes in water level are necessary in order to form the ultimate failure mechanism of the ice cover. As in the case of determining the bearing capacity of the ice cover, the nominal values for Young's modulus and flexural strength decrease considerably with the duration of loading. If the formation of the ultimate failure mechanism takes several hours, the proper value for the nominal Young's modulus may be on the order of 1000 MPa (150,000 lb/in²) and for flexural strength about 7 MPa (1000 lb/in²).

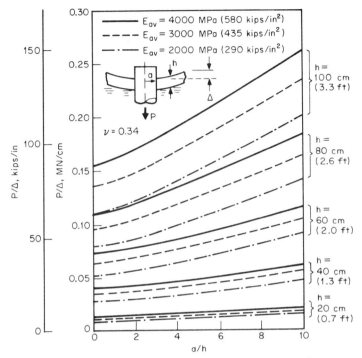

Figure 3.48 Uplift force on isolated cylindrical structure for a unit rise in water level. *(Adapted from Kerr, 1975.)*

Large and constantly occurring water-level fluctuations may create weak failure zones in the vicinity of structures. This naturally eliminates the possibility of very large vertical ice forces.

The shear strength or adhesion of the ice cover may also sometimes limit the maximum vertical load on the structure. In rapid loading conditions, a safe estimate for nominal shear strength of lake ice may be about 0.8 MPa (120 lb/in^2). However, the formation of a circumferential crack in the vicinity of the structure with the application of a very slow shear deformation rate may reduce the nominal shear strength of the ice cover to 0.2 MPa (30 lb/in^2) or less (Frederking, 1979).

The ice adhesion on certain plastics is much lower than the shear strength of the ice. The Finnish Board of Navigation has measured field ice adhesion values in the range of 0.012 to 0.04 MPa (2 to 6 lb/in^2) with an average value of 0.02 MPa (3 lb/in^2) between low-salinity sea ice and polyethylene tubes [$d = 0.5$ m (20 in)] filled with expanded polystyrene (Alaluusua, 1980). The ice thickness was between 0.1 and 0.4 m (4 and 16 in), and the average ice temperature varied between -1.3 and

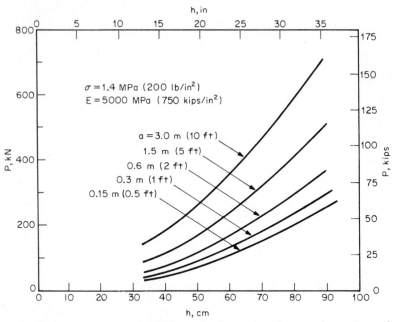

Figure 3.49 Vertical ice force corresponding to cracking in the vicinity of structure as a function of ice thickness. *(Wortley, 1978.)*

$-5.2\,^{\circ}$C (30 and 23$\,^{\circ}$F). The force required to pull concrete-filled tubes out from the ice was found to be more than twice the force required for those filled with polystyrene. This is obviously due to the thermal conductivity of concrete. The ice is thicker in the vicinity of a tube filled with concrete, and the average temperature of the contact surface is lower.

The thermal conductivity of structures may also force bending cracks to form in thinner ice at some distance from the contact surface. The effective structural radius of narrow steel or concrete piles may be about twice the radius of the pile. This should be taken into account when ice forces are estimated.

Vertical ice forces on a straight wall can be calculated using formulas for the bearing capacity of the ice cover subjected to strip loading. In this case, the load is half the bearing capacity because the ice sheet is semi-infinite. If w represents a small change in the water level, the corresponding load can be approximated from

$$p = \sqrt{2}\,kLw \qquad (3.24)$$

and the corresponding nominal bending stress near the surface of the structure is

$$\sigma_f = \frac{3\sqrt{2}\,Lp}{h^2} \tag{3.25}$$

When the flexural strength is exceeded, a crack forms near the wall. The crack may still carry shear. In this case the load and the maximum stress are governed by

$$p = \frac{1}{\sqrt{2}}\,kLw \tag{3.26}$$

$$\sigma_f = \frac{2.7\,pL}{h^2} \tag{3.27}$$

until a crack forms at a distance $1.1L$ from the wall, creating the ultimate failure mechanism. The effects of temperature and the rate of water-level change on the strength and Young's modulus of the ice should also be considered in this case.

The simple formulas given to estimate vertical ice forces on isolated circular structures and straight walls are valid when there are no severe disturbances in the stress field of the ice cover. Usually the distance from an isolated structure to a restraint should be at least on the order of $3L$, and the distance between channel walls should be greater than $5L$. In many cases these requirements are not fulfilled. The design loads may, however, be reduced significantly if the group action of structures is considered.

If the water-level fluctuations are so small that no cracking occurs in the ice cover, the group action of individual piles can be analyzed using the solutions given in Kerr (1978). More general cases can be solved by computer. The water level is increased step by step. When the flexural strength of ice is exceeded, cracks are simulated by hinges until the failure mechanism has been created. If the water-level fluctuations are very large, such an analysis may give even higher vertical ice forces than those for isolated structures. However, the development of high force values is in this case generally avoided due to water penetrating to the surface, which eliminates a large part of the buoyancy (Eranti, 1978).

The vertical loads for rectangular structures and groups of piles are often estimated by using the theory for circular structures with equivalent diameter $d = \sqrt{ab}$, where a and b are the sides of the rectangle. If the structure is long, the sides may be treated as straight walls and the ends as semicircles. This same philosophy may be useful for estimating the vertical loads on pile rows supporting a dock or a platform. If the piles are relatively close to each other, the row can be treated as a wall and the pile load is approximated by $p = bp$, where b is the distance between piles and p is the load per unit length of wall. A slight addition to the load is required due to the buoyancy of ice between the pile rows. However, the

Figure 3.50 Thick ice buildup on piles in harbor of Anchorage, Alaska. *(Courtesy of TAMS Engineers.)*

piles at the end of the dock and at the corners of platforms are subjected to larger forces.

Sometimes ice formations attached to structures cause problems connected directly to their weight, the falling of ice pieces, or to operational aspects. The buildup of ice on piles reaches sizable dimensions in harbors, where tidal water-level changes are very large (Fig. 3.50). The piles are strained by the weight of the ice when the water level is high. Nonvertical piles are especially vulnerable to damage. This kind of ice formation should also be considered in earthquake design (Peyton, 1968; Perdichizzi and Yasuda, 1978).

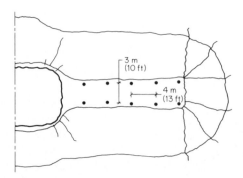

Figure 3.51 Piling of pier and assumed yield lines.

EXAMPLE 3.1 Find design values for vertical ice forces on the piles of the pier shown in Fig. 3.51. Relatively large and rapid water-level fluctuations are assumed to occur, but the ice cover is not assumed to be weakened by an ice failure zone. The design values are: ice strength 0.8 MPa (120 lb/in²), Young's modulus 2000 MPa (300,000 lb/in²), Poisson ratio 0.33, and ice thickness 0.6 m (2 ft). The piles have a radius of 0.15 m (6 in).

The characteristic length of the ice cover [Eq. (3.12)] is given by

$$L = \left[\frac{2000(0.6)^3}{12(0.0098)(1 - 0.33^2)} \right]^{1/4} = 8.0 \text{ m (26 ft)}$$

The maximum ice force on a straight wall is, from Eq. (3.27),

$$p = \frac{0.8(0.6)^2}{2.7(8.0)} = 0.0133 \text{ MN/m (0.91 kip/ft)}$$

and the corresponding change in water level according to Eq. (3.26) is

$$w = \frac{\sqrt{2}(0.0133)}{0.0098(8.0)} = 0.24 \text{ m (10 in)}$$

We may treat the end of the pier as a cylindrical structure with a diameter of about 3.4 m (11 ft). Using, for example, Fig. 3.36 with $a/L = 1.7/8.0 = 0.21$, we get

$$P = 1.56(0.8)(0.6)^2 = 0.45 \text{ MN (100 kips)}$$

The corresponding deflection is roughly estimated by (Fig. 3.34)

$$w = \frac{0.45(1.7)}{2\pi(0.0098)(8)^2} = 0.2 \text{ m (8 in)}$$

If we treat the ice cover between the piles as a rigid plate, the maximum vertical load on an ordinary pile in a row can be computed:

$$P = 4(0.0133) + 4(1.7)(0.24)(0.0098)$$
$$= 0.054 + 0.016 = 0.07 \text{ MN/pile (16 kips/pile)}$$

It is interesting to note that if the ice cover had thickened to 1 m (3 ft) between piles because of flooding and if it were totally supported by piles (water-level drop is more than 0.9 m), the vertical load would be

$$P = 4(1.7)(1)(0.0092) = 0.062 \text{ MN/pile (14 kips/pile)}$$

The maximum vertical ice force on the piles at the end of the pier is

$$P = \frac{0.45}{4} + \frac{0.07}{2} = 0.15 \text{ MN/pile (33 kips/pile)}$$

(two piles and only half a cylinder plus a portion of the straight wall load). The shear stress at the surface of the pile is given by

$$\tau = \frac{0.15}{2\pi(0.15)(0.6)} = 0.27 \text{ MPa (40 lb/in}^2)$$

Shear failure does not necessarily result from shear stress of this magnitude, but ice adhesion to certain plastics is lower. Use of plastic covers on wooden piles should be considered here, because the large uplift forces cannot be accommodated easily in the design.

3.5.2 Thermal Ice Pressures

The ice cover tends to expand and contract along with temperature fluctuations. Dams, reservoir walls, and other structures that restrict these lateral ice movements are subjected to thermal ice forces (Fig. 3.52). Because high tensile stresses in the ice cover are eliminated by cracking, only compressive forces are generally considered in structural design. The magnitude of the thermal ice pressure depends on a number

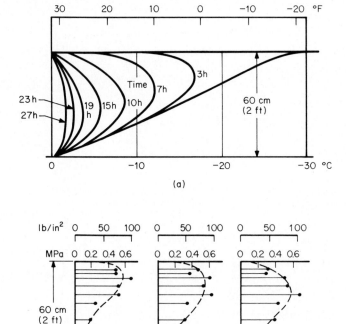

Figure 3.52 Measured ice temperatures (*a*) and pressures (*b*) in ice sheet. If thermal expansion of concrete tank had been eliminated, pressures would have been 25% higher. (*Löfquist, 1954.*)

of factors, including the initial temperature distribution in the ice cover, its rate of change, the viscoelastic properties of ice, the existence of cracks in the ice cover, and the degree of confinement of the ice sheet.

The development of thermal ice pressures has been analyzed in Bergedahl (1978) and Drouin and Michel (1971). The approach of the latter authors is reviewed briefly in the following. The surface temperature of the ice cover is assumed to be equal to the air temperature. The ice temperature is initially assumed to increase linearly from the surface to the ice-water interface, where it is $0\,°C$ ($32\,°F$). If the air temperature is subjected to fluctuation,

$$T_a = \frac{T_0}{2} + \frac{T_0}{2} \cos \frac{2\pi t}{t_0}, \qquad 0 < t < \frac{t_0}{2} \tag{3.28}$$

with T_0 being the initial air temperature and t_0 the period of the cycle, and the theory of a semi-infinite body is applied, the temperature at any depth z of the cover is given by

$$T(z, t) = \frac{T_0}{2}\left[1 + \exp\left(-z\,\sqrt{\frac{\omega}{2d}}\right) \cos\left(\omega t - z\,\sqrt{\frac{\omega}{2d}}\right) \right] \tag{3.29}$$

where

$$\omega = \frac{2\pi}{t_0}$$

and d is the thermal diffusivity of ice, assumed to be $1.16 \times 10^{-6}\ \text{m}^2/\text{s}$ ($1.25 \times 10^{-5}\ \text{ft}^2/\text{s}$).

Differentiating the equation with respect to time gives

$$\frac{\partial T}{\partial t} = \dot{T}(z, t) = -\frac{\pi T_0}{t_0}\left[\exp\left(-z\,\sqrt{\frac{\omega}{2d}}\right) \sin\left(\omega t - z\,\sqrt{\frac{\omega}{2d}}\right) \right] \tag{3.30}$$

On the other hand, strains are related to temperature change by

$$\dot{\epsilon} = \alpha \dot{T} \tag{3.31}$$

The coefficient of thermal expansion of ice decreases slightly with decreasing temperature and is expressed in the form

$$\alpha = (5.4 + 0.018T) \times 10^{-5} \qquad [°C^{-1}] \tag{3.32}$$

When the strain rates at different levels of the ice cover and the rheological model for ice are known, the thermal thrust can be computed numerically step by step. The rheological model of Drouin and Michel (1971) is of the following form:

$$\frac{d\sigma}{dt} = \dot{\epsilon} E_a - 2b\beta E_a \left[\left(\frac{n_0}{\beta} + \dot{\epsilon}t\right) - \frac{\sigma(t)}{E_a}\left(\frac{\sigma(t)}{2p}\right)^m \right] \tag{3.33}$$

where E_a = apparent elastic modulus
$\quad n_0$ = initial number of dislocations
$\quad \beta$ = rate of multiplication of dislocations
$\quad b$ = Burgers vector
$\quad p, m$ = constants

The factors E_a, n_0, β, and p were determined based on uniaxial constant-strain-rate tests. Computed pressure values for columnar grained S1 ice are given in Fig. 3.53.

The temperature profiles given by Eq. (3.5) are reasonably accurate only if the ice cover is thick and the period of the thermal cycle is long. Accurate computation of the thermal profile of the ice cover and its rate of change is complicated. In addition to the air temperature history, factors such as wind speed, cloudiness, solar radiation, vapor pressure, existence of snow, and ice thickness should be considered (Fig. 3.54). Different aspects of thermal profile calculations are discussed thoroughly in Bergedahl (1978).

The rheological model for ice represents another important element in the thermal pressure computations. Leppävuori (1977) computed thermal ice pressures with an ice model of the form

$$\epsilon = a + bt^n \tag{3.34}$$

where the values for factors a, b, and n depend on temperature, stress, and ice type. Some results for a theoretical ice sheet at uniform temperatures are given in Table 3.2. The results illustrate clearly that natural ice

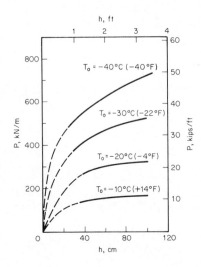

Figure 3.53 Maximum ice thrust for S1 ice, assuming 30-h sinusoidal temperature rise to 0°C (32°F). *(Adapted from Drouin and Michel, 1971.)*

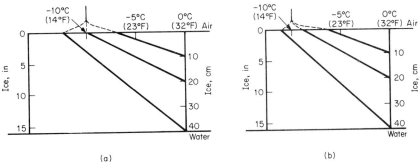

Figure 3.54 Effect of ice thickness and weather conditions on temperatures of snow-free ice cover at night. (*a*) Clear sky. (*b*) Overcast. Air temperature $-10\,°$C ($14\,°$F); wind speed 2 m/s (5 mi/h); vapor pressure 300 Pa (6.3 lb/ft^2). (*Bergedahl, 1978.*)

gives much lower ice pressures than laboratory-made ice. The degree of confinement is also important for the development of thermal ice pressures. Highest pressures are created when the lateral ice movements are completely restricted in all directions. However, if the structures are allowed to deform, much lower ice pressures can be expected.

Only a few attempts have been made to measure thermal ice thrusts in nature. The results of Monfore's (1954) measurements at several reservoirs in the mountains of Colorado are given in Table 3.3. Ice was not covered with snow and had a thickness of about 0.5 m (20 in) when maximum ice forces were measured. According to these measurements, the nature of the shore has great significance with regard to the value of thrust. Flat shores do not appear to restrain the movement of the ice cover as effectively as do steep rocky shores.

The design values for thermal ice thrust commonly used in North America, Scandinavia, and the U.S.S.R. vary between 0.05 and 0.30 MN/m (4 and 20 kips/ft) depending on local meteorological conditions, the degree of restraint to thermal expansion, and the flexibility of structures. The design values as well as the results from Monfore's measurements are considerably lower than theoretical estimates. Some of the reasons for this are listed below.

1. The rheological models for ice are usually based on small-scale experiments using new laboratory ice with regular crystallographic structure. Natural ice contains irregularities and impurities and creates smaller pressures.

2. Existing cracks in the ice cover delay the buildup of pressures.

TABLE 3.2 Pressures of an Ice Sheet Subjected to Uniform Linear Temperature Rise to 0°C (32°F) Computed Using Rheological Models for Natural Saimaa Channel Ice (S) and for Laboratory-Made S2 Ice (L)°

Situation	Ice thickness, m (ft)	Ice model	Initial temperature T, °C (°F)	Warming time t, h	Radius of structure R, m (ft)	Wall thickness, cm (in)	p_{max}, MPa (lb/in²)
Uniaxially constrained ice plate	No effect	S	−20 (−4)	12	No effect	No effect	0.37 (53.7)
		S	−20 (−4)	24			0.30 (43.5)
		S	−10 (+14)	12			0.18 (26.1)
		S	−10 (+14)	24			0.15 (21.7)
		L	−20 (−4)	12			0.59 (85.6)
		L	−20 (−4)	24			0.42 (60.9)
		L	−10 (+14)	12			0.33 (47.9)
		L	−10 (+14)	24			0.24 (34.8)
Ice movement totally constrained, as in a cylindrical or rectangular reservoir with rigid walls	No effect	S	−20 (−4)	12	No effect	No effect	0.65 (94.3)
		S	−20 (−4)	24			0.53 (76.9)
		S	−10 (+14)	12			0.31 (45.0)
		S	−10 (+14)	24			0.26 (37.7)
		L	−20 (−4)	12			1.08 (15.7)
Reservoir with steel walls	0.6 (2.0)	S	−20 (−4)	12	5 (16)	1 (0.4)	0.23 (33.4)
	0.6 (2.0)	S	−20 (−4)	12	10 (33)	1 (0.4)	0.16 (23.9)
	0.6 (2.0)	S	−20 (−4)	12	20 (66)	1 (0.4)	0.11 (16.0)

0.6 (2.0)	L	−20 (−4)	12	5 (16)	1 (0.4)	0.31 (45.0)
0.6 (2.0)	L	−20 (−4)	12	10 (33)	1 (0.4)	0.20 (29.0)
0.6 (2.0)	L	−20 (−4)	12	20 (66)	1 (0.4)	0.13 (18.9)
0.6 (2.0)	S	−20 (−4)	12	10 (33)	0.5 (0.2)	0.09 (13.1)
0.6 (2.0)	S	−20 (−4)	12	10 (33)	3.0 (1.2)	0.35 (50.8)
0.6 (2.0)	L	−20 (−4)	12	10 (33)	0.5 (0.2)	0.10 (14.5)
0.6 (2.0)	L	−20 (−4)	12	10 (33)	3.0 (1.2)	0.53 (76.9)
1.2 (4.0)	S	−20 (−4)	12	10 (33)	1 (0.4)	0.15 (21.8)
0.1 (2.0)	S	−20 (−4)	12	10 (33)	1 (0.4)	0.46 (66.7)
Reservoir frozen along walls 1 (3.3)	S	−20 (−4)	12	5 (16)	1 (0.4)	0.06 (8.70)
1 (3.3)	L	−20 (−4)	12	5 (16)	1 (0.4)	0.12 (17.4)
1 (3.3)	S	−20 (−4)	12	5 (16)	Rigid	0.09 (13.1)
1 (3.3)	L	−20 (−4)	12	5 (16)	Rigid	0.15 (21.8)
3 (10)	S	−20 (−4)	12	10 (33)	1 (0.4)	0.08 (11.6)
3 (10)	L	−20 (−4)	12	10 (33)	1 (0.4)	0.10 (14.5)
3 (10)	S	−20 (−4)	12	10 (33)	Rigid	0.15 (21.8)
3 (10)	L	−20 (−4)	12	10 (33)	Rigid	0.24 (34.8)
6 (20)	S	−20 (−4)	12	10 (33)	1 (0.4)	0.13 (18.9)
6 (20)	L	−20 (−4)	12	10 (33)	1 (0.4)	0.17 (24.7)
6 (20)	S	−20 (−4)	12	10 (33)	Rigid	0.40 (58.0)
6 (20)	L	−20 (−4)	12	10 (33)	Rigid	0.65 (94.6)

* The Poisson ratio was assumed to be $0.5 + 0.01T$.

SOURCE: Leppävuori (1977).

TABLE 3.3 Maximum Thermal Ice Loads in Mountain Reservoirs in Colorado

Reservoir	Type of shore	Year of measurement	Maximum temperature in January/February, °C (°F)	Minimum temperature in January/February, °C (°F)	Maximum load, kN/m (kips/ft)
Eleven Mile	Steep and rocky	1948	+11 (52)	−43 (−45)	234 (16)
		1949	+11 (51)	−34 (−29)	205 (14)
		1950	+12 (54)	−27 (−16)	292 (20)
Antero	Flat and sandy	1951	+19 (66)	−46 (−50)	53 (3.6)
Shadow Mountain	Flat and sandy	1951	+11 (51)	−41 (−41)	85 (6)
Evergreen	Moderately steep	1951	+22 (72)	−36 (−33)	137 (9)
Tarryall	Steep and rocky	1951	+16 (60)	−34 (−30)	249 (17)

SOURCE: Monfore (1954).

3. Snow cover, even thin, eliminates the possibility of large temperature fluctuations in the ice cover.

4. Cases where structures are rigid and the ice cover is fully confined are rare.

In most cases maximum ice thrusts occur in early winter. After the first long cold period the ice cover may have grown substantially in thickness, but it may still be bare and relatively free of cracks. Engineering judgment is needed to select the design value for the ice pressure. High loads may be exerted on rigid structures such as dams. However, the risk of sudden catastrophic failure is small, because the deformations of ice are limited and the pressures decrease as the structure gives.

3.5.3   External Forces on Ice Cover

Ice fields and individual ice floes may be subjected to the action of several external forces, including water and wind friction, wave force, the Coriolis force, the gravitational component, and the hydrodynamic force at the frontal edge of the ice cover. In general the frictional components are most important in practical applications. An understanding of these forces is necessary, not only because they may give direct estimates of the ice forces acting on structures, but also because they are basic elements in the models simulating the dynamics of ice fields and ice floes.

The air friction force exerted on the ice cover is given by

$$\tau_a = \rho_a C_a(z)[v_a(z)]^2 \qquad (3.35)$$

with

$$C(z) = k^2 \left(\ln \frac{z + z_0}{z_0} \right)^{-2} \qquad (3.36)$$

where ρ_a = density of air
$C_a(z)$ = wind drag coefficient specified at height z
$v_a(z)$ = wind velocity at height z above ice surface relative to ice velocity
k = von Karman's constant, considered to be 0.4
z_0 = roughness length of interface

In large and deep water bodies, such as seas and big lakes, the effect of the bottom topography on the characteristics of the surface flow can be neglected. Equation (3.35) can also be used to compute water friction:

$$\tau_w = \rho_w C_w(z)[v_w(z)]^2 \qquad (3.37)$$

TABLE 3.4   Roughness and Drag Coefficients for Ice Cover According to Various Measurements

Source	Description	Roughness length z_0, cm (in)		Drag coefficient C_a, 10 m (33 ft) $\times 10^3$ C_w, 0.5 m (1.6 ft) $\times 10^3$
Banke and Smith (1971), air	Broken ice, 30 cm (1 ft) high over 25% of area	0.23 − 0.57	(0.091 − 0.22)	2.27 − 2.87
Banke and Smith (1973), air	Snow covered	0.0023 − 0.285	(0.00091 − 0.11)	0.95 − 2.38
Banke and Smith (1976), air	Flat ice	0.0017 − 0.03	(0.00067 − 0.012)	0.97 − 1.48
	Hummocked ice	0.0011 − 0.32	(0.00043 − 0.13)	0.85 − 2.47
	Covered with slushy snow	0.0036 − 0.33	(0.0014 − 0.13)	1.02 − 2.49
Johannessen (1970), water	Very rough	8.1 − 10.0	(3.2 − 3.9)	41 − 50
	Rough	1.5 − 4.7	(0.59 − 1.9)	12.8 − 26.6
	Smooth	0.2 − 1.4	(0.079 − 0.55)	5.2 − 12.3

Some measured roughness lengths and drag coefficients are given in Table 3.4.

In the case of rivers and channels, the computation practice for water friction differs from that for large water bodies. Michel (1966) has estimated the water friction force by

$$\tau_w = 1.26 \, \frac{\gamma_w V_0^2 n_1^{3/2} n^{1/2}}{(H_0)^{1/3}} \tag{3.38}$$

with

$$n = \left(\frac{n_1^{3/2} + n_2^{3/2}}{2} \right)^{2/3} \tag{3.39}$$

where γ_w is the unit weight of water.

Equation (3.38) is not dimensionally correct. It requires the use of metric units. (For imperial units the multiplier is 0.57.) The notations are given in Fig. 3.55.

Values for Manning's roughness coefficient n_1 for ice covers vary considerably. Smooth river ice covers and thin slush accumulations may have roughness coefficients n_1 between 0.005 and 0.02, and thick ice accumulations consisting of solid floes, for example during spring breakup, may have values of n_1 up to 0.10 (Michel, 1971). For values of the bottom roughness coefficient n_2, the reader is referred to Ven Te Chow (1959).

The ice cover has a downstream component of weight due to the hydraulic slope, given simply by

$$q = \gamma_i h (1 - e) \sin \alpha \tag{3.40}$$

where γ_i = unit weight of ice
e = porosity of accumulation
α = hydraulic slope

The hydrodynamic force on the frontal edge of the ice cover is given by (Michel, 1971)

$$p = \gamma_w H \left(1 - \frac{H_0}{H} \right)^2 \frac{V_0^2}{2g} \tag{3.41}$$

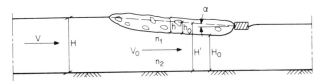

Figure 3.55 Notation for computing external forces on ice accumulation.

When the external forces on the ice floe, ice accumulation, or ice cover have been determined, estimates can be made on the forces transmitted to structures and other objects restricting its movement. Michel (1966) has studied the forces that an ice accumulation exerts on an ice boom using grain elevator theory. The total load on the boom is given by

$$P = B^2(\tau_a + \tau_w + q)\left(\tan\frac{\psi}{6} + \frac{2f}{3B}\right) + pB\exp\left(-\frac{\alpha x}{B}\right)$$
$$+ \frac{(\tau_a + \tau_w + q)B^2}{\alpha}\left[1 - \exp\left(-\frac{\alpha x}{B}\right)\right] \quad (3.42)$$

where $\alpha = \sin 2\psi \tan^2(45° - \psi/2)$
$\quad B =$ width of river
$\quad x =$ length of ice accumulation
$\quad f =$ deflection of boom
$\quad \psi =$ angle of friction between shore and accumulation

Equation (3.42) can also be used to estimate pressures in the accumulation. It can be readily seen that the load does not approach infinity as the length of accumulation increases, because the external forces are transmitted to the shore due to friction and arching effects. Michel's estimate for the angle of friction ψ between the shore and the accumulation is between 15 and 30°. For these values the coefficient α varies very little. If the deflection of the boom is small, the total load caused by a long ice accumulation can be estimated by

$$P = 3.6B^2(\tau_a + \tau_w + q) \quad (3.43)$$

However, there are certain conditions in which the pressures given by Eq. (3.43) do not develop. For shallow rivers with low discharge, the accumulation may thicken to the bottom of the river under the action of high wind drag. When the discharge is considerable, the hydraulic limit corresponding to the accumulation may become critical with a ratio of $h_0/H' = 0.4$. The accumulation then cannot thicken further because the stability limit is reached.

The limiting thrust values for different river widths and depths are conveniently determined by using Fig. 3.56. In zone A the accumulation is grounded for wind speeds of 30 m/s (67 mi/h) and the maximum value of thrust is given by the border between zones A and B. In zone B the discharge is small and wind drag is the dominant factor. The border between zones B and C gives the hydraulic limit.

EXAMPLE 3.2 An ice boom has been constructed across a 300-m (1000-ft)-wide and 10-m (33-ft)-deep river. The average velocity of flow is 0.5 m/s (1.6 ft/s) under the ice cover, the design wind velocity is 20 m/s (47 mi/h), the roughness coefficient for the riverbed and the ice cover is assumed to be

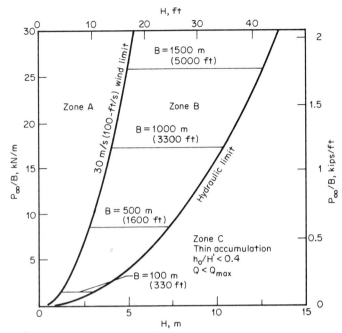

Figure 3.56 Ice thrust on a boom. *(Michel, 1966.)*

$n = 0.05$, and the hydraulic slope is 1.0×10^{-4}. Determine the ice thrust on the boom during the ice-formation period.

Assuming that the ice accumulation is about 2.5 m (8 ft) thick, the water drag [Eq. (3.38)] is

$$\tau_w = 1.26 \frac{9.81(0.5)^2(0.05)^2}{7.5^{1/3}} \approx 4.0 \times 10^{-3} \text{ kPa } (0.084 \text{ lb/ft}^2)$$

The wind drag is given by Eq. (3.35). Assuming that the drag coefficient $C_a = 2.0 \times 10^{-3}$ for slushy ice cover, we get

$$\tau_a = 1.3(2.0)(10^{-3})(20^2) = 1.0 \text{ Pa}$$
$$= 1.0 \times 10^{-3} \text{ kPa } (0.021 \text{ lb/ft}^2)$$

The gravity component of the ice accumulation, assuming an effective unit weight of 6 kN/m³ (38 lb/ft³), according to Eq. (3.40), is

$$q = 6(2.5)(1.0)(10^{-4}) = 1.5 \times 10^{-3} \text{ kPa } (0.031 \text{ lb/ft}^2)$$

The total thrust on the boom for long accumulation is [Eq. (3.43)]

$$p = 3.6(300)(4.0 + 1.0 + 1.5)(10^{-3})$$
$$\approx 7.0 \text{ kN/m } (480 \text{ lb/ft})$$

We can now observe that the result is well within the stable area shown in Fig. 3.56.

The result given by Eq. (3.42) is for ice accumulations in straight rivers and channels. When the degree of irregularity increases, the situation becomes more complicated. For the rivers that do not have straight banks but have relatively constant width, Eq. (3.42) should yield conservative predictions. When the river width increases upstream from the boom, estimates of the force can be made by considering the area of ice accumulation held back before freezing stabilizes the situation (Fig. 3.57).

Equation (3.42) does not necessarily give the worst loading condition on an ice boom or other structures restricting the movement of ice. The impact of large ice floes may cause very high local loads. The structure alone or together with the ice accumulation in front of it must be able to

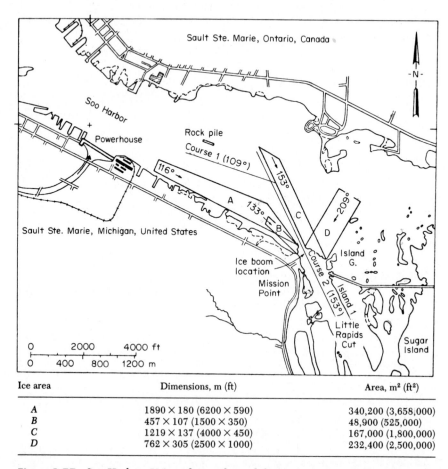

Ice area	Dimensions, m (ft)	Area, m² (ft²)
A	1890 × 180 (6200 × 590)	340,200 (3,658,000)
B	457 × 107 (1500 × 350)	48,900 (525,000)
C	1219 × 137 (4000 × 450)	167,000 (1,800,000)
D	762 × 305 (2500 × 1000)	232,400 (2,500,000)

Figure 3.57 Soo Harbor, U.S. and Canada, and design ice areas. *(Perham, 1977.)*

deform and dissipate the kinetic energy of the incoming floe or to sustain the force corresponding to failure of the ice floe. Ships may also induce stresses in ice accumulation and hence to the supporting structures (Perham, 1977). Furthermore, in restricted channels, ships may cause thickening of the accumulation, thereby increasing the local flow velocity. Finally, in the spring melting may occur first along shores, and the support for the accumulation may be lost. As a result large ice areas may be left supported only by the ice boom.

3.5.4 Static Ice Forces on Isolated Structures Due to Slow Horizontal Movements of the Ice Cover

When the ice cover moves, forces are exerted on fixed hydraulic structures such as lighthouses, linemarks, bridge piers, and artificial islands. In the areas of land-fast ice, movements of strong and thick ice covers are generally slow and limited. Ductile failure of ice is in many cases the governing loading condition. In this case the force can be estimated from

$$F = kmnf\sigma_c Dh \qquad (3.44)$$

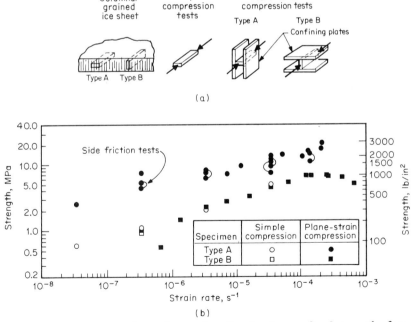

Figure 3.58 Strain-rate dependence of confined and unconfined strength of columnar grained S2 ice at $-10°C$ ($14°F$). *(Frederking, 1977.)*

where k = contact coefficient
$\quad m$ = shape factor
$\quad n$ = aspect ratio factor
$\quad f$ = strength factor
$\quad \sigma_c$ = uniaxial compressive strength of ice in the direction of loading and corresponding to the relevant strain rate
$\quad D$ = width of structure
$\quad h$ = thickness of ice cover

It is not clear which indentation velocity corresponds to the strain rate $\dot{\epsilon}$ in the uniaxial compression test. Korzhavin (1962) has suggested the formula

$$\dot{\epsilon} = \frac{v}{2D} \tag{3.45}$$

where v is the ice indentation velocity and D the width of the structure. Rough estimates of the ice movements can be obtained by estimating deformations of the ice cover under external loads and thermal fluctuations. Local experience and statistical analysis of actual field measurements may provide better results.

The strength factor takes into account the fact that the loading condition is not uniaxial in indentation. The failure zone is more or less laterally confined by the surrounding ice sheet. Frederking (1977) found that columnar grained ice samples were two to five times stronger when

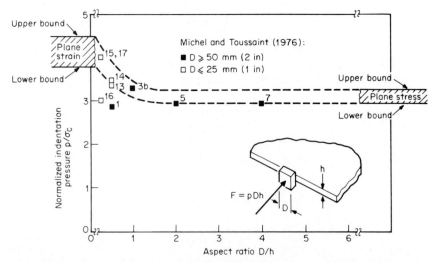

Figure 3.59 Aspect ratio effect based on plastic limit analysis for S2 ice at $-10°C$ $(14°F)$ compared with experimental data. *(Ralston, 1978.)*

lateral deformation was prevented than in normal uniaxial tests (Fig. 3.58). However, the strength of snow ice samples was almost independent of the degree of lateral restraint. Ralston (1978) used uniaxial and confined laboratory test results to determine coefficients of a yield function in limit analysis. He reported that the strength coefficient f is about 3 for columnar grained ice and large aspect ratios D/h (Fig. 3.59). Michel and Toussaint (1976) determined empirically that a value of 2.97 for f gave reasonable agreement between indentation and uniaxial tests for columnar grained laboratory S2 ice over the entire strain-rate range at $-10°C$ ($14°F$). They used $\dot{\epsilon} = v/4D$ as the strain rate.

The contact coefficient k is assumed to be 1.0 for good contact and initial failure. In theory the contact coefficient may be larger than 1.0 for frozen-in condition. For continuous ductile failure the contact coefficient may be somewhat below 1.0, especially at higher loading rates.

The shape of the structure has a certain effect on the failure mode and therefore on the load. Values for the shape factor suggested by Korzhavin (1962) are $m = 1.0$ for a flat indentor, $m = 0.9$ for a round indentor, and $m = 0.85\sqrt{\sin \beta}$ for a wedge-shaped indentor when the front angle 2β is between 60 and 120°.

The aspect ratio h/D does not appear to be a very important factor in ice failure in the ductile range, as shown in Fig. 3.59. The aspect ratio effect for cold ice is in this case about 1.3 for a very narrow structure. Reinicke (1979) has reported that the aspect ratio effect might be larger, up to 2.3 for warm ice and a narrow structure. The result for warm ice is uncertain because confined strength data were not available. However, field tests by Rantamäki (1972) do lend some support to the obtained aspect ratio effect (Fig. 3.60).

Some results of laboratory and in situ indentation tests are shown in Fig. 3.60. The specimens in the laboratory tests were 5- to 10-cm (2- to 4-in)-thick S2 ice sheets at $-10°C$ ($14°F$). In Croasdale's (1974) measurements for arctic sea ice the thickness of ice was 1.0 to 1.5 m (3.3 to 5 ft), and the lowest average ice temperature was about $-10°C$ ($14°F$). The measurements by Croasdale et al. (1976) were for lake ice with surface temperatures between -2.2 and $-17.8°C$ (28 and 0°F) and ice thicknesses of 0.74 to 0.9 m (2.4 to 3 ft). Rantamäki's (1972) measurements were for warm sea ice with average temperatures between 0 and $-2°C$ (32 and 28°F). The salinity of the ice was close to 0.0% and the ice thickness ranged from 0.67 to 0.94 m (2.2 to 3.1 ft).

We can readily see from Fig. 3.60 that field measurements give much lower indentation pressures than those obtained for laboratory S2 ice, although the difference is smaller for purely ductile indentation than for indentation in the transition zone. However, even in a purely ductile failure, the difference can only partially be explained by the shape factor and the possible effects of temperature, salinity, and air content on the

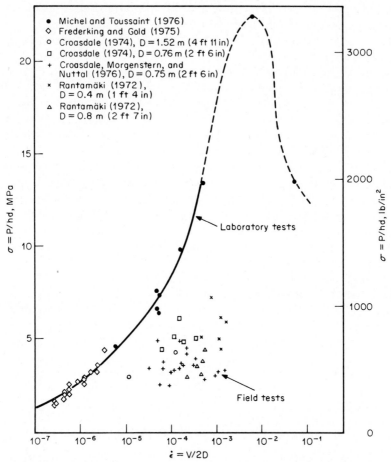

Figure 3.60 Comparison of ice pressures in laboratory and field indentation tests. The initial contact between ice and indentor was good in all measurements.

strength of ice. The temperature and the crystal structure vary at different depths of the ice cover, which is layered due to the uneven freezing process. Layer boundaries are usually weak areas in the ice cover, initiating cracking and failure planes.

These factors suggest that the failure mechanism is quite complicated. It differs from the idealized mechanisms used in the analysis, and it is also different from the failure mechanism of laboratory-grown ice, not only in brittle and transition zones, but also in the ductile zone.

Equation (3.44) can nevertheless be used to estimate the ice forces in nature. However, the value of the coefficient f depends on the type of structure, the temperature of the ice cover, and the strain rate. The theoretical value of 3 for nonoriented columnar grained S2 ice can be used as an upper limit.

Ice samples should be taken from different levels of the ice cover for the uniaxial laboratory tests. The samples should be large when compared to the crystal size. The design ice temperatures are usually high because of the insulative effect of the snow cover. Even in very cold arctic environments design temperatures below $-10\,°C$ ($14\,°F$) are seldom necessary.

Buckling of the ice sheet may reduce the ice force, especially if the ice sheet is thin and the structure is wide. Two theoretical results for the buckling load in an indentation situation are shown in Fig. 3.61. The results for frictionless boundary conditions provide a reasonable upper limit for the buckling load because of the imperfections of the material properties, the load eccentricity effects, and nonlinear ice behavior. The observed buckling values have usually been lower than the ones based on the theory of elasticity (Wang, 1978).

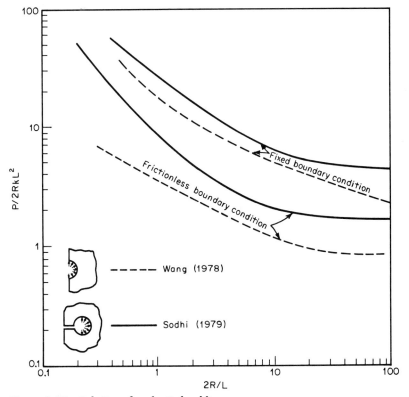

Figure 3.61 Solutions for elastic buckling pressure.

The static ice loads on bridge piers in the direction perpendicular to the shores is often neglected in structural design. Bridge piers are usually quite rigid. However, in large rivers the ice movements result from the freezing of cracks, and thermal expansion may be considerable, as reported in Gold (1966). One failure has been reported in Larsson (1974). In Scandinavia the design ice loads for this case range from 200 to 300 kN/m (14 to 21 kips/ft). For some extreme cases higher design loads may be required.

If the ice movements are not large, the ice loads applied to flexible structures are not necessarily very critical. The design ice load of a Finnish linemark shown in Fig. 3.62 is only 70 kN (15.7 kips). This kind of mark is used in Finnish lakes. The philosophy behind the design is based on the flexibility of the structure and the viscoelastic properties of ice. This essentially calls for the structure to sustain the winter ice movements by deforming elastically. If larger ice movements should occur in the spring, the ice is typically very weak and fails against the structure.

In conclusion, the importance of slow ice movements in the design of hydraulic structures depends not only on the maximum rate of ice movement, but also on the extent of movement and the flexibility of structures. In small lakes and rivers and well-protected harbors, horizontal ice movements are small and structures are likely to be flexible enough so that this loading condition is not critical. However, in the land-fast ice zone at some distance from the shoreline, the flexibility of the structure is less important and the strain rate is the dominant factor in the design.

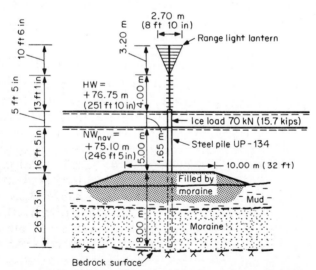

Figure 3.62 Linemark used in Finnish inland waterways. *(Rekonen, 1973.)*

EXAMPLE 3.3 Ice movements of 0.4-m/h (16-in/h) magnitude are known to occur in midwinter in an area of land-fast low-salinity sea ice with design thickness 1 m (3.3 ft) and design temperature $-3\,°\mathrm{C}$ ($27\,°\mathrm{F}$). Estimate the ice force on a lighthouse with a 6-m (20-ft) diameter under these circumstances.

The strain rate is, from Eq. (3.45),

$$\dot{\epsilon} = \frac{0.4}{3600(2)(6)} \approx 1.0 \times 10^{-5}\ \mathrm{s}^{-1}$$

The one-dimensional design strength for these conditions is about 0.7 MPa (100 lb/in²), the strength factor f for this ice, temperature, and strain rate is assumed to be 2, and the shape factor m is 0.9. Because ice is practically in continuous horizontal and vertical movement, frozen-in conditions or even initial failure with perfect contact are not likely to occur. We select a contact coefficient of 0.8. With the aspect ratio factor n close to 1.0, Eq. (3.44) gives

$$P = 0.8(0.9)(1.0)(2)(0.7)(6)(1) = 6\ \mathrm{MN}\ (1350\ \mathrm{kips})$$

In order to calculate the buckling force, we need the characteristic length of ice L. With Young's modulus $E = 4000$ MPa (580,000 lb/in²),

$$L = \left[\frac{4000(1)^3}{12(0.0098)(1-0.33^2)} \right]^{1/4} = 13.9\ \mathrm{m}\ (45\ \mathrm{ft})$$

With $D/L = 0.43$ we get, for a semi-infinite ice sheet with frictionless boundary (Fig. 3.61),

$$P = 5.5(6)(0.0098)(13.9)^2 = 62\ \mathrm{MN}\ (14{,}000\ \mathrm{kips})$$

Buckling is thus not likely to occur and ductile indentation governs the design.

3.5.5 Dynamic Ice Forces on Isolated Narrow Vertical Structures

Drifting ice floes and ice fields may move with considerable speed under the action of environmental forces. Major ice movements typically occur in the fall when the thin ice cover is easily broken and in the spring during ice breakup. In some arctic sea areas strong ice fields are also in motion during winter. The velocity of drifting ice is usually sufficiently high so that ice fails in a brittle manner against a hydraulic structure. The associated forces are generally large and have an uneven, dynamic character.

It may take an ice field with an area of several square kilometers before external forces acting on the field can reach the value of the ice force. However, much smaller ice floes are also able to produce high dynamic ice forces. The kinetic energy of an impacting ice floe is often sufficient to cause full penetration and failure along the entire width of a structure.

Design force Ice indentation failure in the transition or brittle zone is a complicated process, which includes some local crushing, cracking that originates at layer boundaries and stress concentration points, and ice

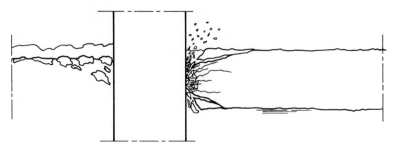

Figure 3.63 Ice crushing against structure.

wedges pealing off from the failure area (Fig. 3.63). As pointed out earlier, a thin ice cover may also fail by buckling. Small ice floes may split into pieces before the full crushing load is reached (Fig. 3.64).

An equation similar to the one used for ductile indentation may also be used to estimate the crushing force. However, the aspect ratio factor n and the contact factor k are often combined into an indentation factor I. The equation for the ice force is given by

$$F = mI\sigma_0 Dh \tag{3.46}$$

Figure 3.64 Ice floe failing against bridge pier by splitting. *(Courtesy of F. D. Haynes.)*

The values for the shape factor m are the same as for ductile indentation. The indentation factor suggested in Afanasev (1972) seems to be in reasonable agreement with laboratory experiments. It is given by

$$I = \begin{cases} \sqrt{5\dfrac{h}{D} + 1} & 6 > \dfrac{D}{h} > 1 \\[2ex] 4 - 1.55\dfrac{D}{h} & \dfrac{D}{h} \leq 1 \end{cases} \tag{3.47}$$

For aspect ratios greater than 6, the indentation factor may become smaller than unity since the ice failure occurs at totally different phases in different zones along the perimeter of the structure.

The nominal strength value σ_0 in Eq. (3.46), unlike the uniaxial strength of ice, seems to depend only slightly on the average temperature of the ice. This is because the better contact for warm and ductile ice eliminates the strength difference that exists with cold and brittle ice. The indentation strength depends somewhat on the strain rate. It reaches the maximum value at the transition zone with $10^{-4} < \dot{\epsilon} < 10^{-2}$, depending on the temperature and other factors. A suitable design value for maximum nominal ice strength σ_0 is in the range of 1.5 to 2.5 MPa (200 to 350 lb/in^2). If ice failure at the critical strain rate and with perfect contact can be anticipated (for example, in earthquake conditions or when a major ice feature impacts a partly consolidated rubble surrounding the structure) the nominal strength values are higher, maybe in the range of 3.0 to 4.0 MPa (400 to 600 lb/in^2). On the other hand, during spring breakup, ice is often very warm and partly deteriorated, and the corresponding design strength should then be low.

The reduction of river ice strength prior to ice breakup has been taken into consideration in most design recommendations. For example, Korzhavin (1962) recommended reducing the ice strength values to about half their maximum values if the ice temperature is at the melting point during ice breakup. Further reduction of the ice force is recommended if the ice floes are small. The Canadian Standards Association's bridge code (1974) states the ice force against a bridge pier can be computed from

$$P = C_n pbh \tag{3.48}$$

where the factor C_n depends on the slope angle α of the frontal edge of the pier. For a vertical edge $C_n = 1.0$; for $75° > \alpha > 60°$, $C_n = 0.75$; and for $60° > \alpha > 45°$, $C_n = 0.5$. The effective ice pressure depends on local ice conditions during river breakup:

$p = 0.7$ MPa (100 lb/in^2) when breakup occurs at melting temperatures and ice is disintegrated

$p = 1.4$ MPa (200 lb/in^2) when breakup occurs at melting temperatures and ice is moving in large internally sound pieces

$p = 2.1$ MPa (300 lb/in^2) when the ice sheet moves as a whole during breakup or large sound ice sheets may hit the pier

$p = 2.8$ MPa (400 lb/in^2) when breakup or major ice movement may occur with ice temperatures well below the melting point

This recommendation does not include the effect of the aspect ratio D/h and should be used with care for very slender structures.

The ice force may also have a transverse component acting on a bridge pier during breakup. This component is often assumed to be about 20% of the force given by Eq. (3.46). It may be even larger for a triangular leading edge and cause significant torsional moments.

Dynamic ice-structure interaction   The design based on the probable maximum value of the ice force has worked reasonably well in the case of traditional massive and rigid hydraulic structures. It has been possible to neglect the dynamic nature of the load because these structures are not sensitive to fatigue and deformations are small. Furthermore the nature of the ice-structure interaction is governed only by the behavior of the ice. The nature of ice force fluctuations is quite random, and the possibility of severe dynamic magnification is small.

In recent years slender and light designs have also been proposed and adopted for ice-covered waters (Fig. 3.65). In this case the behavior of both the ice and the structure contributes to the ice-structure interaction process. The possibility for dynamic magnification exists (Määttänen, 1977), and the response of the structure should also be considered in the serviceability and fatigue design.

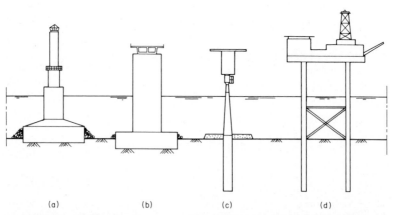

(a)   (b)   (c)   (d)

Figure 3.65   Hydraulic structures subjected to ice forces. (*a*) Traditional reinforced concrete lighthouse. (*b*) Massive bridge pier. (*c*) Modern steel lighthouse. (*d*) Offshore oil-drilling platform.

One approach to the ice-structure interaction problem is to divide the process into two basic phases. In the indentation phase, the structure penetrates into the ice sheet until a random strain-rate-dependent failure strength is reached. Assuming that the penetration depends linearly on the nominal ice pressure, we have

$$[m]\{\ddot{x}\} + [c]\{\dot{x}\} + [k]\{x\} = \{P(t)\}$$

$$x_c(t) = \int_0^t v_{ice}(t)\, dt - \sum_j (P_j + F_j) - \frac{P_c(t)}{mhD} A \tag{3.49}$$

where $[m]$ = mass matrix where hydraulic effects have been included by using the approximate added mass concept
$[c]$ = damping matrix
$[k]$ = stiffness matrix
$\{P(t)\}$ = force vector
$\{x\}$ = nodal point displacement vector
x_c = displacement of contact point
v_{ice} = ice velocity
P_j, F_j = lengths of jth indentation and failed zones
A = constant giving relation between pressure and penetration

and the dot represents time derivative. The first equation is the equation of motion of the structure, while the second equation states that the penetration of the structure into the ice is proportional to the force or pressure. If we solve $P_c(t)$ in the second equation and substitute it into the first, the following results:

$$[m]\{\ddot{x}\} + [c]\{\dot{x}\} + ([k] + [k_{cc}])\{x\} = \{P_0(t)\} + \{P_c(t)\} \tag{3.50}$$

In the force vector $\{P_0(t)\}$ the component with subscript c is zero; in the vector $\{P_c(t)\}$ all components are zero but the one with subscript c, corresponding to the lateral degree of freedom at the contact point between ice and structure. This component is given by

$$P_c(t) = \left[\int_0^t v_{ice}(t)\, dt - \sum_j (P_j + F_j) \right] \frac{mDh}{A} \tag{3.51}$$

Similarly, in matrix k_{cc} all elements are zero except for

$$k_{cc} = \frac{mDh}{A} \tag{3.52}$$

When the critical ice force P_{cr} is reached, ice is assumed to fail in front of the structure within an area with finite length given by

$$F = \frac{P_{cr}}{mDh} B \tag{3.53}$$

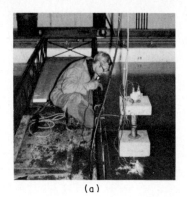

(a)

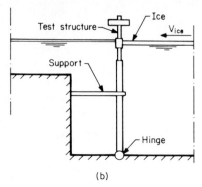

(b)

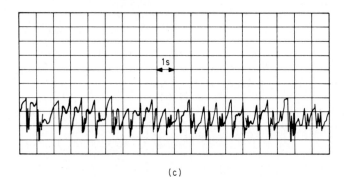

(c)

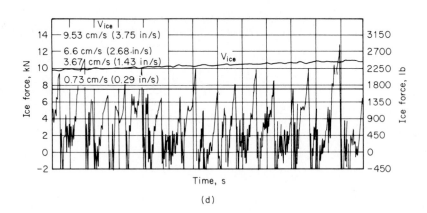

(d)

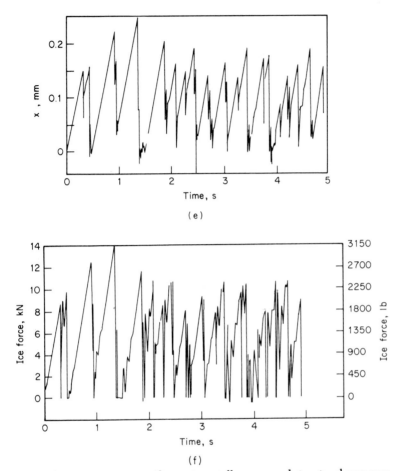

Figure 3.66 Comparison of experimentally measured structural response and computed results using dynamic ice-structure interaction analysis. (*a*), (*b*) Test arrangement. (*c*) Measured support reaction. (*d*) Measured ice force. (*e*) Computed support reaction. (*f*) Computed ice force. (*Eranti et al., 1981.*)

The equation of motion as defined by Eq. (3.49) is now

$$[m]\{\ddot{x}\} + [c]\{\dot{x}\} + [k]\{x\} = \{P_0(t)\} + \{P_r(t)\} \qquad (3.54)$$

where P_r is the small resistance of ice in the failed zone.

Equations (3.50) and (3.54) can be solved conveniently with the aid of mode shape analysis with numerical evaluation of the Duhamel integral by eliminating rotational degrees of freedom using static condensation. A computer program for this purpose is given in Eranti and Lee (1981). Measured and computed results are compared in Fig. 3.66.

In the analysis, the values of constants A and B are the two basic elements. Eranti (1979) found that for indentation the constant A was the order of $0.003h$ [m^3/MN] ($1.4 \times 10^{-4}h$, ft^3/kip) in laboratory indentation tests with natural lake and sea ice plates. This means that the penetration predicted by elastic analysis considerably underestimates the real value because of local crushing and plastic deformations. The values for constant B were approximated to be on the order of $0.006h$ [m^3/MN] ($2.9 \times 10^{-4}h$, ft^3/kip). The tests were pursued with circular indentors, the ice temperature was close to $-10°$C ($14°$F), and the aspect ratios were in the range of $0.7 < D/h < 1.3$. Direct or indirect full-scale field measurements are needed to verify these values for structural design purposes.

The ice strength in relatively stationary failures was found to be log-normally distributed with variance δ^2 on the order of 0.06. For good-quality natural ice, the design value for maximum average strength σ_0 could be up to 1.0 MPa (145 lb/in^2). If necessary, one may further define the dependence of ice strength on loading rate.

Although this analysis is based on some rough assumptions, it provides a rational approach for the design of slender hydraulic structures. The amplitude of vibrations and the possible dynamic magnification effects can be found by increasing the ice velocity slowly over the entire possible velocity range. When the extent of ice movements can be estimated over the design life of the structure, the fatigue design can also be pursued in a proper manner. If the aspect ratio D/h is larger than 6, the analysis can be made by assuming that the ice-structure interaction occurs in independent zones following the outlined mechanism.

3.5.6 Ice Forces on Sloping Structures

In some cases it may be feasible to use a sloping structure in order to reduce the magnitude of ice forces. In theory the mode of ice failure will be bending instead of crushing for slope angles α less than about $70°$. Ice may also fail by shear when impacting a sloping structure, especially in spring when it is partly deteriorated.

The mechanism of ice failure against a sloping structure is illustrated in Fig. 3.67. Horizontal and vertical force components form as some local crushing occurs at the contact point where the ice sheet impacts the structure. The maximum load is reached when ice fails by bending at some distance from the structure. Subsequent failures require additional force to push the ice forward and to remove it from the face of the structure.

Ralston (1977) used plastic limit analysis to estimate the maximum ice force against a conical structure. The results are expressed by

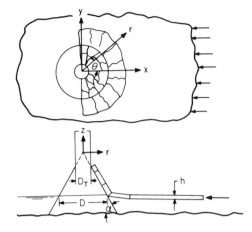

Figure 3.67 Failure mechanism of ice against a sloping structure. *(Ralston, 1977.)*

$$H = [A_1\sigma_f h^2 + A_2\rho_w ghD^2 + A_3\rho_w gh(D^2 - D_T^2)]A_4 \qquad (3.55)$$
$$V = B_1 H + B_2\rho_w gh(D^2 - D_T^2) \qquad\qquad\qquad (3.56)$$

where H = horizontal force
$\quad\quad V$ = vertical force
$\quad\quad D$ = diameter of structure at waterline
$\quad\quad D_T$ = top diameter
$\quad\quad \rho_w g$ = unit weight of water

The dimensionless coefficients A_1, A_2, A_3, A_4, B_1, and B_2 are given in Fig. 3.68 as functions of the appropriate parameters. The effects of both the ice sheet breaking and the broken ice sliding over the surface of the cone are included in Ralston's theory. The first two terms in Eq. (3.55) represent the ice-breaking force. The third term is due to removing broken ice pieces, and it becomes important as the width of the structure increases.

There are no reliable field measurements of the actual ice forces acting on conical structures. Therefore some conservative measures are needed when estimating the design ice force. Ralston's equation seems to be in reasonable agreement with model tests (Kry, 1980). The plastic limit theory and the assumption of instantaneous bending failure along the entire width of the structure are on the safe side. Reasonable values for the design ice force should be obtained by assuming the flexural strength of ice σ_f to be on the order of 1.0 MPa (145 lb/in^2) and taking the value of the kinetic coefficient of friction μ as 0.15 between ice and steel, and 0.25 between ice and concrete. When ice has started to deteriorate or the salinity of ice is high, the flexural strength value can be reduced.

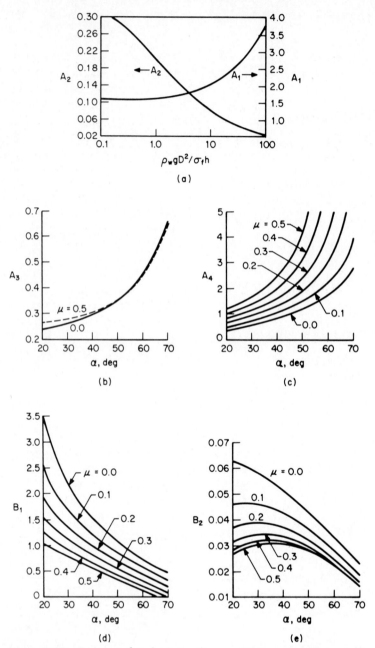

Figure 3.68 Constants for plastic ice-force analysis. μ— coefficient of friction. *(Ralston, 1977.)*

Sloping structures have been used as bridge piers and lighthouses, especially in areas where heavy ice conditions develop during the spring breakup. However, the advantages of the reduced vertical force and the stabilizing horizontal component may be partially lost if an adfreeze bond exists between ice and structure when the ice begins to move. The application of heat at the ice contact zone should be considered in areas where ice movements occur also in the winter.

Proposals have been made to use sloping structures also as arctic off-shore oil-drilling platforms. These structures would have great lateral dimensions at the waterline compared with the ice thickness. Practice has shown that especially ice ridges, but also level ice, tend to form a rubble in front of even narrow conical structures. Careful model tests are required to verify that ice will actually clear away and that the desirable failure mode can be maintained.

EXAMPLE 3.4 Estimate the ice force against a conical lighthouse with $D = 16$ m, $D_r = 4$ m, and cone angle $\alpha = 50°$. The ice thickness is 1 m (39 in), the flexural strength 0.8 MPa (116 lb/in^2), and the friction coefficient between ice and concrete $\mu = 0.25$. Bending failure of ice is assumed.

With $\rho_w gD^2/\sigma_f h = 0.0098(16)^2/0.8(1.0) = 3.1$ in Fig. 3.68, we get $A_1 = 1.7$ and $A_2 = 0.14$. The remaining constants are $A_3 = 0.35$, $A_4 = 2.2$, $B_1 = 0.6$, and $B_2 = 0.031$. The horizontal component is given by Eq. (3.55),

$$H = [1.7(0.8)(1.0)^2 + 0.14(0.0098)(1.0)(16)^2$$
$$+ 0.35(0.0098)(1.0)(16^2 - 4^2)](2.2) = 5.5 \text{ MN (1240 kips)}$$

of which 3.8 MN (850 kips) arise from breaking the ice sheet and the rest from broken ice pieces sliding over the surface. The vertical component is, from Eq. (3.56),

$$V = 0.6(5.5) + 0.031(0.0098)(1.0)(16^2 - 4^2)$$
$$= 3.4 \text{ MN (760 kips)}$$

The horizontal force is only about one-fourth the force corresponding to the crushing failure against a structure of the same diameter at waterline. In Example 3.2 we computed a typical air drag force, with winds averaging 20 m/s (47 mi/h), to about 1 Pa (0.021 lb/in^2). The ice floe must thus have a size of at least 5.5 km^2 (2.1 mi^2) in order to overcome the resistance of the structure. Of course a smaller drifting floe with kinetic energy may also create the same load.

3.5.7 Pressure Ridges

Pressure ridges in drifting ice fields represent in many cases the most severe loading condition in the design of offshore structures (Fig. 3.69). First-year pressure ridges are usually assumed to be largely unconsoli-dated. Ice blocks are held together mostly by buoyancy and friction,

Figure 3.69 Pressure ridge crushing against lighthouse. *(Courtesy of M. Määttänen.)*

although in some cases the blocks may be frozen together loosely. Multi-year ridges develop in polar regions. These ridges are often massive solidly frozen accumulations of broken ice blocks with few cavities and voids.

The knowledge of the strength and failure characteristics of different types of ridges is very limited. The design load is usually given in the form $F_r = AP$, where P is the force corresponding to the failure of the level ice. The values of the constant A range typically from 1.5 to 3.0 for first-year ridges, depending on local experience, ridge size, and the degree of ridge consolidation (Fig. 3.70).

Dolgopolov et al. (1975) estimated the maximum force on a cylindrical structure penetrating through a first-year pressure ridge by

$$F_r = F_{cr} + F_p \tag{3.57}$$

Here F_{cr} corresponds to the force needed to crush the sail and the frozen upper layer of the ridge. This force can be approximated by using Eq.

Design ridge

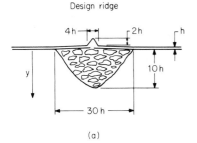

(a)

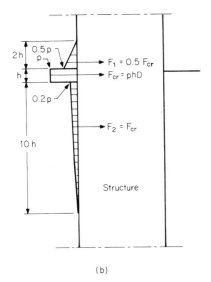

(b)

Figure 3.70 Example of design ice pressure distribution corresponding to interaction between first-year ridge and vertical structure. *(Määttänen et al., 1984.)*

(3.46). One should keep in mind that in certain conditions the thickness of the solidly frozen ice layer in the ridge may be considerably larger than that of the surrounding ice sheet. This thickness exceeded 6 m (20 ft) in one measurement at the Beaufort Sea (McGonical and Wright, 1982). The fast freezing rate can be explained by the low water content of the rubble and by the possible flooding sequences at the surface. The role of force F_p is to push aside the underwater portion of the ice ridge. Dolgopolov et al. (1975) estimated this force by

$$F_p = \mu H_e D (0.5 H_e \mu \gamma_b + 2c) \left(1 + \frac{2H}{3D} \right) \qquad (3.58)$$

where $\mu = \tan(45° + \phi/2)$
ϕ = angle of internal friction in mass of ice blocks
H = thickness of underwater portion of ridge

D = design pier width within boundaries of underwater portion of ridge

H_e = design thickness of underwater portion of ridge; $H < H_e < H + D/2$, because the ridge thickens during the interaction with the structure

γ_b = ice buoyancy

c = cohesion between ice blocks

Internal friction and cohesion between ice blocks are key parameters in Eq. (3.58). These values appear to depend strongly on the ice thickness and other characteristics of the ridge. The value of 30° is commonly used as a first estimate for internal friction. Hudson (1983) has concluded that cohesion in an ice rubble may reach 35 kPa (5 lb/in²) if the ice blocks are very thick.

EXAMPLE 3.5 Estimate the force that a pressure ridge with an underwater portion thickness of 12 m (40 ft) exerts on an 8-m (27-ft)-wide cylindrical structure. The angle of friction during penetration is assumed to be 30°, the average cohesion between blocks is 10 kPa (1.5 lb/in²), ice buoyancy including porosity is 0.7 kN/m³ (4.5 lb/ft³), and the thickness of the level ice is 1 m (3.3 ft).

Assuming that the thickness of the solidly frozen layer in the ridge does not increase considerably compared to the level ice thickness, we get from Eq. (3.46) with indentation factor $I = 1.0$

$$F_{cr} = 0.9(1.0)(1.5)(1.0)(8.0) = 10.8 \text{ MN (2400 kips)}$$

The contribution of the underwater portion of the ridge is given by Eq. (3.58). Assuming that the effective height of the ridge is 14 m (46 ft), we get

$$F_p = \tan(60°)(14)(8)[0.5(14)\tan(60°)(0.7) + 2(10)][1 + (\tfrac{2}{3})(\tfrac{12}{8})]$$
$$= 11{,}000 \text{ kN} = 11 \text{ MN (2500 kips)}$$

In this case the total design force is 22 MN (4900 kips), and the ratio between ridge load and the load corresponding to the crushing of the level ice sheet is close to 2.

When a conical structure penetrates through a first-year pressure ridge, the frozen surface layer of the ridge may fail by bending and the loose underwater blocks can move more easily in both the horizontal and the vertical directions than in the case of a cylindrical structure. This is why ridge forces are considered to be lower against conical structures than against equivalent cylindrical structures (Lavonie, 1966; Dolgopolov et al., 1975) provided that ice does not form a rubble in front of the structure.

A multiyear ridge that impacts a vertical structure may fail by crushing, shearing, or in-plane bending, depending on the width of the structure and on the dimensions and the strength of the ridge (Prodanovic,

1981). The ridge may also come to rest after impacting the structure if the ice behind the ridge cannot sustain the driving force but begins to fail. In each case the loads are high. The average multiyear ridge ice strength naturally varies a lot but may be comparable to the strength of first-year sea ice (Cox et al., 1984). The preferable mode of a multiyear ridge failure might appear to be bending. The analysis by Croasdale (1980) based on the theory of a beam on an elastic foundation could be used as a first estimate for ridge loads against a conical structure.

Small loose ridges in some temperate seas do not have great influence on the design of offshore structures. In exceptional conditions in arctic seas, however, the height of first-year ridges may reach 50 m (160 ft) and that of multiyear ridges 30 m (100 ft). Large ridges pose a major hazard not only to offshore structures but also to seabed installations. Conservatism is needed in estimating loads, because the failure modes and the design parameters of different ridge types are not yet properly known, for example, in a head-on collision.

3.5.8 Ice Forces against Wide Structures

The interaction between moving ice fields and wide structures differs in many respects from that between ice and narrow structures. The crushing failure is nonsimultaneous with irregular contact, which means a considerable reduction in the maximum effective stress (total force divided by structural width). Failing ice tends to form a rubble in front of a wide structure instead of clearing away, and the rubble formation process involves a change in the ice failure mode. Finally in case of a very wide structure, ridge formation in the ice field may limit the driving force below the value required to crush the ice against the structure. While Eq. (3.44) appears to work reasonably well in the case of ductile ice failure against a wide structure (for example, an artificial island in the land-fast ice zone), new approaches are required when wide structures are subjected to loads from moving ice fields.

Multizonal ice failure Kry (1978) has used the concept of zonal ice failure to estimate the forces connected to ice crushing against wide structures. The principle of his approach is illustrated in Fig. 3.71. It is assumed that as ice moves past the structure, effective stresses in each failure zone develop independently of those in adjacent zones. As the number of failure zones increases, the extent of the total effective stress fluctuations decreases.

A statistical analysis of multizonal ice failure is presented in Kry (1979). Assume that the local stress in one zone has a log-normal probability distribution with median $\bar{\sigma}_1$ and geometric standard deviation σ_{g1}.

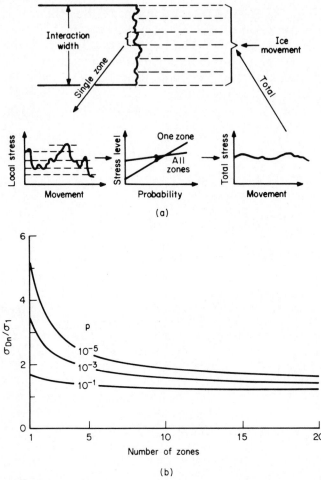

Figure 3.71 Principle of zonal ice failure approach in estimating ice forces on wide structures. *(Kry, 1979.)*

A particular local stress σ_{D1} has an instantaneous probability of exceedance P, given by

$$\sigma_{D1} = \bar{\sigma}_1 \sigma_{g1}^y \tag{3.59}$$

where y is related to P through the probability integral.

If the risk of exceedance of a particular local stress over the life of the structure is a unique function of the instantaneous probability of exceedance of stress and independent of the number of zones, the appropriate design stress σ_{Dn} is given by

$$\frac{\sigma_{Dn}}{\bar{\sigma}_1} = \frac{\exp(\tfrac{1}{2}\ln^2\sigma_{g1})\exp\{y[\ln(1 - 1/n + (1/n)\exp(\ln^2\sigma_{g1}))]^{1/2}\}}{[1 - 1/n + (1/n)\exp(\ln^2\sigma_{g1})]^{1/2}}$$

$$(3.60)$$

where n is the number of failure zones and y corresponds to a particular exceedance risk P. The width of a single failure zone is indicated to be four to five times the ice thickness or greater. When the extent of ice movement is known, it is in principle possible to calculate the effective design stress. Only a few failure zones are required to provide potential for design stress reductions on the order of 50%.

Rubble formation and the associated forces When ice begins to fail against a wide structure, it will not generally clear away. Broken ice accumulates and eventually forms a rubble in front of the structure. Rubble formation is a complicated process, which involves ice failure by bending and ice penetration into the rubble. Some local crushing and buckling may also occur. There are two principal approaches to estimate the ice forces involved in the process. One considers multimodal and multizonal ice failure and the other is based on an energy method coupled with observed modes of rubble building.

Assur (1971) assumed that the ice sheet fails by bending, buckling, or crushing against the rubble, depending on the ice properties and the angle θ of the rubble block in contact with the ice sheet (Fig. 3.72). The horizontal force per unit width F_h is then given by

$$F_h = \begin{cases} F_v \tan(\theta + \phi) & 0 \leq \theta \leq \theta_c \\ F_c & \theta_c \leq \theta \leq \pi/2 \end{cases} \qquad (3.61)$$

where F_v = vertical force per unit width required to fail ice sheet in bending

F_c = force per unit width required to fail ice by buckling or crushing

ϕ = angle of friction between ice sheet and ice blocks

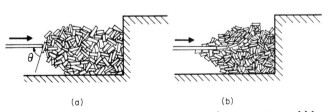

(a)   (b)

Figure 3.72 Principles to determine ice forces against rubble. (a) Ice fails against rubble. (b) Ice penetrates through rubble.

The critical angle θ_c where bending does not occur is determined by F_c/F_v as

$$\theta_c = \tan^{-1}\left(\frac{F_c}{F_v}\right) - \phi \tag{3.62}$$

Kry (1980) assumed a random distribution of block angles. Because the force required to fail the ice in bending is much less than the buckling or crushing force, the average force F_h is proportional to the buckling or crushing fraction and can be approximated by

$$F_h = \frac{2\phi}{\pi} F_c \tag{3.63}$$

An alternative approach is based on estimates to add gravitational potential energy to the rubble pile in order to overcome friction and push the leading edge of the ice sheet through the rubble. The force required to add gravitational potential energy into a floating ridge (Fig. 3.73), after Parmerter and Coon (1973), is simply

$$F_E = \tfrac{1}{2}\rho_i g H_s h \tag{3.64}$$

where ρ_i = density of ice
  g = acceleration of gravity
  H_s = ridge sail height
  h = ice thickness

Kry (1977) showed that the force required to generate potential energy in a grounded rubble pile (Fig. 3.73) with a triangular sail is given by

$$F_E = \frac{1}{2}\rho_i g H_s h \left\{ \frac{[(\rho_w - \rho_i)/\rho_i](H_{km}/H_s)^2 + 1}{H_{km}/H_s + 1} \right\} \tag{3.65}$$

where ρ_w is the density of water and H_{km} is the water depth. The friction

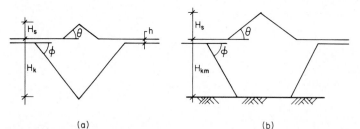

(a)   (b)

Figure 3.73 Models for (a) floating ridge and (b) grounded rubble pile.

resistance as the ice sheet penetrates into the rubble (Fig. 3.72) is given by the friction components of sail and keel, or for grounded rubble roughly by

$$F_f = \rho_i g(1 - e)f\phi H_s^2 \cot \theta \qquad (3.66)$$

where ϕ = coefficient of friction between ice blocks
e = porosity of sail
f = fraction of sail mass that acts to create frictional force
θ = slope angle of sail or pileup

According to Kovacs and Sodhi (1979) the value of the coefficient f lies generally between 0.1 and 0.5. The passive pressure on the leading edge of the ice sheet is estimated in Allyn and Charpentier (1982) as

$$F_p = K_p \rho_i g(1 - e)H_s h \qquad (3.67)$$

where K_p is the passive pressure coefficient.

It is obvious that all the described factors contribute to the total force, although not simultaneously with full effect. Estimates for the rubble-building forces are quite low compared to the ice-crushing forces. Based on a comprehensive review of predictions and models, Vivitrat and Kreider (1981) concluded that forces in the range of 150 to 750 kN/m (10 to 50 kips/ft) were possible for 2-m (6-ft)-thick ice.

The total force that is transferred through the rubble to the structure depends on the dimensions of the rubble and on the resistance provided by the grounding of the rubble (Fig. 3.74). Allyn and Wasilewski (1979) have estimated the total force as

$$F = p_x h D \left[1 + 2(a - 1) \frac{l_r}{D} \tan \alpha + 2a \frac{l_r}{D} \tan \phi \right] \qquad (3.68)$$

where p_x = rubble-building pressure
h = ice thickness
D = rubble width
a = confinement ratio, $= p_y/p_x$
l_r = rubble length
α = angle of rubble wedge
ϕ = ice-rubble dynamic friction coefficient

For example, with $a = 0.5$, $\alpha = 5°$, and $\phi = 20°$, the total force against a fully developed rubble becomes $F \approx 2.5 p_x h D$. The resistance provided by rubble grounding can be estimated from

$$R = \int_A g(1 - e)[H_s \rho_i - H_{km}(\rho_w - \rho_i)] \tan \phi \, dA \qquad (3.69)$$

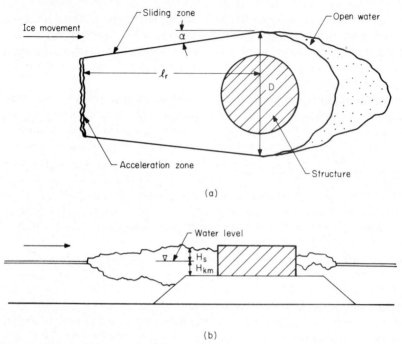

Figure 3.74 Rubble growth and grounding.

where A is the area of the grounded rubble and ϕ the angle of friction between the rubble and the sea bottom.

The ice forces exerted on structures via a rubble can be computed based on experience gathered from rubble formations around similar structures. It is also possible to simulate the rubble formation process, including its sliding resistance, stability, and forces exerted on the structure, using data on ice velocities and movement directions, as shown in Allyn and Charpentier (1982). An extensive grounded rubble tends to form around island and caisson-type structures with gentle slopes or shallow berms in dynamic ice environments. The influence of such a rubble is often positive since it protects the structure from very heavy ice forces and may provide additional sliding resistance.

3.5.9 Icebergs, Ice Islands, and Other Multiyear Ice Features

Icebergs, ice islands, and other exceptionally thick multiyear ice features pose the ultimate hazard for offshore structures and seabed installations in certain sea areas such as the North Atlantic and the Beaufort Sea. The size, shape, kinetic energy, and occurrence frequency of these ice features are key factors in the design. Offshore structures can be designed to withstand the impact of specified ice features. They can be protected by arrangements that dissipate the kinetic energy of the approaching ice feature. In some cases it is possible to remove the structure or to tow the ice feature away from its route if collision seems probable.

The symmetric collision between a structure and an ice feature can be analyzed using the momentum equation

$$F \, dt = mv_1 - mv_2 \qquad (3.70)$$

where the force F corresponds to ice failure, usually by crushing. Cammaert and Tsinker (1981) have given some simple examples of such computations. The hydrodynamic effects are included in the mass of the ice feature as an added quantity. On the other hand, the collision is never fully symmetric, and some impact energy may be lost if the ice feature begins to rotate (Bruun and Johannesson, 1971).

The kinetic energy of a large ice feature may be over 1000 MJ $(7.4 \times 10^8 \text{ ft} \cdot \text{lb})$. It is sufficient to create a large penetration and contact area. The ice force may reach an extremely high value, and a rigid structure that could survive the collision would probably be very expensive. Sacrificial structures that are able to dissipate kinetic energy have been suggested for protection in deep waters. In shallow waters rubble fields around structures may provide protection for a large part of the year and earth berms can be used as a permanent protection.

The dissipation of kinetic energy during the interaction between the seabed and an ice feature includes components from plowing action, bottom friction, and change in potential energy (the center of gravity of the feature tends to rise). Plowing action appears to be the largest component in most cases. Chari and Muthukrishnaiah (1978) have studied the case of an iceberg impacting a slope of soft sediment (Figs. 3.75 and 3.76) and derived an equation for frontal soil resistance of the form

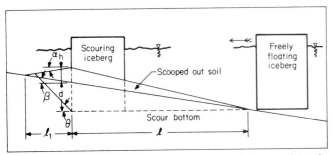

Figure 3.75 Concept of iceberg scouring. (*Chari and Muthukrishnaiah, 1978.*)

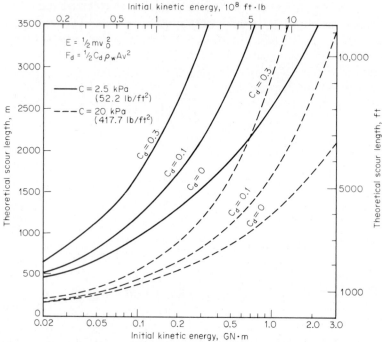

Figure 3.76 Theoretical estimates for scour lengths. C—cohesion of soil; γ—unit weight of soil, 15 kN/m³ (96 lb/ft³); E—initial kinetic energy of iceberg; m—mass of iceberg; v_0—initial velocity of iceberg, 0.3 m/s (1 ft/s); F_d—current drag force; C_d—drag coefficient; ρ_w—water density; A—cross-sectional area on which drag force acts; v—differential velocity between iceberg and water current; B—width of scour, 30m (98 ft); β—angle of slope, 1 in 1000. *(Chari and Muthukrishnaiah, 1978.)*

$$P = \frac{\gamma'(h+d)^2}{2}B + \frac{2\tau dB}{\sin 2\theta} + \frac{\tau d^2}{\sqrt{2}}(1 + \cos\theta) \qquad (3.71)$$

where P = frontal soil resistance

γ' = submerged unit weight of soil

h = height of gaged soil in front of iceberg

d = depth of gage

B = width of gage track

θ = angle between inclined shear plane and horizontal plane

α = angle made by scoured soil in front of iceberg with the horizontal

β = slope of ocean floor

τ = shear strength of soil

The total length of the scour or penetration L can be computed by equating the kinetic energy of the ice feature and the work done by external forces with the work done by soil resistance,

$$\int_0^L P\,dx = \frac{1}{2}\,mv_0^2 + \int_0^L F\,dx \qquad (3.72)$$

where P = total soil resistance
  x = penetration
  m = mass of ice feature
  v_0 = initial velocity of feature
  F = external forces, which may include wind drag, current drag (both skin and form drag), and pack-ice driving force

The pack-ice driving force may become a significant factor in Eq. (3.72) in the winter. Croasdale and Marcellus (1981) have suggested that in this case the driving force could correspond to the ridge buildup forces if the ice feature has very large lateral dimensions. Estimates for these forces range from 10 to 100 kN/m (0.7 to 7 kips/ft) for arctic pack ice on the geophysical scale. The values are an order of magnitude smaller than the ones corresponding to crushing or rubble formation forces.

After the kinetic energy of the ice feature has been dissipated, the ice feature may stay grounded in front of the structure. Drifting pack ice will then form a rubble, as shown in Fig. 3.77. If the ice feature has large lateral dimensions, the steady forces concentrated by the ice feature onto the structure may also be very high.

3.5.10 Local Ice Pressures

The instantaneous local ice pressures may be much higher than the average design pressure. This consideration is important in the design of

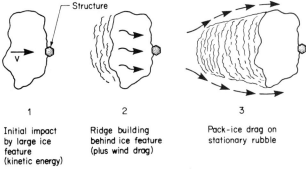

1
Initial impact
by large ice
feature
(kinetic energy)

2
Ridge building
behind ice feature
(plus wind drag)

3
Pack-ice drag on
stationary rubble

Figure 3.77 Stages of limit driving force–ice structure interaction. (*Croasdale and Marcellus, 1981.*)

ice-resistant walls for hydraulic structures. High local ice pressures and
dynamic friction may also create abrasion wear on the wall surface.

The design value of local ice pressure depends on the loading area and
on the thickness, structure, and temperature of the ice. Afanasev's formula [Eq. (3.46)] has been used to estimate also the local peak ice pressures for first-year ice over the practical design area in the form

$$p = \sqrt{\frac{5h^2}{A} + 1}\, \sigma_0 \qquad (3.73)$$

where A is the loaded area and h the ice thickness. A typical value for σ_0
for first-year sea ice could be 1.5 to 2 MPa (200 to 300 lb/in²). Peak
pressures up to 10 MPa (1500 lb/in²) have been measured occasionally
over a small area [200-mm (8-in) diameter] on the hull of an icebreaker
(Riska et al., 1983).

The uniaxial compressive strength values for multiyear ice are generally higher than those for first-year ice. A combination of worst grain
orientation and good confinement results in "hard spots," giving very
high local ice pressures. Typical criteria for local multiyear ice loads
range from 10 to 20 MPa (1500 to 3000 lb/in²) over an area of 1 m²
(10 ft²), depending on the structure and on the design ice feature.

The selection of the critical loading area depends on the structure
type. Concrete walls designed to resist ice impacts are typically over
0.5 m (1.6 ft) thick, with support spacings of about 6 m (20 ft). The
critical loading area for design against moment, shear, or punching shear
is larger than 2 m² (20 ft²). An ice-resistant steel wall, on the other hand,
may have rib spacings of about 0.3 m (12 in). Although the flexibility of
the plating may cause some redistribution of the loading (Varsta, 1983),
the outer plate and the framing have to be designed against very high
local ice loads concentrated on a small area. Composite steel and concrete structures fall between these two extremes.

The zone of solid ice impacts should naturally be designed against the
highest local ice pressures. However, loose blocks can also create high
local pressures during ridge impacts and the rubble-formation process. A
50% reduction in the local ice pressure criteria could be used as a first
design estimate for the zone of loose block impacts.

Narrow structures located in a dynamic ice environment are most
likely to be frequently subjected to high local ice pressures, while wider
structures tend to be protected by rubble fields most of the time, especially in shallow waters. Heavy and frequent ice impacts set strict requirements not only on the strength of the structure, but also on the
durability of its surface (Fig. 3.78). Coatings have a tendency to wear off
and poor-quality concrete wears out at a considerable rate under the
combined effect of ice abrasion and freeze-thaw cycles. Heavy mainte-

Figure 3.78 Effect of ice abrasion at the surface of a concrete lighthouse, observed best when contrasted to position of loading cell. *(Courtesy of M. Määttänen.)*

nance and repair efforts are often required to keep the structures in service.

3.6 Ice Control and Structural Design

3.6.1 Ice Control in Rivers

The purpose of ice control is to modify the existing ice conditions or to predict and minimize ice problems that may result from engineering activities. The most severe problem connected with ice conditions in rivers is flooding. During the fall and winter flooding is usually a result of

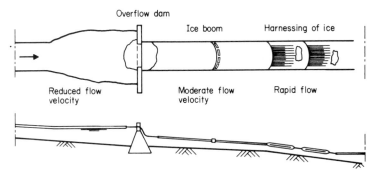

Figure 3.79 Different methods to help formation of ice cover.

Figure 3.80 Lake Erie ice boom at entrance of Niagara River, with a piece of ice riding over boom. *(Courtesy of New York Power Authority.)*

extensive formation of frazil ice. In the spring ice jams that form during the ice breakup may cause severe flooding.

The most severe fall and winter floodings occur in rivers with long reaches of rapid flow. The best way to prevent excessive frazil formation, ice jams, and flooding is to facilitate the formation of the insulative ice cover as rapidly and as smoothly as possible. Manipulation of flow velocities by channel modifications or discharge control is one possibility. If rapid flow velocities cannot be avoided, the area of rapid flow and free water surface exposed to freezing temperatures can be minimized by using an overflow dam (Fig. 3.79).

An ice boom can be used to stop ice transport and initiate ice-cover formation at reaches of moderate flow velocities (Fig. 3.80). The booms are usually made of timber, but steel booms and polystyrene-filled reinforced concrete booms can also be used. The boom has to be designed to have adequate stability and buoyancy under heavy loading conditions.

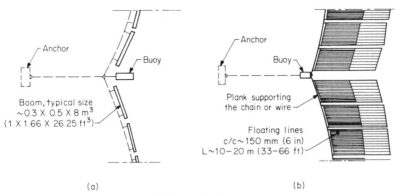

(a)   (b)

Figure 3.81   (a) Ice boom and (b) frazil collector structure.

Otherwise it may roll or sag, allowing the ice to ride over. For an analysis
the reader is referred to Perham (1978).

If the flow velocities are very high, natural formation of an ice cover
becomes very difficult. In that case the frazil collector system innovated
by Perham (1980) may be applicable. The principle of the system is
shown in Fig. 3.81. A set of lines attached to a support chain floats at
water surface. These lines gather frazil effectively during the ice-formation period, forming the initial ice cover locally. The ice barrier then acts
like an ice boom and does not fail easily because the lines reinforce it.

Figure 3.82   Severe flooding caused by ice jams. *(Courtesy of Pressfoto Oy.)*

Occasionally rigid structures, such as the Montreal ice control structure (Donelly, 1966), might be applicable to facilitate the formation of the ice cover. Explosives and icebreakers are used to break up ice jams. However, in severe cases the use of explosives has often been found ineffective.

The formation of ice jams in the spring during ice breakup is strongly dependent on the local character of the river. Ice jams occur typically just below rapidly flowing river reaches where the ice cover is still strong and thick because of frazil deposits. Bridge piers, islands, sharp bends, narrowings, and other channel irregularities may also stop the smooth running of ice and cause jamming. A potential for the formation of dry ice jams exists especially at sites where the channel widens and becomes shallow and rocky (Fig. 3.82).

One way to eliminate ice jams and other problems connected with ice breakup is to stop the ice run and let ice melt in place. This can be done by using ice booms or other structures that keep the ice in place or by changing the hydraulic gradient. However, these measures are seldom taken only because of flood control. Channel modifications at critical frazil-producing and ice-jamming reaches represent the most potential permanent solution for spring as well as for fall and winter flooding. A temporary solution is to break the ice cover at critical areas in advance. Blasting is an effective method (Frankenstein and Smith, 1970), but unfortunately counteractive measures are usually taken only after jamming has occurred and severe flooding begun. It is much easier to blast the ice cover at critical reaches before the breakup, for example, using

Figure 3.83 Explosives have been used with variable results to break up ice jams. *(Courtesy of Pressfoto Oy.)*

line explosion techniques, than to try to break a massive ice jam (Fig. 3.83).

Recently icebreakers and air-cushion vehicles have also been used to break the ice cover and help the ice breakup. Air-cushion vehicles are either run at high speed, taking advantage of the hydrodynamic wave effect, or at low speed so that the air penetrates under the ice cover, eliminating the fluid support on ice.

3.6.2   Ice Problems in Power Generation

Ice control is an important consideration in the generation of hydroelectric power in cold regions. Ice jams upstream from a hydroelectric power plant may cause a considerable temporary decrease in discharge. Jamming downstream from a plant results in an increase in water level and thus a reduction in power generation. Especially during ice-formation and breakup periods there is also a danger of frazil ice or ice floes clogging the water intakes and stopping the power generation. This is a problem concerning not only intakes for hydroelectric power generation, but also intakes for municipal water supply and other power-generation and industrial purposes (Fig. 3.84).

A smooth formation process of a stable ice cover is desirable for hydroelectric power generation. The hydrodynamic losses and the frazil problems can thus be minimized. The methods described in Sec. 3.6.1, including discharge control in particular, can be used. A large deep pond with slow water velocities, which is often created as a result of the construction of a dam, may be very beneficial. A stable ice cover forms into the pond early in the winter. If frazil forms upstream from a pond of adequate size, it is likely to deposit under the stable ice cover and will not reach the intakes. Frazil problems, if any, are thus limited to the initial ice-formation period (Tesaker, 1975).

Ice spillways and ice booms can be used to guide and control the ice transport and accumulation during ice-formation and breakup periods. The required width of the spillways is usually only a fraction of the total river width, and the depth is controlled so that the thickest ice floes can pass but the water discharge is minimized. Calkins and Ashton (1976) have summarized some experiences of the passage of ice through these spillways. In rivers where the ice cover mostly thaws in place (such as in those where the power potential has been fully utilized), the opening of ice spillways is seldom needed. An example of ice flushing in the other extreme case, with very severe ice conditions during the entire winter, is described in Mariusson et al. (1975).

If the water velocities are high, as in power channels, ice booms and discharge control can be used to facilitate the creation of a solid ice

Figure 3.84   Ice jammed in front of discharge control structure at Niagara River. *(Courtesy of New York Power Authority.)*

cover. The production of frazil can be eliminated at an early stage of the winter, and the risk of ice breakup and jamming due to stormy weather conditions and water-level fluctuations can also be reduced. The Beauharnois Canal, Quebec, shown in Fig. 3.85, represents an example of this kind of ice control. Here the formation of the initial ice cover is further assisted by breaking ice at Lake St. Francis and floating it in the canal behind the booms. In the spring the booms maintain the stability of the

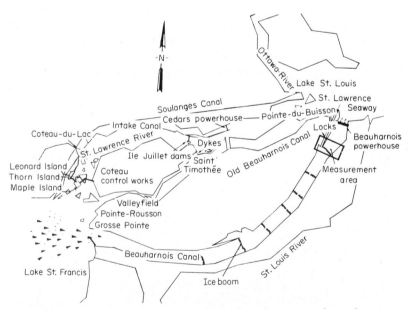

Figure 3.85   Map of Beauharnois Canal, Quebec, showing ice boom locations, powerhouses, and locks. *(Perham 1976.)*

canal ice cover, which remains in place for weeks after shipping commences in early spring. For a further discussion of this and other applications of ice boom techniques and their economical benefits the reader is referred to Perham (1976).

Large and abrupt changes in water discharge should be avoided from the viewpoint of ice engineering. The stability of an ice accumulation may be lost during the ice-formation period. Later the ice cover may be broken because of water-level changes. Another possible result is flooding of the ice surface and accelerated thickness growth of ice. A desparate effort to correct the mistake is shown in Fig. 3.86. The hydroelectric power plant was used for peak energy consumption during the daytime. The ice-covered channel had not been able to carry the abruptly increased flow and as a result daily flooding of the ice cover had occurred. The thickness growth of the ice cover had reached two to three times the normal growth rate. In order to help the ice breakup and prevent excessive flooding in the spring, mechanical methods were used to break the ice cover, and dust that absorbs solar radiation was spread on the ice to weaken it. Experience showed that surface flooding of the ice cover can be kept under control with very smooth and careful changes in the discharge rate.

Frazil ice in its newly formed active state is a major concern in hydroelectric power generation as well as in other water intake operations in cold regions. It can be found at all depths in turbulent water in rivers and also in considerable depths in lakes and seas. However, it only appears during a very limited time period, in the beginning of the ice-formation process, except in the vicinity of rapidly flowing frazil-generating open river reaches. Moderate amounts of frazil ice seem to be able to pass the intake canals and the turbines. The problem is that active frazil tends to stick on trash racks and block the flow at that point.

Figure 3.86 Mechanical weakening of ice to assist spring breakup. (*Courtesy of E. Kuusisto.*)

Figure 3.87 Icebreaker trying to break loose ice jammed in front of water intakes at Niagara River. Aboveground portions of intakes can be seen in background. *(Courtesy of New York Power Authority.)*

Electric heating is a commonly used method to prevent frazil from adfreezing on the rack bars. The energy required to maintain the surface of a cylindrical bar at a certain temperature, as given in Logan (1974), is

$$E = 1.78\Delta T\sqrt{vd} \tag{3.74}$$

where E = power requirement, kW/m [Btu/(h·ft)]
ΔT = temperature difference between supercooled water and the surface of the bar, °C (°F)
v = water velocity, m/s (ft/s)
d = bar diameter, m (in)

(If U.S. customary units are used in Eq. (3.74), the multiplier is 90.7.)

The design values for required temperature differences are commonly on the order of 1°C (1.8°F). One of the many possible electric heating systems is described in Gevay and Erith (1979).

In the case of nuclear, fossil fuel, and coal-fired generating stations the frazil problems at intakes can be conveniently avoided by recirculating a portion of the cooling water through the intake. These and other possible methods to prevent frazil problems at intakes, including air bubbler systems, insulation techniques, and coating materials, are further discussed in Michel (1971) and Wahanik (1978).

Water intakes can also be blocked mechanically by inactive frazil floes, thick ice accumulations, and ice jams (Fig. 3.87). The removal of trash racks for the period of initial ice cover formation is sometimes used to avoid the problems of passive as well as active frazil. The solid frazil floes are concentrated at the upper part of the flow, and thus the removal of the upper trash racks may prevent the clogging caused by frazil floes. Ice spillways, control of the thickness and stability of the ice cover, and proper selection of the location of the intake are the best measures to prevent clogging of the intake by ice accumulations or jams.

3.6.3   Some Aspects of Harbor Design

Winter navigation is not very common under severe ice conditions. Engineers involved in harbor design have been most concerned with ice forces. However, as the need for year-round navigation increases in cold regions, more attention will be given to the control of ice conditions in harbors, including drifting ice and accelerated growth of the constantly broken brash ice cover.

Design against ice forces   As pointed out in Sec. 3.5, the ice forces on structures depend on local ice conditions and the characteristics of the structures. Harbors are often located in well-protected areas. Horizontal ice movements are small. Lateral ice loads that may be of importance in structural design can be expected only on rigid structures. However, structures may be subjected to heavy ice impacts in rivers and areas of strong tidal currents. Similar impacts also occur in harbors that are operating during winter as ice floes get squeezed between ships and structures.

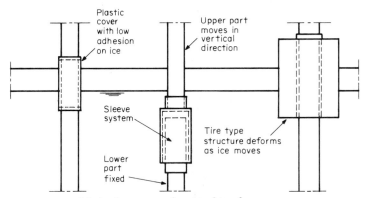

Figure 3.88   Methods to control vertical ice forces.

The vertical uplift on piles is often difficult to handle economically. Sometimes the deadweight of the structure is increased artificially or piles are used in groups in order to resist uplift. It is good engineering practice to keep some margin between the superstructure and the highest assumed ice level. Otherwise the ice rubble that often forms around the pile may damage the superstructure.

Some techniques to reduce or eliminate uplift forces are shown in Fig. 3.88. The purpose of using certain types of plastic tubes to cover piles is to force the failure to occur at the interface between the ice and the plastic before other failure modes become critical. The sleeved pile system (Wortley, 1978) eliminates the uplift problem totally. However, in that case the structural system has to be very flexible, because the piles may break through the ice at different moments when the dock is in the lifted stage. A sleeving system that would allow vertical ice movement without transferring any forces onto the structures would be a better approach. Unfortunately it is very difficult to design such a system to accommodate large water-level fluctuations.

Problems related to moderate lateral and vertical ice movements have been avoided by using floating pontoon structures, such as the pier shown in Fig. 3.89. Some structural failures have been experienced at the branch and at the fingers of the system. Special attention should be given to the lateral flexibility of the branch. Fingers of the floating type should be lifted up before the winter.

Another solution to eliminate vertical and also horizontal ice forces in case of slow horizontal ice movements is the air bubbler system (Fig.

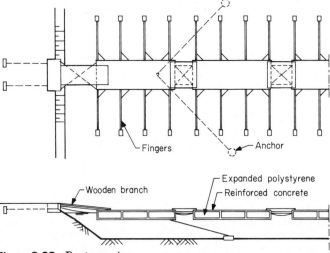

Figure 3.89 Pontoon pier.

Figure 3.90 Air bubbler system in operation. *(Courtesy of J. L. Zabilansky.)*

3.90). The principle is shown in Fig. 3.91. A compressor delivers air through the orifices into the water. The initial momentum of the air jet is soon dissipated and buoyancy of the air occurs. As the bubbles rise, they entrain water into the rising plume from the lateral direction. The buoyant plume spreads near the surface as it either impinges on the ice cover or encounters the free air surface. The melting of ice is primarily a result of the convective heat losses of the spreading water flow. Thus the temperature of the water flow has to be above freezing for the ice to melt.

An analysis of a line-source and a point-source bubbler system suppressing ice formation has been undertaken by Ashton (1974, 1979). There are five major steps of analysis: diffuser-line analysis, induced-plume analysis, heat-transfer analysis, ice-melting analysis, and thermal-reserve analysis. The diffuser-line analysis is not within the scope of this

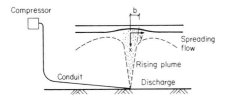

Figure 3.91 Schematic representation of air bubbler system.

publication. We only note that as the air flows through the orifices, it expands and the temperature cools down from its original level, which is usually close to the water temperature. If the pressure is not properly induced, icing may occur at the orifices. Typical orifice diameters are on

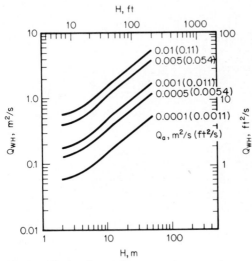

Figure 3.92 Induced flow at surface as a function of diffuser submergence depth and air discharge rate for line-source bubbler system. *(Ashton, 1974.)*

the order of 1 mm (0.04 in), and the outlet pressures may be as high as 3.5×10^5 Pa (50 lb/in^2), depending on the depth of submergence. Even distribution of air through all orifices and proper maintenance of the system are essential for successful operation. For further details of diffuser-line design, the reader is referred to Ashton (1978, 1974).

The effectiveness of the bubbler system depends partly on its submergence depth. The induced water flow at the surface as a function of diffuser submergence and air discharge rate is given in Fig. 3.92 for a line-source bubbler system and in Fig. 3.96 for a point-source system. The standard deviations of the vertical velocity profiles at the surface (width b) for the systems are given in Figs. 3.93 and 3.97.

The water bodies in which the bubbler systems are installed are seldom isothermal. The temperature of the flow at the surface T_{WH} compared to the freezing point T_m can be computed by

$$T_{WH} - T_m = \frac{1}{Q_{WH}} \int_0^H [T_W(x) - T_m] \frac{dQ_W(x)}{dx}\, dx \qquad (3.75)$$

where $Q_W(x)$ is the induced flow at depth x and Q_{WH} is the induced flow at the surface. However, the thermal stratification of the water body may be quite mixed because of currents and also because of the operation of the bubbler system itself. Conservative estimates of the average temperature of the water body are often used as the temperature of the flow.

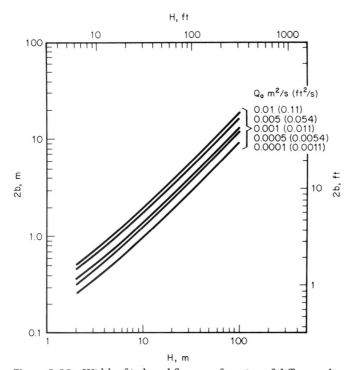

Figure 3.93 Width of induced flow as a function of diffuser submergence depth and air discharge rate for line-source bubbler system. *(Ashton, 1974.)*

The heat-transfer rates at the reference distance b for different diffuser submergence depths and air discharge rates for line- and point-source bubbler systems are given in Figs. 3.94 and 3.98. The variations for the heat-transfer coefficients h for a lateral distance $y > b$ are also shown in Figs. 3.95 and 3.99. For $y < b$ the values for the heat-transfer coefficient h are somewhat higher than at distance b. The heat-transfer rate to the undersurface of the ice is given by

$$q_W = h_y (T_{WH} - T_m) \tag{3.76}$$

where the temperature of impinging flow T_{WH} is slowly decreasing as the distance y increases because thermal energy is lost into air or to melt the ice cover.

The actual heat balance at the water-ice interface is given by

$$q_i - q_W = \rho_i \lambda \frac{d\eta_i}{dt} \tag{3.77}$$

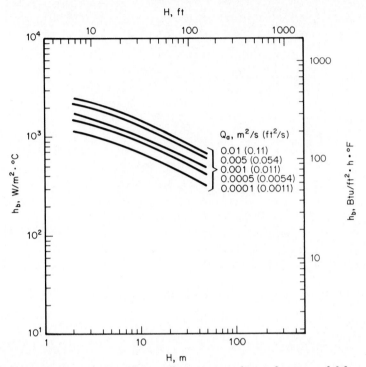

Figure 3.94 Heat-transfer coefficient at $y = b$ as a function of diffuser submergence depth and air discharge rate for line-source bubbler system. *(Ashton, 1974.)*

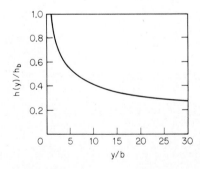

Figure 3.95 Variation of heat-transfer coefficient with distance outward from impingement region for line-source bubbler system. *(Ashton, 1974.)*

where q_i = rate of heat conduction through ice
  ρ_i = mass density of ice
  λ = heat of fusion of ice, = 334 kJ/kg (144 Btu/lb)
$d\eta_i/dt$ = rate of change in ice thickness

The rate of heat conduction through the ice is

$$q_i = \frac{T_m - T_a}{\eta_i/k_i + \eta_s/k_s + 1/h_a} \qquad (3.78)$$

where k_i and k_s are the thermal conductivities of snow and ice, and $1/h_a$ represents the thermal resistance of the air boundary layer [Eq. (3.5)].

Finally, the thermal reserve of the water body should be sufficient to cause melting of the ice cover. The total thermal reserve of a closed water body is

$$Q_{tot} = V\rho_w C_p (T_W - T_m) \qquad (3.79)$$

where C_p = specific heat capacity of water, = 4187 J/(kg·°C) [1000 Btu/(lb·°F)]
  T_W = average water temperature
  V = volume of water body

In practice, the air discharge should be adjusted to climatic conditions so that the loss of thermal capacity can be minimized.

Ashton (1979) has presented computer programs to simulate the changes in the thickness profile of ice under given meteorological conditions for both line-source and point-source systems. The simulations seem to give results in reasonable agreement with actual field measurements, especially for relatively small values of lateral distance ($y \leqslant 15b$). Some uncertainties still exist. For example, it is not clear whether the simulation gives good results for deep water [depth over 10 m (30 ft)]. Furthermore, it is also difficult to analyze how disturbing factors such as structures or other air discharge sources in the vicinity affect the characteristics of the flow and the results. Finally the mechanism of entrainment of far-field warm water into the system is not well understood. In any case, field measurements of the water temperatures at the site are required before the air-bubbler system can be installed. The minimum average water temperatures may be surprisingly low, in seawater often practically at the freezing point.

EXAMPLE 3.6 The piles of the pier in Example 3.1 are protected against ice forces by an air-bubbler system. An orifice with air discharge Q_a = 0.0005 m³/s (0.018 ft³/s) is located at both sides of each pile. The water depth is 5 m (16 ft), average water temperature is 0.2°C (32.36°F), and the wind velocity is 7 m/s (16 mi/h). Find the steady-state condition at temperatures of −10 and −20°C (14 and −4°F) with no snow.

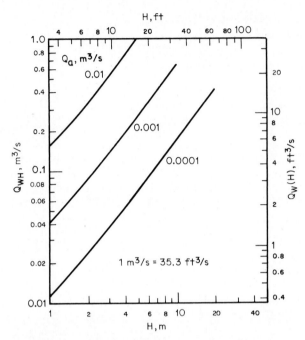

Figure 3.96 Induced flow of water at impingement as a function of orifice submergence depth and air discharge rate for point-source bubbler system. *(Ashton, 1979.)*

Since the distance between the piles and the air sources is large compared to the water depth, the point-source analysis seems more appropriate than the line-source analysis. From Figs. 3.96 to 3.98 we get (assuming that the pile cuts the interference between the two adjacent plumes)

$$Q_{WH} = 0.18 \text{ m}^3/\text{s} \ (6.4 \text{ ft}^3/\text{s})$$
$$b = 0.3 \text{ m} \ (1 \text{ ft})$$
$$h_b = 1.8 \times 10^3 \text{ W/(m}^2 \cdot {}^\circ\text{C)} \ [320 \text{ Btu/(ft}^2 \cdot \text{h} \cdot {}^\circ\text{F)}]$$

The heat-transfer rate at a distance of 0.3 m (1 ft) from the centerline of the orifice is thus

$$q_w = 1.8(10^3)(0.2) = 360 \text{ W/m}^2 \ [114 \text{ Btu/(ft}^2 \cdot \text{h)}]$$

Assuming no ice and snow at the surface, we get from Eqs. (3.5) and (3.78),

$$q_i = 10[3.4 + 4.4(7)] = 342 \text{ W/m}^2 \ [108 \text{ Btu/(ft}^2 \cdot \text{h)}]$$

The surface is thus free from ice over a radius of about 0.3 m (1 ft) above each orifice. The ice begins to thicken rapidly outside this area. At the distance of 1.5 m (5 ft) from the centerline the heat-transfer rate from a single orifice would be (Fig. 3.99)

$$q_w = 0.5(360) = 180 \text{ W/m}^2 \ [57 \text{ Btu/(ft}^2 \cdot \text{h)}]$$

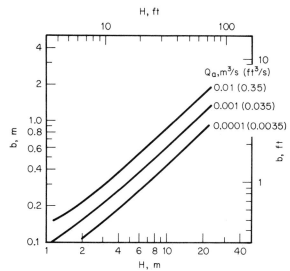

Figure 3.97 Diameter of impinging plume as a function of orifice submergence depth and air discharge rate for point-source bubbler system. *(Ashton, 1979.)*

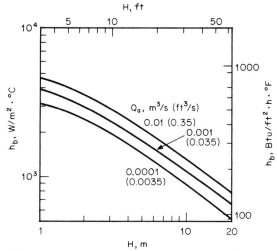

Figure 3.98 Heat-transfer coefficient at distance b from centerline as a function of orifice submergence depth and air discharge rate for point-source bubbler system. *(Ashton, 1979.)*

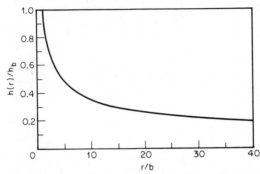

Figure 3.99 Variation of heat-transfer coefficient with radial distance from centerline of impingement for point-source bubbler system. *(Ashton, 1979.)*

The equilibrium ice thickness can be solved from Eq. (3.78) with $q_i = q_w$,

$$q_i = \frac{10}{\eta_i/2.24 + 1/[3.4 + 4.4(7)]} = 180$$

$$\eta_i = 0.06 \text{ m (2.4 in)}$$

This estimate is conservative because of the combined effect of air sources. On the other hand, at a temperature of $-20°C$ $(-4°F)$ and a distance of 0.3 m (1 ft) we also get

$$q_i = \frac{20}{\eta_i/2.24 + 1/[3.4 + 4.4(7)]} = 360$$

$$\eta_i = 0.06 \text{ m (2.4 in)}$$

Thus the pile has to be able to resist vertical ice forces corresponding to the shear capacity of an ice sheet with estimated thickness of about 6 cm (2.4 in) if $-20°C$ $(-4°F)$ is taken as the design air temperature.

If we assume that 10% of the total heat capacity of the flow is used to melt ice, the thermal energy to be replaced by currents or bubbler-induced circulation per orifice is

$$Q = 0.1(0.18)(0.2)(10^3)(4187) = 1.5 \times 10^4 \text{ W } (5 \times 10^4 \text{ Btu/h})$$

In small closed water bodies the thermal reserve given by Eq. (3.79) is soon exhausted.

Drifting ice   Drifting ice floes may interfere with harbor operations in large rivers and in coastal areas with strong tidal currents. Berthing and mooring may be hazardous in a dynamic ice environment. The floes may cause navigational problems to approaching and departing vessels. Finally, impacts by major floes may damage the ships.

Proper site selection for the harbor based on a careful survey of the flow characteristics and the ice environment is naturally essential in order to manage this kind of problem. Operational limitations, such as the use of protected sides of piers during difficult periods, provide one solution. Protective structures such as breakwaters, ice booms, or ice deflectors offer an alternative approach. Model tests may prove useful in studying changes in the flow characteristics and the dynamics of the ice environment.

A good example of structural protection of berthing places is given in Fig. 3.100. In this case cylindrical concrete caissons stop major ice floes and form protective ice accumulation upstream. The berthing area is free of drifting ice while the natural flow characteristics have not been extensively disturbed by the protection (Saunders and Timascheff, 1973).

In exposed coastal harbors ice may pile up against harbor structures during major ice movement events, as reported in Kovacs (1983) and Bruun and Johannesson (1971). The pileups may occur during storm surges as ice fields may push hundreds of meters ashore over gently sloping terrain. Breaking of the ice cover by heavy winter navigation may also contribute to ice movements and accumulation in the harbor area.

Large ice forces will naturally be exerted on structures during such events. However, the events are also hazardous to winter operations. Heavy accumulation will stop the traffic and may even crush vessels. Breakwaters can be used as protection, but they have to be sufficiently high and heavily armored to avoid ice override and armor layer damage.

Accelerated ice growth   Accelerated ice growth is a severe problem in protected harbors with high winter traffic. The heat losses from the

Figure 3.100   Oil terminal wharf at St. Romuald, Quebec, protected from drifting ice by concrete cells. *(Courtesy of F. Saunders.)*

(a)

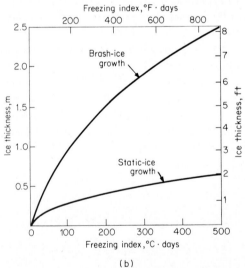

(b)

Figure 3.101 Heat losses of broken ice cover are large, and as a result the ice growth rate is greatly increased. (*a*) Helsinki harbor during winter. *(Courtesy of Helsinki Port Authority.)* (*b*) Comparison of growth rates for static and brash ice.

constantly broken ice cover are large, with plenty of unfrozen water near the surface. As a result the growth of brash ice is typically three to five times larger than the normal static ice growth. Maneuvering of ships becomes difficult and even impossible in thick brash ice.

Michel and Berenger (1975) have developed an algorithm to estimate the growth of brash ice. They showed that while static ice growth is proportional to the square root of the freezing index, the growth of brash ice is almost a linear function of the freezing index when the ice cover is broken daily (Fig. 3.101). A simplification of the algorithm (Vance, 1980) can be presented in the form

$$h_i = (1 - \beta)(h_{i-1} + \alpha\sqrt{\Sigma F_i}) + \beta\alpha\sqrt{\Sigma F_i} \qquad (3.80)$$

where h_i = ice thickness after time step i corresponding to ice-breaking period
β = percentage of open water (typically 0.25)
α = coefficient from Eq. (3.1)
ΣF_i = freezing index up to time step i

Some typical methods used to manage brash ice problems in harbor operations are illustrated in Fig. 3.102. Approach and berthing operations have proven to be especially difficult. The most common method is to use a harbor icebreaker to break the brash ice cover in advance. The berthing vessel then uses its own propeller flows to flush away loose ice blocks between itself and the dock. Human intervention and the propeller flows of a harbor icebreaker are often needed to assist in this operation, which under difficult conditions may take hours. Harbor cranes are occasionally used to remove ice from the berthing area in emergency situations.

Erosion caused by the propeller flows is a serious problem, which has even led to failures of harbor structures. Heavy protective concrete plates are often required. Ice accumulations or ice collars, which form as ice slush is repeatedly squeezed against structures, as in ferry wharves, represent another problem. Such formations may become quite extensive and concentrated, especially under the action of tidal fluctuations, and may prevent the vessels from reaching the wharf. Heating the structures is one way to eliminate the problem.

The applicability of air bubbling in the control of brash ice conditions depends on the available thermal reserves. If there is a sufficient continuous supply of water that is at least 0.1 °C (0.2 °F) above the freezing point, the condition for successful bubbling exists. Such conditions may for example exist in straits and river mouths, where currents may carry thermal energy. In this case bubbling can keep the brash ice growth within tolerable limits and will prevent the ice blocks from freezing together, which means a significant relief for operations. However, the

(a)

(b)

Figure 3.102 Methods to manage brash ice problems in Finnish harbors. (*a*) Use of propeller flows and human labor to remove ice floes from berthing place. (*b*) Mechanical removal of brash ice using harbor cranes. (*c*) Ice management with air bubbler system. (*d*) Utilization of thermal effluents. (*Courtesy of M. Penttinen.*)

risk of propellers getting jammed on the bubbler lines should be considered in the design.

Thermal effluents are in some cases readily available for ice control in harbors from power plants, industry, or water-treatment plants. A thermal discharge of 10 MW (34×10^6 Btu/h) typically maintains an open water area of 1 ha (2.5 acres) if the freezing index is on the order of

(c)

(d)

$800\,°\text{C}\cdot\text{days}$ $(1400\,°\text{F}\cdot\text{days})$. However, from the navigational point of view it is not necessary to maintain a totally ice-free area. Greatly improved efficiency can be achieved if the distribution of thermal effluents is directed to the critical areas and controlled so that a thin protective brash ice cover is maintained.

Year-round operation of heavily trafficked harbors with the average freezing index exceeding $1000\,°\text{C}\cdot\text{days}$ $(1800\,°\text{F}\cdot\text{days})$ becomes extremely difficult without active ice-control methods. Different ice-control methods for arctic harbors, including thermal energy, mechanical ice removal, insulative foams, and harbor enclosures, have been discussed in Cammaert et al. (1979). One potential solution is represented in Fig. 3.103. In this case warm water is distributed to critical areas and air bubbling is used to circulate the thermal energy to the underside of the brash ice cover. When a ship approaches, the air is directed to air

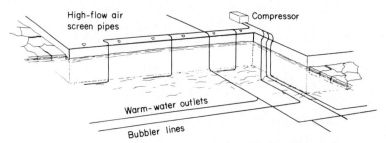

Figure 3.103 Combination of ice-control methods. *(Eranti, Leppänen, and Penttinen, 1983.)*

screen lines and pumped at a rate of 1 to 5 m³/(min·m) [10 to 50 ft³/(min·ft)] to create a water flow that pushes the loose ice blocks temporarily away from the berthing area. Naturally ordinary flow developers can also be used instead of the high-flow air screen.

3.6.4 Some Engineering Problems Related to Winter Navigation

An attempt to extend the winter navigation season or maintain year-round navigation in ice-covered waters requires a solution to ice navigation problems but often also to problems related to ice control and structural design. Instability of the broken ice cover, accelerated growth of brash ice, and the operation of locks represent the most severe problem areas for inland waterways. Ships getting pinned between the land-fast ice and the pack ice and design of navigation aids that can survive in the dynamic ice environment are difficulties in large lakes and in sea areas.

Year-round navigation in inland waterways often becomes very difficult when the average freezing index exceeds 500°C·days (900°F·days). The first problems appear in locks. Traffic is disturbed by ice buildup along the lock walls, floes drifting into the locks, and ice jams forming below the tail gates.

Ice collars form on lock walls by compaction and adhesion of ice debris during the passage of ships and by direct freezing of water on a cold surface. The ice buildup will eventually prevent the passage of large beam ships. Possible solutions to this problem include built-in wall heating or ice removal using mechanical contact tools, hot steam, or high-pressure water jets (Calkins and Mellor, 1975).

The locks are usually kept full of water between long intervals of ship passage during freezing conditions in order to minimize the icing of lock walls. However, this exposes one side of the lower gates to freezing ambient temperatures while the other side is in contact with water. Ice formation will occur and may prevent operation of the gate machinery or full opening of the gate unless proper insulation and heating measures have been pursued.

Floating ice tends to obstruct the movement of vessels in lock areas, interferes with gate operations, and partly causes the formation of ice collars on the lock walls. Air bubbler systems, high-flow air screens, and surface current generation can be used to prevent ice formation and to remove ice floes in the vicinity of locks. Webb and Blair (1975) have described ice-flushing systems that produce a surface current capable of flushing accumulated ice out of the lock, and compressed-air systems that remove the floes interfering with lock openings. However, if the ice cover is solidly frozen, these methods are inefficient. External heat may be required to prevent the ice floes from freezing together if, for example, a high-flow air screen is used to stop ice floes from drifting into the lock along with the vessel. Some ice control methods used in locks are illustrated in Fig. 3.104.

Accelerated growth of brash ice also rapidly develops into a problem that causes slowdown and total shutdown of traffic in canals and narrow rivers. Methods to control the situation including air bubbling, melting by external heat, and mechanical removal of ice (Vance, 1980) have been considered. Bubbling has been found inefficient because the thermal reserves of a small water body are soon exhausted.

The thermal energy requirements to keep long canal stretches operational appear to be too large to be feasible. Also mechanical ice removal methods appear to have only limited applicability. However, the thermal reserves of the water body have generally increased considerably in ice-covered canals by spring as a result of thermal heat flux from the ground. Air bubbling has been used successfully to melt holes into the ice cover in the Saimaa Canal, Finland (Eranti, Penttinen, and Rekonen, 1983). This has helped ice breaking at critical canal stretches and made an early opening date possible.

In rapidly flowing rivers instability of the broken ice cover often causes ice jamming, flooding, and interference with ship traffic. The use of ice booms to stabilize the ice cover is illustrated in Figs. 3.57 and 3.85. The arrangements in the Lac St. Pierre, Quebec, provide an example of the use of artificial islands and lighthouses together with ice booms to stabilize the ice cover on a large scale (Fig. 3.105). Before the ice control structures were installed, the ice cover broke loose, plugged the navigation channel, and even caused flooding several times per winter. Ice interference with the vessel track decreased considerably, the danger of winter flooding was reduced, and year-round navigation became possible as a result of ice control (Danys, 1978).

(a)

(b)

Figure 3.104 Winter lock operations at Saimaa Canal, Finland. (*a*) Ship entering a lock. (*b*) High-flow air screen is used to prevent ice blocks from drifting into lock. (*c*) Insulation of tail gate. (*d*) Mechanical ice removal. (*Courtesy of M. Penttinen.*)

(c)

(d)

Similar solutions could also be applied in areas where ships get pinned between land-fast ice and pack ice under severe weather conditions and even icebreaker-assisted convoys have difficulties. Artificial islands, navigation aids, and pack-ice barriers (Potter et al., 1982) could be used to stabilize the ice cover outside the navigation canal.

One important problem related to navigation is to design inexpensive navigation aids that can survive the winter ice conditions. Some typical designs are shown in Fig. 3.106. It is interesting to note how the cost of the navigation aids increases rapidly as the ice conditions get more severe. Linemarks, beacons, and light buoys (Fig. 3.106a, b, and c) used in

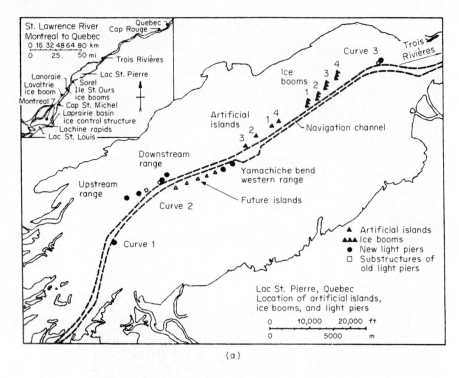

(a)

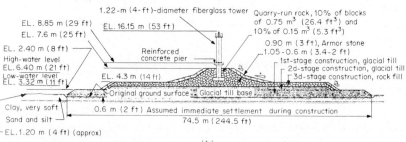

(b)

Figure 3.105 Ice-control arrangements at Lac St. Pierre, Quebec. (a) General plan. (b) Cross section of artificial island. *(Danys, 1978.)*

protected areas with a stable ice cover do not represent very significant investments. However, in a dynamic ice environment innovation and good engineering judgment are required in order to achieve a reliable design with reasonable cost.

The prestressed polyethylene ice buoy filled with polystyrene (Figs. 3.106d and 3.107) is an example of this kind of innovative engineering. The tube easily slips through the ice in case of vertical or lateral ice movements because of low adhesion between ice and polyethylene. The older design of steel (Fig. 3.106e) is not as reliable in this respect. It is removed from its position every now and then because of ice movements. Wider dredging is also required because the buoy has a free-floating area in the lateral direction.

Slender designs (Fig. 3.106f, g, and h) are replacing traditional massive concrete lighthouses in heavy dynamic ice conditions. At the early development stage steel lighthouses (Fig. 3.106g) experienced serviceability problems and even structural failures because of ice-induced vibrations. A special shock-absorber system was designed to cut down vibrations (Määttänen, 1975).

3.6.5 Ice and Offshore Hydrocarbon Development

The most intensive ice-engineering activities are currently concentrated around offshore oil drilling. Exploration programs are under way at several ice-infested sea areas, and production started already in the 1960s in moderate ice conditions in Cook Inlet, Alaska. The ice conditions at the Beaufort Sea with thick first-year ice and different forms of multiyear ice belong to the harshest class. That is why the ice-engineering approach adopted for this area provides an especially interesting example for discussion.

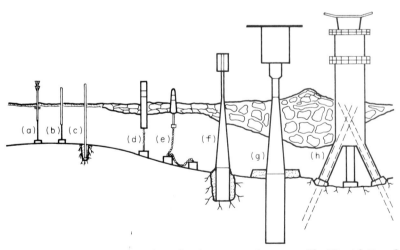

Figure 3.106 Navigation aids with relative cost classes used by Finnish Board of Navigation. (a), (b), (c) Stable ice cover; relative cost range 100 to 1000. (d), (e) Ice cover occasionally broken and unstable; relative cost 10,000. (f), (g), (h) Severe ice conditions; relative cost range 100,000 to 1,000,000.

Figure 3.107 Prestressed plastic buoy in ice. *(Courtesy of Wiik & Höglund Ltd.)*

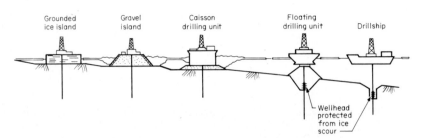

Figure 3.108 Typical drilling schemes used at Beaufort Sea.

Some typical exploration drilling schemes used in the Beaufort Sea are illustrated in Fig. 3.108. Removable caisson units, gravel islands, and ice islands are used for exploration in shallow waters and floating units with seasonal operation restrictions in water depths exceeding 30 m (100 ft). Gravel islands, caisson islands, and gravity structures are considered feasible also as production platforms.

Experience with the performance of structures in heavy ice conditions has been very limited. There are severe environmental concerns with regard to accidents in arctic offshore hydrocarbon development. Finally, ice is a major factor contributing to the extremely high development costs at the Beaufort Sea. That is why a thorough ice-engineering approach plays an essential role in the development and design of offshore structures. This approach includes three major elements:

1. Survey of the ice environment

2. Development of the ice design criteria, considering ice processes, ice forces, and ice defense

3. Monitoring the validity of design assumptions and criteria during the service stage of the structure

Naturally the extent of each measure depends on the type of structure and on the environment where it will be located, as discussed in a publication by the American Petroleum Institute (1982).

Survey of ice conditions   The purpose of an areal ice survey is to define the operational environment and to form a basis for developing design criteria for offshore structures. An ice survey is systematic and time-consuming work. Some typical elements of an areal ice survey are illustrated in Fig. 3.109.

The general mapping of ice conditions is based on satellite photographs, ice surveillance flights, and field observations. It produces periodical ice maps, which give the areal distributions of different ice forms, ice concentrations, and average ice thicknesses. In addition observations are registered on factors such as ice movements, ridge frequencies and heights, occurrence and nature of multiyear ice, and extent of ice rubble features and ride-ups.

General ice mapping provides quite rapidly information on seasonal ice coverage variations and local ice conditions, which is required in planning offshore operations such as construction, transportation, and logistic support. Statistical methods are used to define the extreme ice features and ice events that govern the design of offshore structures. It may take at least 10 years before estimates can be made in a statistically reliable manner, depending on the type of structure and ice environ-

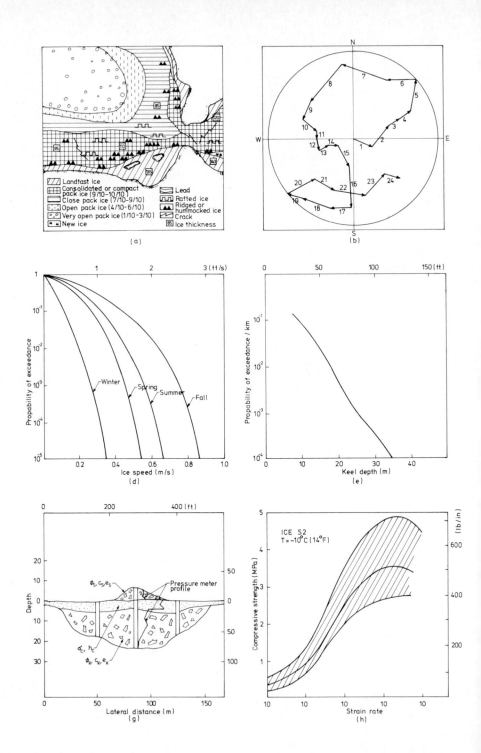

(a)

(b)

(d)

(e)

(g)

(h)

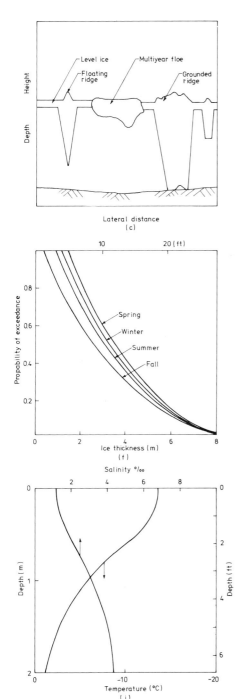

Figure 3.109 Typical results of areal ice survey. (*a*) Ice maps. (*b*) Ice-movement plots. (*c*) Ice profiles. (*d*) Ice movement. (*e*) First-year features. (*f*) Multiyear features. (*g*) Ice coring and in situ testing. (*h*) Mechanical properties of ice. (*i*) Salinity and temperature profiles.

ment. In some cases offshore structures are designed against events with return periods of 1000 years. Statistical approaches have to be supplemented with theoretical worst-case scenarios and their physical limitations for reasonable results. Recorded historical extreme events in similar conditions may also prove helpful.

Studies of the physical and mechanical properties of the ice cover are an essential part of the ice survey. Grain structure, porosity, seasonal variation of temperature and salinity profiles, and mechanical properties of different ice types are basic parameters that influence the design criteria. Recently a great deal of attention has also been paid to the friction and cohesion parameters and to the consolidation of ice ridges and rubble formations. Different kinds of field tests may provide valuable information in addition to results obtained from traditional sampling and laboratory experiments.

Ice-related design considerations   Ice forces, ice processes, and ice defense or control are closely associated with one another in the design of ice-resistant offshore structures. The design of Beaufort Sea structures requires an analysis of several different loading schemes corresponding to extreme ice formation, midwinter ice breakup, and summer ice conditions, as illustrated in Fig. 3.110. Ice forces were discussed in Sec. 3.5, and one should only note that the probable extent of extreme ice-structure interaction and the severity of structural failure should have an influence on the design force level. However, ice processes and ice defense also have an influence on the ice forces and on the general design and need further addressing in this context.

The ice processes to be considered in offshore operations include ice ride-up, ice pileup, rubble formation, and ice scouring. Ice ride-up is an extremely hazardous event for operations and structures on artificial islands and gently sloping beaches (Fig. 3.111). Kovacs and Sodhi (1979) and Kovacs (1983) have described events where ice has been pushed over 100 m (300 ft) inland and over cliffs up to 10 m (30 ft) high. Ice ride-up is typically associated with the ice-formation period or a storm surge. Forces required to create ice ride-up are small and can be estimated by

$$F = \rho_i ghL \, (\sin \alpha + \mu \cos \alpha) \qquad (3.81)$$

where ρ_i = mass density of ice
$\quad g$ = acceleration of gravity
$\quad h$ = ice thickness
$\quad L$ = length of ice ride-up on slope
$\quad \mu$ = coefficient of friction
$\quad \alpha$ = slope angle from the horizontal

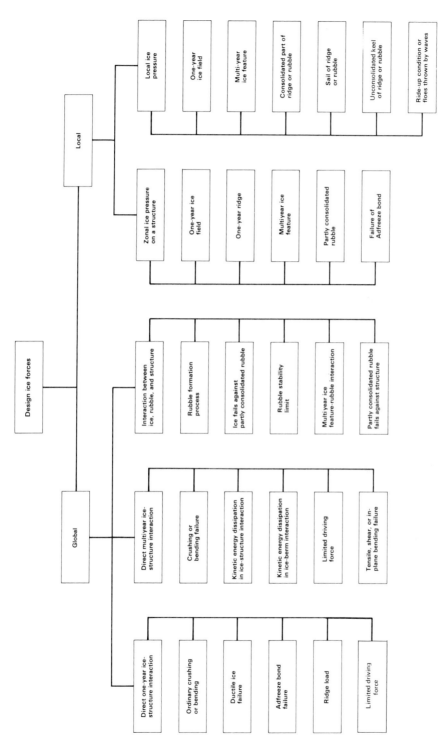

Figure 3.110 Different ice loading cases to be considered in design of arctic offshore structures.

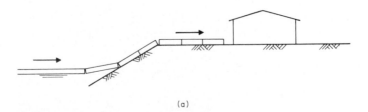

(a)

(b)

Figure 3.111 (a) Mechanism of ice ride-up. (b) Example of ride-up event. (*Courtesy of J. L. Wuebben.*)

A pileup process instead of ride-up is triggered near the waterline on rough and uneven beaches. Up to 10-m (30-ft)-high ice piles are commonly found at the shoreline, and the formation will grow seaward as grounded or floating rubble if the ice movement continues. The model by Allyn and Charpentier (1982) may prove valuable in predicting the growth, grounding, extreme extent, and consolidation of a rubble around an offshore structure (Fig. 3.112). This information is needed to estimate ice forces and to evaluate the feasibility of winter navigation around the structure.

Ice scouring is a major hazard to offshore pipelines and other seabed installations. The deepest furrows are created by multiyear ice features, but scouring by ordinary first-year ridges may also be quite substantial. The depth of ice scouring and the disruption of the soil can be estimated from ice keel or ice feature thickness statistics combined with plowing analysis, as discussed in Sec. 3.5.9. An alternative approach is to gather statistical data on scour depths at different water depths. Methods to determine pipeline burial depths are discussed in Pilkington and Marcellus (1981). Their suggestion to bury pipelines up to 5 m (16 ft) into the seabed in the Canadian Beaufort Sea in water depths less than 55 m (180 ft) underlines the economic significance of this problem.

Ice defense contains both structural and operational aspects, and it is an essential part of many designs. Possible ice defense measures include:

Structural defense against ice ride-up

Active countermeasures against high ice forces

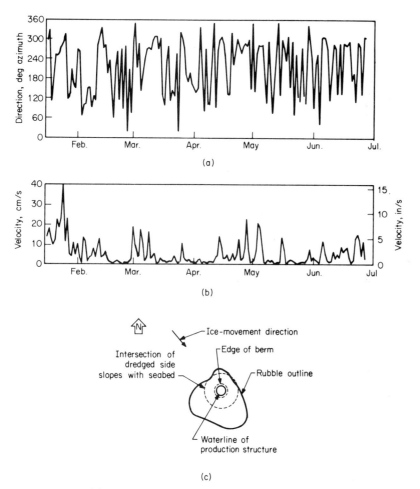

Figure 3.112 (a) Ice-movement direction, (b) velocity plots, and (c) computed rubble outline plot for deep-water structure. *(Allyn and Charpentier, 1982, reprinted by permission of Offshore Technology Conference.)*

Monitoring and predicting extreme ice events for evacuation and shutdown procedures

Active countermeasures against exceptional ice features or ice events

Countermeasures against difficult operational conditions

Vertical or steep and uneven slopes with adequate freeboard are used for protection against ice ride-up. The idea is to trigger rubble formation or ice pileup processes instead of ice ride-up (Fig. 3.113).

Active countermeasures against high ice forces include manipulation of ice contact and artificial buildup of ice rubble. Adfreeze loading conditions can be avoided by mechanical methods or by heating the contact surface between the ice and the structure. Grounded ice islands in the stable ice zone are protected by maintaining a moat between ice cover and island (Fig. 3.108). Rubble buildup can be increased by seawater spraying to improve rubble grounding and structural stability and to provide better protection against major multiyear ice features (kinetic energy dissipated by the rubble).

Continuous monitoring of ice behavior and structural response, prediction of changes in meteorological and ice conditions, and surveillance of exceptional ice features are also part of the ice defense system. If extreme ice events are predicted or collision with an exceptional ice feature seems possible, preparation for evacuation and shutdown operations can be made in advance to minimize the possible damage. Active attempts can also be made to build up protection, break dangerous formations, or tow major ice features away from their course.

Difficult ice conditions may cause severe problems for different operations such as navigation, offshore construction, or drilling from floating platforms. Icebreakers are commonly used as a countermeasure. However, operational aspects should also be considered in the design of

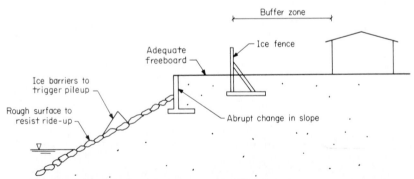

Figure 3.113 Possible ice ride-up protection features.

Figure 3.114 Arctic production and loading atoll scheme ensuring year-round operation.

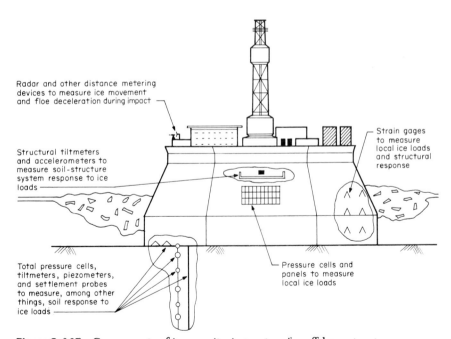

Radar and other distance metering
devices to measure ice movement
and floe deceleration during impact

Strain gages
to measure
local ice loads
and structural
response

Structural tiltmeters
and accelerometers to
measure soil-structure
system response to ice
loads

Total pressure cells,
tiltmeters, piezometers,
and settlement probes
to measure, among other
things, soil response to
ice loads

Pressure cells and
panels to measure
local ice loads

Figure 3.115 Components of ice-monitoring system for offshore structure.

(a)

(b)

(c)

(d)

Figure 3.116 Examples of fixed drilling platforms used in Beaufort Sea. (a) Grounded ice island. (Courtesy of G. Cox.) (b) Artificial gravel island. (Courtesy of Esso Resources Canada Ltd.) (c) Artificial island retained by monolithic caisson unit. (Courtesy of Esso Resources Canada Ltd.) (d) Monolithic caisson drilling unit. (Courtesy of Canmar Ltd.)

offshore structures. One example is shown in Fig. 3.114. The harbor area of the hydrocarbon production and loading atoll is protected from drifting ice by the structure itself. Brash ice growth in the harbor area can be controlled using methods described in Sec. 3.6.3. At least one entrance remains open during major rubble-formation events around the structure.

Offshore structures are commonly equipped with sophisticated instrumentation systems, which monitor their behavior under environmental loads. These systems serve as an early warning channel and also provide information on the validity of design criteria and applied theories. Some components of an ice-monitoring system are shown in Fig. 3.115. Ice forces can be estimated from pressures in the ice cover, deceleration of impacting floes, pressure sensors and panels at the surface of the structure, structural reactions, and soil-structure interaction. All methods are inaccurate, but reliable results can be obtained by comparing information from different measurements. Ice features, processes, and dynamics around the structure are mostly studied with optical and mechanical equipment. This kind of monitoring and field experience will form a sound basis for the development of design criteria and statistical design methods.

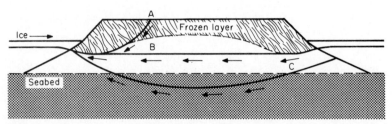

Figure 3.117 Possible island failure modes. *A*—edge failure; *B*—failure through island fill; *C*—failure through seabed. *(de Jong et al., 1975.)*

Structures for ice-covered waters The role of ice should not be over-emphasized in the design of offshore structures. The simplicity, effectiveness, and reliability of construction and installation operations are extremely important factors in ice environments. Also other offshore design criteria, such as earthquakes, wave loads, bottom scour, wave overtopping, and icing, should be given full attention. As a matter of fact, wave loads are larger than ice loads for many types of deep-water structures, even in relatively severe ice conditions. Massive ice-resistant structures very often also interfere with waves, causing surprisingly severe wave run-up, scouring, and icing problems.

Some fixed drilling platforms used in the Beaufort Sea are shown in Fig. 3.116. The designs are of the massive type. The overall stability against ice forces is one principal design consideration. There are several methods to analyze foundation behavior and capacity.

Some typical failure planes for a limit-state analysis are shown in Fig. 3.117. Local edge failure would require the failure of the frozen soil layer in the winter, while failure through the seabed is possible in weak soil conditions. The sliding resistance along the horizontal plane through the fill can be estimated simply from

$$R = (\rho_1 V_1 + \rho_2 V_2)g \tan \phi + Ac \qquad (3.82)$$

where V_1 = volume of fill above water level
V_2 = volume of fill between water level and sliding plane
ρ_1 = density of fill
ρ_2 = buoyancy density of fill
ϕ = angle of friction of fill or seabed at horizontal failure plane
A = area of failure surface
c = cohesion of seabed at failure plane

The factor of safety against sliding is obtained simply by dividing the sliding resistance by the design ice force. Replacing the top layer of weak

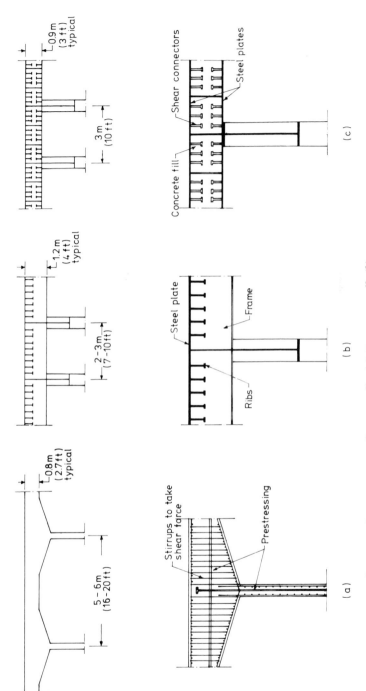

Figure 3.118 Typical structural arrangements for ice walls. (*a*) Concrete wall. (*b*) Steel wall. (*c*) Composite wall.

183

Figure 3.119 Ice-resistant monopod-type platform operated by Union Oil Company at Cook Inlet, Alaska.

sea-bottom soil by better material is a common practice to reduce settlements and improve sliding resistance.

EXAMPLE 3.7 Estimate the factor of safety against fill failure for a gravel island constructed into a land-fast ice zone. The island has a freeboard of 5 m (16 ft), a diameter at the waterline of 120 m (400 ft), a fill density of 1800 kg/m^3 (110 lb/ft^3), and an angle of friction of the fill of 30°. The design ice thickness in midwinter is 2.0 m (7 ft) and the maximum ice-movement rate, 1 m/h (3 ft/h). The partly consolidated rubble field around the island is assumed to increase the total force on the island by 20%.

From Eq. (3.45) we get

$$\dot{\epsilon} = \frac{1}{2(120)(3600)} = 1.2 \times 10^{-6}\,\text{s}^{-1}$$

The uniaxial strength of cold arctic columnar grained ice at this strain rate may be on the order 0.5 MPa (70 lb/in^2). Assuming conservatively that the strength factor equals 3 and the contact coefficient k is 0.8, we get from Eq. (3.44)

$$F = 1.2(0.8)(0.9)(3.0)(0.5)(120)(2.0) = 310 \text{ MN } (70{,}000 \text{ kips})$$

At breakup conditions with a nominal ice strength of 1.5 MPa (220 lb/in^2), an effective ice thickness of 1.8 m (6 ft), and an indentation factor $I = 0.6$, we get from Eq. (3.46)

$$F = 1.2(0.9)(0.6)(1.5)(120)(1.8) = 210 \text{ MN } (47{,}000 \text{ kips})$$

The breakup load is smaller than the midwinter load with the above assumptions.

If the island slope is 1/3, the fill buoyant density 1000 kg/m^3 (62 lb/ft^3), and the location of the critical shear plane 3 m (10 ft) below the mean water level, the island resistance against fill failure is, from Eq. (3.82),

$$R = \left\{ 1800 \frac{\pi(5)}{3} [60^2 + 60(45) + 45^2] \right.$$
$$\left. + 1000 \frac{\pi(3)}{3} [69 + 69(60) + 60] \right\} (9.8)(\tan 30)$$
$$= 670 \times 10^6 \text{ N} = 670 \text{ MN } (150{,}000 \text{ kips})$$

The factor of safety is then about 2.1.

Heavy ice walls are typical for caisson and gravity structures designed for ice conditions. Exceptionally strong structures are necessary to resist high local ice impacts. Some typical structural arrangements are shown in Fig. 3.118.

In areas with no multiyear ice, massive caisson and gravity structures may be replaced by more slender designs, especially in deeper waters where ordinary oceanographic criteria govern. However, the existence of ice does have an impact on the design, ruling out structural elements that are too slender, such as those in ordinary steel jackets. Simple monopod or multileg structures should be favored in deep waters (Fig. 3.119), while caisson and island-type structures are still applicable in shallow waters.

CHAPTER

4

FROST

4.1 Seasonal Frost and Permafrost

The existence and the degree of frost are closely connected with the climate. The requirement for substantial frost penetration [about 0.3 m (1 ft)] is usually met when the mean freezing index exceeds 50°C·days (90°F·days). If the mean annual temperature is lower than 0°C (32°F), some frost may exist throughout the year. This is regarded as permafrost. The permafrost is first sporadic or discontinuous, both laterally and vertically, but becomes thicker and continuous further north as the mean annual temperature approaches -10°C (14°F). The existence of frost in the northern hemisphere is illustrated in Fig. 4.1.

The thermal regime of the ground depends on its surface temperature, thermal properties, and geothermal gradient. The makeup of seasonal frost and permafrost areas is illustrated in Fig. 4.2. At the top there is the active layer where seasonal freezing and thawing occur. This layer is very important in cold region geotechnical engineering and foundation design. Below the active layer there is thawed ground or permafrost, depending on the mean ground surface temperature. The mean surface temperature is related to the mean air temperature, but it may be several

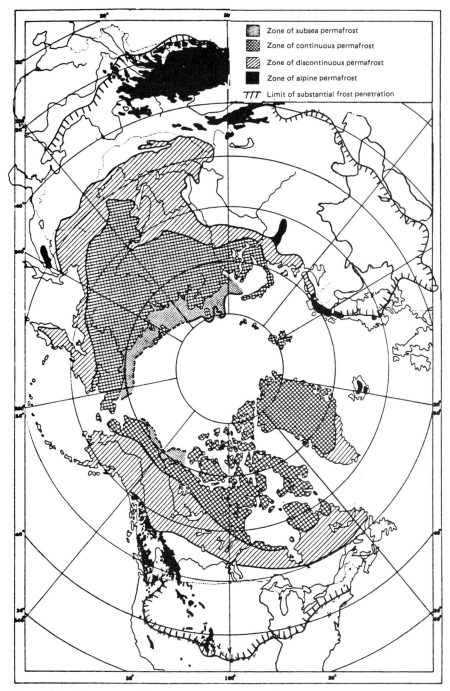

Figure 4.1 Existence of frost in the northern hemisphere. [*Péwé, 1982* (permafrost); *Burdick et al., 1978* (seasonal frost).]

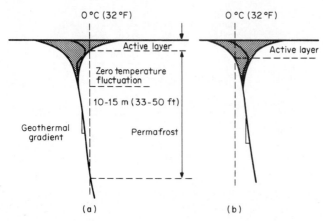

Figure 4.2   Thermal regime of ground in *(a)* permafrost and *(b)* seasonal frost areas.

degrees warmer due to the seasonal snow cover and the conditions of the energy exchange process at the surface. The seasonal ground temperature variation decreases with increasing depth and becomes insignificant for engineering purposes at depths of 5 to 15 m (15 to 50 ft). In greater depths the ground thermal regime is governed by the mean surface temperature and the geothermal flux. The geothermal gradient is typically on the order of 2 to 3 °C per 100 m (1 to 1.5 °F per 100 ft), but varies depending on the type of soil and long-term climatological changes. In discontinuous and marginal permafrost areas, layers of talik, that is, unfrozen ground, may exist within the permafrost or between active layer and permafrost.

4.1.1   Frozen Soil Classification

All ground material below 0 °C (32 °F) is generally classified as frozen, although the bulk of the freezing of soil water may occur significantly below this temperature in fine-grained soils. The freezing of soil alters its texture and mechanical properties with the exception of dry and coarse mineral soils and rock. In the classification of frozen soils the conventional soil classification is combined with a description of segregated ice in the frozen soil. If a substantial ice stratum exists, it is described separately. The description and classification of frozen soils by Linell and Kaplar (1966) is given in Table 4.1.

4.1.2   Frost Features

There are various types of frost features and processes, especially in permafrost areas, that have to be considered in engineering design and

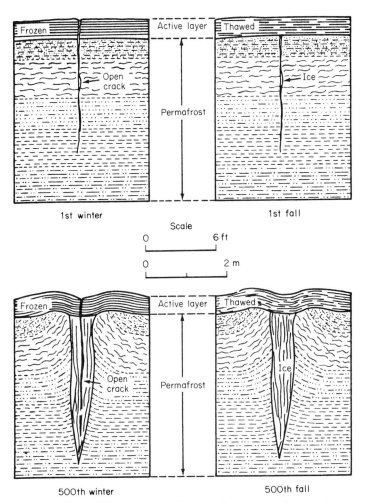

Figure 4.3 Schematic drawing showing the growth of ice wedges. *(Lachenbruch, 1962.)*

construction. Ice wedge polygons are maybe the most important and widespread terrain features in permafrost areas. They are easily detected from the air in open tundra but may be well masked by vegetation and a deep active layer in the discontinuous permafrost zone.

Ice wedges are formed in the following fashion. Frozen ground cracks due to thermal contraction during winter and the cracks become filled with snow, ice, and water. The ground then warms up and expands during summer and some of the soil material is replaced by ice. In the subsequent winter cracks are formed at the same locations, and the process repeats itself season after season (Fig. 4.3). Polygon ground is not

TABLE 4.1   Classification of Frozen Soils

I. *Description of soil phase* (independent of frozen state)[a]

Major group		Subgroup		Field identification
Description	Designation	Description	Designation	

II. *Description of frozen soil*

Major group		Subgroup		Field identification
Segregated ice is not visible by eye.[b]	N	Poorly bonded or friable	Nf	Identify by visual examination. To termine presence of excess ice, procedure given below[c] and hand m nifying lens as necessary. For soils fully saturated, estimate degree of saturation: medium, low. Note p ence of crystals, or of ice coat around larger particles.
		Well bonded — No excess ice	Nb — n	
		Well bonded — Excess ice	Nb — e	
Segregated ice is visible by eye. Ice 25 mm (1 in) or less in thickness.[b]	V	Individual ice crystals or inclusions	Vx	For ice phase, record the followin *applicable:* Location Size Orientation Shape Thickness Pattern of Length arrangement Spacing Hardness ⎤ Structure ⎬ per III below Color ⎦ Estimate volume of visible segreg ice present as percentage of total s ple volume.
		Ice coatings on particles	Vc	
		Random or irregularly oriented ice formations	Vr	
		Stratified or distinctly oriented ice formations	Vs	

III. *Description of substantial ice strata*

Major group		Subgroup		Field identification
Ice greater than 25 mm (1 in) in thickness.	ICE	Ice with soil inclusions	Ice + soil type	Designate material as *ice*[d] and use scriptive terms as follows, usually item from each group, as applicab
		Ice without soil inclusions	Ice	Hardness: Hard, Soft (of mass, not individual crystals) — Structure: Clear, Cloudy, Porous, Candled, Granular, Stratified — Color (Examples): Colorless, Gray, Blue — Adm tur (Exam Con few silt i sic

Classify soil phase by the unified soil classification

Pertinent properties of frozen materials which may be measured by physical tests to supplement field identification	Guide for construction on soils subject to freezing and thawing	
	Thaw characteristics	Criteria
-place temperature ensity and void ratio 　a. In frozen state 　b. After thawing in place ater content (total H_2O, cluding ice) 　a. Average 　b. Distribution	Usually thaw-stable	The potential intensity of ice segregation in a soil is dependent to a large degree on its void sizes and *for pavement design purposes* may be expressed as an empirical function of grain size as follows: 　Most inorganic soils containing 3 percent or more of grains finer than 0.02 mm (0.0008 in) in diameter by weight are frost-susceptible for pavement design purposes. Gravels, well-graded sands, and silty sands, especially those approaching the theoretical maximum density curve, which contain 1½ to 3 percent finer by weight than 0.02 mm (0.0008 in) size should be subjected to a standard laboratory frost susceptibility test to evaluate actual behavior during freezing. Uniform sandy soils may have as high as 10 percent of grains finer than 0.02 mm (0.0008 in) by weight without being frost-susceptible. However, their tendency to occur interbedded with other soils usually makes it impractical to consider them separately.
rength 　a. Compressive 　b. Tensile 　c. Shear 　d. Adfreeze astic properties astic properties nermal properties e crystal structure (using tical instruments) 　a. Orientation of axes 　b. Crystal size 　c. Crystal shape 　d. Pattern of arrangement	Usually thaw-unstable	Soils classed as frost-susceptible under the above pavement design criteria are likely to develop significant ice segregation and frost heave if frozen at normal rates with free water readily available. Soils so frozen will fall into the *thaw-unstable* category. However, they may also be classed as *thaw-stable* if frozen with insufficient water to permit ice segregation. 　Soils classed as non-frost-susceptible under the above criteria usually occur without significant ice segregation and are usually *thaw-stable* for pavement applications. However, the criteria are not exact and may be inadequate for some structure applications: exceptions may also result from minor soil variations. 　In permafrost areas, ice wedges, pockets, veins, or other ice bodies may be found whose mode of origin is different from that described above. Such ice may be the result of long-time surface expansion and contraction phenomena or may be glacial or other ice which has been buried under a protective earth cover.
me as Part II above, as plicable, with special emphasis n ice crystal structure.		

TABLE 4.1 *(Continued)*

Definitions

Ice coatings on particles are discernible layers of ice found on or below the larger soil particles in a frozen soil mass. They are sometimes associated with hoarfrost crystals, which have grown into voids produced by the freezing action.

Ice crystal is a very small individual ice particle visible in the face of a soil mass. Crystals may be present alone or in a combination with other ice formations.

Clear ice is transparent and contains only a moderate number of air bubbles.[e]

Cloudy ice is relatively opaque due to entrained air bubbles or other reasons, but is essentially sound and nonpervious.[e]

Porus ice contains numerous voids, usually interconnected and usually resulting from melting at air bubbles or along crystal interfaces from the presence of salt or other materials in the water, or from the freezing of saturated snow. Though porous, the mass retains its structural unity.

Candled ice is ice that has rotted or otherwise formed into long columnar crystals, very loosely bonded together.

Granular ice is composed of coarse, more or less equidimensional ice crystals weakly bonded together.

Ice lenses are lenticular ice formations in soil occurring essentially parallel to each other, generally normal to the direction of heat loss and commonly in repeated layers.

Ice segregation is the growth of ice as distinct lenses, layers, veins, and masses in soils, commonly but not always oriented normal to direction of heat loss.

Well bonded signifies that the soil particles are strongly held together by the ice and that the frozen soil possesses relatively high resistance to chipping or breaking.

Poorly bonded signifies that the soil particles are weakly held together by the ice and that the frozen soil consequently has poor resistance to chipping or breaking.

Friable denotes extremely weak bond between soil particles. Material is easily broken up.

Thaw-stable frozen soils do not, on thawing, show loss of strength below normal long-time thawed values nor produce detrimental settlement.

Thaw-unstable frozen soils show, on thawing, significant loss of strength below normal long-time thawed values and/or significant settlement as a direct result of the melting of the excess ice in the soil.

[a] When rock is encountered, standard rock classification terminology should be used.

[b] Frozen soils in the N group may, on close examination, indicate presence of ice within the voids of the material by crystalline reflections or by a sheen on fractured or trimmed surfaces. However, the impression to the unaided eye is that none of the frozen water occupies space in excess of the original voids in the soil. The opposite is true of frozen soils in the V group.

[c] When visual methods may be inadequate, a simple field test to aid evaluation of volume of excess ice can be made by placing some frozen soil in a small jar, allowing it to melt and observing the quantity of supernatant water as a percentage of total volume.

[d] Where special forms of ice, such as hoarfrost, can be distinguished, more explicit description should be given.

[e] Observer should be careful to avoid being misled by surface scratches or frost coating on the ice.

NOTES: The letter symbols shown are to be affixed to the Unified Soil Classification letter designations, or may be used in conjunction with graphic symbols, in exploration logs or geological profiles. Example — a lean clay with essentially horizontal ice lenses.

The descriptive name of the frozen soil type and a complete description of the frozen material are the fundamental elements of this classification scheme. Additional descriptive data should be added where necessary. The letter symbols are secondary and are intended only for convenience in preparing graphical presentations. Since it is frequently impractical to describe ice formations in frozen soils by means of words alone, sketches and photographs should be used where appropriate, to supplement descriptions.

The abbreviation nfs is commonly used to designate non-frost-susceptible materials on exploration logs and drawings.

SOURCE: Linell and Kaplar (1966).

a very desirable base for construction because of its sensitivity to thermoerosion. Other types of patterned ground resulting from very slow frost processes in the active layer include circles and nets in coarse or rocky ground and mudboils in fine-grained ground.

There are also separate frost features that do not create any kind of pattern on the ground. These include

Frost mounds, that is, small ice-cored structures, which may form and disappear in one season

Pingos, that is, up to 50-m (160-ft)-high hills, which form generally from the thaw bulbs beneath former lakes as water trapped between permafrost and the freezing front begins to freeze and expand

Palsas, that is, ice-cored structures up to 7 m (23 ft) high, covered usually with organic matter and commonly found in the discontinuous and sporadic permafrost zone

The frost features are illustrated in Fig. 4.4.

Thermokarst, slides and flows, and aufeis are among those frost processes that are of special concern in engineering. Thermokarst is the result of thawing of ice-rich permafrost. The change in the ground thermal regime may be caused by disturbance of the vegetation cover, for example by fire or surface traffic or by alteration of the ground drainage patterns due to construction (Figs. 4.5 and 4.6). Thermokarst topography is very uneven with mounds and sinkholes, beaded streams, and extensive erosion.

Flows and slides are also often caused by changes in the ground thermal regime in natural or cut permafrost slopes. Flows range from solifluction and skinflows, which are limited to the active layer, to the more extensive bimodal and multiple retrogressive flows. Slides through frozen ground are not very common except along some riverbanks where thawing and erosion changes the natural equilibrium conditions.

When the path of water flow is blocked in the winter, water may penetrate to the surface and freeze in subsequent layers. This phenomenon, called icing or aufeis, reaches its most extensive forms in small streams that freeze to the bottom. Although the amount of flow may be limited, the icing may cover considerable areas and attain several meters in thickness by the end of the winter. Aufeis may also be caused by springs or groundwater blocked between the freezing front and permafrost or other impervious materials. The changes in natural drainage patterns caused by road construction often trigger icing formations. Icing occurs commonly both in permafrost and in seasonal frost areas, causing problems especially for road maintenance.

(a)

(b)

(c)

(d)

Figure 4.5 Thermokarst caused by construction that had altered natural drainage patterns.

4.1.3 Frost and Engineering

Frozen ground is generally very strong and provides excellent support for short-term loads. Under long-term loading creep will occur, especially in warm ice-rich frozen ground. Undesirable settlements may thus result under heavy loads and the allowable values for bearing pressure will have to be reduced.

The most important frost problems are, however, not connected directly to frozen ground, but to freezing and thawing of the ground. Only clean coarse-grained mineral soils and ice-free bedrock are not affected by frost action. If soil contains a few percent of grains smaller than 0.02 mm (0.0008 in), it is generally frost-susceptible. Three important factors are related to the behavior of frost-susceptible soils: frost heave, thaw weakening, and thaw settlement.

When a frost-susceptible soil freezes in natural conditions with the presence of moisture, ice segregation generally occurs. Heaving is the result of water migration to the freezing front and ice-lense formation. The severity of heaving has been found to be related to the capillarity of

Figure 4.4 Different types of frost features. *(a)* Ice-wedge polygon formation. *(b)* Large pingo. *(c)* Palsas. *(d)* Frost mound. *(Courtesy of Matti Seppälä.)*

Figure 4.6 Large thermokarst depression. *(Courtesy of Duwayne Anderson.)*

the soil. Frost heave is naturally a factor in pavement design but should also be taken into account in foundation design, because the large forces associated with heave may frost-jack piles and displace foundations.

During thaw the melting of the segregated ice causes settlement and moisture buildup. If there is improper drainage, external loading may cause excessive pore water pressures and greatly reduce the bearing capacity of the soil. As the soil drains and consolidates, it slowly regains its prefreezing strength. Thaw weakening is not necessarily associated with frost heaving. Some fine-grained soils experience strength loss during thawing, even if no ice segregation occurs and the soils actually shrink during freezing. Weight restrictions are often set in the spring for secondary roads to account for thaw weakening of the subgrade.

Some alternatives for foundation design in cold regions are illustrated in Fig. 4.7.

In seasonal frost areas the general design principle is to extend foundations below the frost-affected zone, either directly or through non-frost-

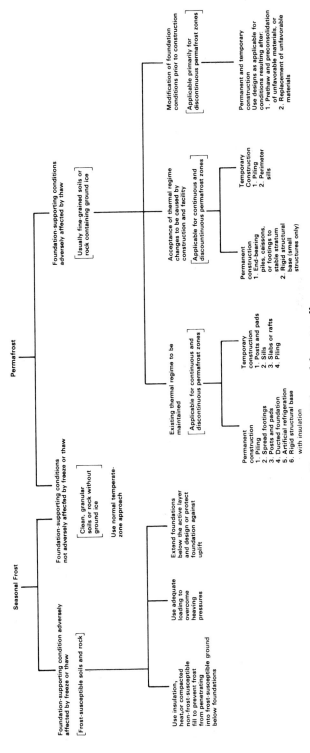

Figure 4.7 Alternatives for foundation design in cold regions. (*Adapted from Linell and Lobacz, 1980.*)

susceptible fill. Structures should also be properly anchored or isolated from frost-jacking and frost pressures. Thermal insulation and heat can be applied to limit frost penetration if deep foundations are not to be used.

In permafrost areas the general design principle is to extend the foundations to the permanently frozen ground and maintain a stable thermal regime in the ground. The most difficult conditions are often met in the discontinuous permafrost zone where permafrost is warm and the thermal balance of the frozen ground may be hard to maintain. Sometimes laborious methods such as melting or removal of permafrost may have to be used.

The role of a proper site or route survey should be emphasized in cold region engineering. Many problems can be avoided if the site or route selection is based on sufficient knowledge of soil and frost conditions and drainage patterns. The final geotechnical design is based on the results of field and laboratory investigations and thermal and geotechnical analyses.

4.2 Properties of Frozen Ground

Soil is generally considered to be a complex material with nonlinear properties depending on its grain size distribution, mineral content, compaction, water content, and other factors. When soil freezes, its complexity is further increased by the presence of ice. The soil matrix becomes stronger as ice fills pores and connects mineral particles together. The viscoelastic and thermal properties of ice add to the properties of the soil. On the other hand, especially in fine-grained soils, part of the bound water in the soil matrix remains unfrozen in subfreezing temperatures due to the effect of surface forces. Hence the interaction of all four components, that is, soil particles, air, water, and ice, seems to contribute to the behavior of frozen soil. In engineering the occurrence of ice seems to be the most significant factor.

Frozen soil may contain ice in varying amounts and forms, as shown in Table 4.1. The friable frozen soils have a very low moisture content, and their mechanical properties are not very much affected by freezing and thawing. The ice-bonded coarse-grained soils with relatively low moisture content behave in a brittle manner under short-term loading until the bond is broken. As the degree of ice saturation and ice segregation increases, the viscoplastic properties of ice play a predominant role in the soil behavior, especially under long-term loading.

The unfrozen water portion may have a considerable effect on the mechanical and thermal properties in fine-grained soils. The unfrozen

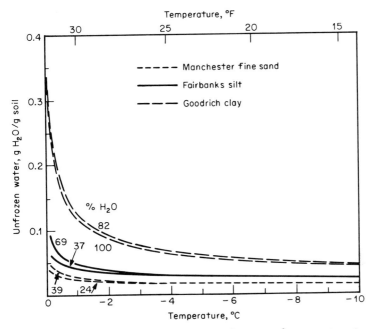

Figure 4.8 Unfrozen water contents as a function of temperature in three soils. *(Anderson et al., 1978.)*

water contents for some fine-grained soils are shown in Fig. 4.8. Anderson et al. (1973) have proposed the following simple power curve to compute the unfrozen water content:

$$w_u = \alpha T^\beta \tag{4.1}$$

where α and β are constants and T is the temperature in degrees Celsius below freezing. Some values for constants α and β are given in Table 4.2. Because water can move via the interfacial water films in partially frozen ground, normal consolidation is possible. Furthermore high stresses may cause some pressure melting at the contact points of soil particles and thus induce additional consolidation. The U.S.S.R. design code SNiP II-18-76 (1977) has categorized coarse and medium-coarse sands if $T > -0.1°C$ (31.8°F), fine and dusty sands if $T > -0.3°C$ (31.5°F), coarse silts if $T > -0.6°C$ (30.9°F), and clays if $T > -1.5°C$ (29.3°F) as plastic frozen because of the viscous properties of partly frozen fine-grained soils. The existence of salts may further lower the temperature at which soils become hard frozen and may have a remarkable effect on the mechanical properties of soils.

TABLE 4.2 Specific Surface Area, Values for Parameters α and β, and Freezing-Point Depression for Six Representative Soils

Soil	Specific surface area, m²/g (ft²/lb)	Experimental values		Freezing-point depression, °C (°F)		
		α	β	$w_w = 0.25$ g H₂O/g soil	$w_w = 0.50$ g H₂O/g soil	$w_w = 1.00$ g H₂O/g soil
Manchester very fine sand	0.016 (78)	0.0346	−0.048	1.00×10^{-18} (1.8×10^{-18})	1.00×10^{-25} (1.8×10^{-25})	1.00×10^{-31} (1.8×10^{-31})
Fairbanks silt	40 (2.0×10^{5})	0.0481	−0.326	6.37×10^{-3} (1.15×10^{-2})	7.60×10^{-4} (1.37×10^{-3})	9.07×10^{-5} (1.63×10^{-4})
Kaolinite	84 (4.1×10^{5})	0.2380	−0.360	8.72×10^{-1} (1.49)	1.27×10^{-1} (2.29×10^{-1})	1.85×10^{-2} (3.33×10^{-2})
Suffield silty clay	140 (6.8×10^{5})	0.1392	−0.315	1.56×10^{-1} (2.81×10^{-1})	1.73×10^{-2} (3.11×10^{-2})	1.91×10^{-3} (3.44×10^{-3})
Hawaiian clay	382 (1.9×10^{6})	0.3242	−0.243	2.91 (5.24)	1.68×10^{-1} (3.02×10^{-1})	9.7×10^{-3} (1.75×10^{-2})
Umiat bentonite	800 (3.9×10^{6})	0.6755	−0.343	18.2 (32.8)	2.41 (4.34)	3.19×10^{-1} (5.74×10^{-1})

SOURCE: Anderson et al. (1973).

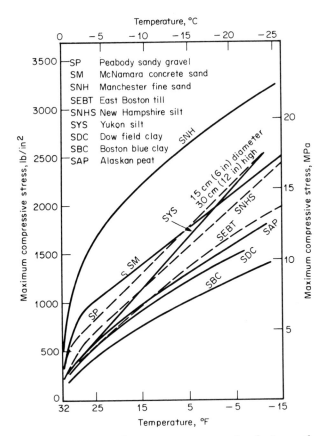

Figure 4.9 Uniaxial compressive strength of nine soil types as a function of temperature. Rate of stress increase 2.8 MPa/min (400 lb/in² · min); specimens 15 cm (6 in) high, 7 cm (2.75 in) diameter, except as noted. SBC, SDC, and SAP undisturbed; others remolded. *(Kaplar, 1954.)*

4.2.1 Strength Properties

Ice may affect the strength properties of frozen soils in several ways. The short-term strength of frozen soil is generally high, but the strength decreases as the duration of loading increases. The strength is also highly dependent on the temperature effects, as shown in Fig. 4.9.

Naturally the amount and the nature of the ice in the soil matrix have a major influence on the properties of soil. There seems to be an optimum ice content in granular soils, where the effects of the ice cementing action, intergranular friction, and dilatancy contribute simultaneously (Fig. 4.10). When the degree of ice saturation decreases, the strength of

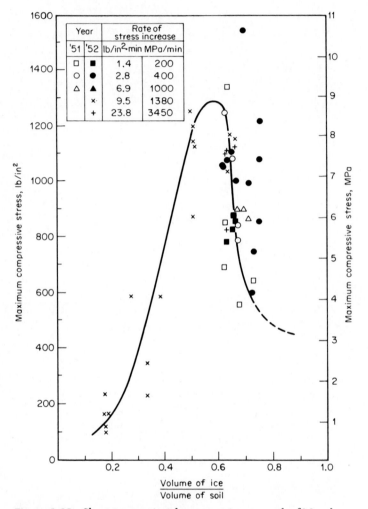

Figure 4.10 Short-term uniaxial compressive strength of Manchester fine sand versus ice content. Test temperature -1.7 to $-0.8°C$ (29.0 to 30.6°F). *(Kaplar, 1971.)*

frozen soil begins to approach that of unfrozen soil. When the degree of ice saturation is large, the strength of ice sets a lower limit also to the strength of frozen soil. The effect of unfrozen water may lower the strength of frozen soil below that of ice in fine-grained soils.

In constant-strain tests the failure of the ice bond occurring at relatively low strain values seems to govern the strength of frozen granular soils under low confining pressures (Sayles, 1973). Under higher confining pressures the ultimate strength is mobilized at larger strains and governed by sliding friction and grain interlocking. If excessive ice is present in the soil matrix, its brittle failure seems to be the governing factor in a short-term strength test. Frozen clays behave in a more plastic manner and their strength is not very sensitive to confining pressures because of the effect of unfrozen water films in the soil matrix.

The shear strength of frozen soil is often approximated based on the conventional Coulomb equation

$$s = c + p \tan \phi \qquad (4.2)$$

where s = shear strength of soil
$\quad c$ = cohesion
$\quad p$ = normal stress
$\quad \phi$ = angle of friction

Mohr's envelopes for a frozen sand illustrating this relationship are given in Fig. 4.11. The apparent angle of friction decreases with increasing confining pressure. The friction term can often be neglected in soils with excessive ice.

The strength of frozen soil also decreases with the duration of loading. This is shown in Fig. 4.12. The long-term strength is only a fraction of the instantaneous strength.

The long-term cohesion of frozen soil may eventually approach the cohesion of unfrozen soil. If the soil contains excessive ice, the long-term cohesion is very low, although the design bearing capacity for the service life of the structure may still have a reasonable value. Some estimates of practical long-term cohesion values are given in Table 4.3.

Vyalov (1963) has suggested that the time dependence of the strength of frozen soil σ_f can be represented by

$$\sigma_f = \frac{\beta}{\ln(t/B)} \qquad (4.3)$$

where t is time and β and B are constants that can be derived by short-term creep tests. Sayles (1973) noted that the relationship also gave acceptable results for frozen sand when effective stress $(\sigma_1 - \sigma_3)$ was used instead of σ_f.

4.2.2 Deformation Properties

The behavior of frozen ground under applied load depends on many factors, but the duration and magnitude of loading, the existence of ice in the soil matrix, and temperature are among the most important ones. When frozen soil is loaded, recoverable elastic deformations occur in-

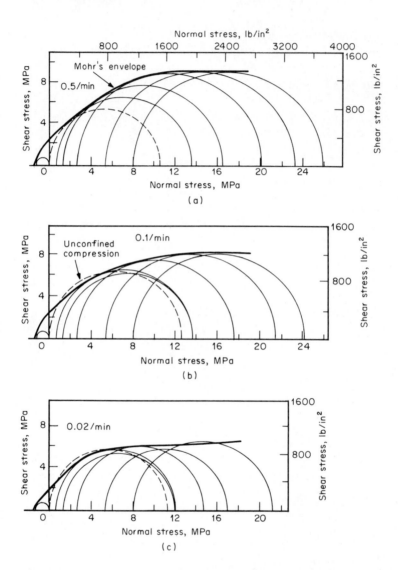

stantaneously. At high stress levels plastic deformation may also occur as a result of collapse or local failures of the soil structure. Creep of soil that may eventually lead to failure governs the deformation behavior in long-term loading. Ordinary consolidation may contribute a significant part of the long-term deformation in fine-grained soils with a large unfrozen water content.

The instantaneous elastic deformations of frozen ground are generally small compared to the long-term deformations. However, the elastic

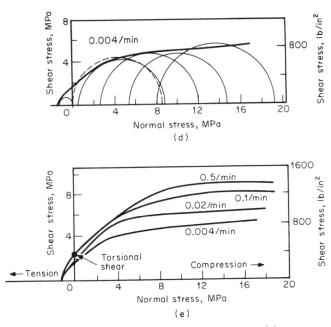

Figure 4.11 Mohr's envelopes for initial failure of frozen saturated Ottawa sand at different strain rates. Temperature −3.85°C (25°F); average dry density 1663 kg/m³ (104 lb/ft³); average void ratio 0.6. (a) – (d) Average applied strain rates per minute 0.5, 0.1, 0.02, and 0.004. (e) Composite Mohr's envelopes. (*Adapted from Sayles, 1973.*)

properties are of importance in the analyses of structures subjected to seismic or other dynamic loads. Some examples of measured values for the modulus of elasticity and the Poisson ratio are given in Fig. 4.13. The velocities of body waves are strongly affected by ground freezing, as shown in Fig. 4.14. This phenomenon can also be used in the mapping of permafrost. Vinson (1978) has presented a thorough review of the dynamic properties of frozen ground.

When a constant load is applied to frozen ground, the instantaneous deformation is followed by creep deformations. Some typical creep curves are illustrated in Fig. 4.15. All the classic stages of creep, primary creep, secondary creep, and tertiary creep can be distinguished in ordinary ice-rich frozen soils at medium stress levels. The transition from primary to tertiary creep may occur without a period of steady-state creep at relatively high stress levels, and it may take a long time before the tertiary creep stage is reached at low stress levels. In ice-poor granular soils the deformation may virtually stop after the elastic deformation

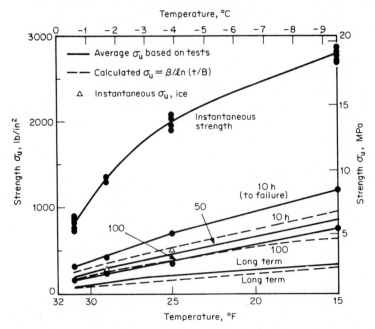

Figure 4.12 Uniaxial compressive strength of Manchester fine sand as a function of time and temperature. *(Adapted from Sayles, 1968.)*

and primary creep at low stress levels or move directly into the tertiary stage at higher stress levels. In fine-grained plastic frozen soils where hydrodynamic consolidation is of significance, the deformation may remain nonlinear with time.

The long-term deformation of frozen soil is strongly affected by stress level and temperature. The deformation under constant load before the tertiary creep stage can be estimated by a simple power expression that adds instantaneous elastic deformation, primary creep, and secondary creep,

$$\epsilon = \frac{\sigma}{E} + \epsilon_p \left(\frac{\sigma}{\sigma_p}\right)^p + \dot{\epsilon}_s \left(\frac{\sigma}{\sigma_s}\right)^s t \qquad (4.4)$$

where ϵ_p and $\dot{\epsilon}_s$ are arbitrary, small values of strain and strain rate; σ_p and σ_s are uniaxial stresses, which cause the primary creep ϵ_p and subsequently the secondary creep rate $\dot{\epsilon}_s$; and p and s are creep constants. The values for σ_p, σ_s, p, and s can be found when experimental stress–primary creep and stress–secondary creep rate curves are drawn to a log-log scale.

TABLE 4.3 Examples of Long-Term Cohesion of Some Frozen Soils

Soil	Moisture content, %	Temperature, °C (°F)	Cohesion, kPa (lb/in²)	Source
Norman Wells silt	42–185	−1 (30.2) −3 (26.6)	86 (12) 110 (16)	McRoberts et al. (1978) McRoberts et al. (1978)
Suffield clay	31–42	−1 (30.2) −3 (26.6)	73 (11) 185 (27)	Sayles and Haines (1974) Sayles and Haines (1974)
Dense varved clay	30–40	−0.35 (31.4) −1.1 (30.0) −4.1 (24.6)	180 (26) 260 (38) 420 (61)	Vyalov (1959) Vyalov (1959) Vyalov (1959)
Heavy, silty, sandy loam	43	−0.35 (31.4) −4.1 (24.6)	75 (11) 200 (29)	Vyalov (1959) Vyalov (1959)

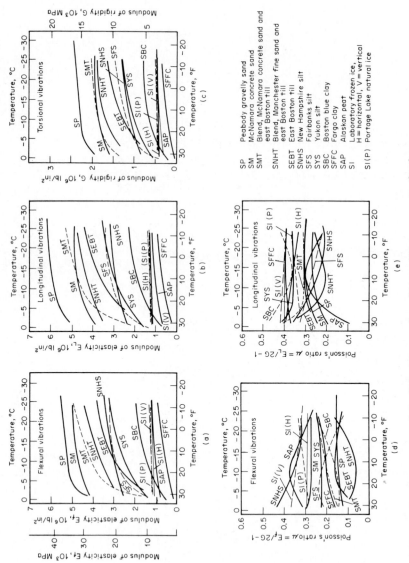

Figure 4.13 Dynamic moduli of elasticity and rigidity and Poisson ratios for frozen soils. *(Kaplar, 1969.)*

SP   Peabody gravelly sand
SM   McNamara concrete sand
SMT   Blend, McNamara concrete sand and
        east Boston till
SNHT   Blend, Manchester fine sand and
        east Boston till
SEBT   East Boston till
SNHS   New Hampshire silt
SFS   Fairbanks silt
SYS   Yukon silt
SBC   Boston blue clay
SFFC   Fargo clay
SAP   Alaskan peat
SI   Laboratory frozen ice,
        H = horizontal, V = vertical
SI(P)   Portage Lake natural ice

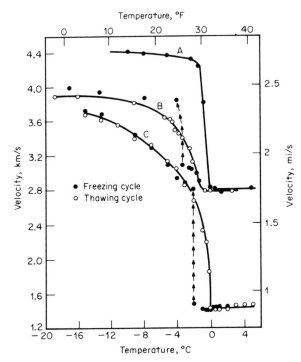

Figure 4.14 Compressional wave velocities versus temperature. A—20–30 Ottawa sand, wet density, $\gamma = 2200$ kg/m³ (137 lb/ft³); B—Hanover silt, $\gamma = 1830$ kg/m³ (114 lb/ft³); C—Goodrich clay, $\gamma = 1800$ kg/m³ (112 lb/ft³), fully saturated. *(Adapted from Nakano and Froula, 1973.)*

In practical applications it is often sufficient to study only primary creep for ice-poor soils or secondary creep when excessive ice is present. Vyalov (1962) has presented an equation for primary creep of the form

$$\epsilon = \left[\frac{\sigma t^{\lambda}}{\omega (T + T_c)^k} \right]^{1/m} \qquad (4.5)$$

where T = temperature below freezing, °C (°F)
T_c = reference temperature, = 1 °C (1 °F)
λ, m, ω, k = material constants

The derivation of these constants from experimental data has been presented in Andersland et al. (1978), for example. Some typical values for the constants are given in Table 4.4. Equation (4.5) is often also presented in nondimensional form,

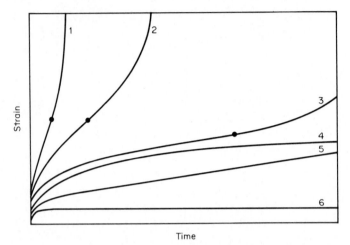

Figure 4.15 Typical creep curves for frozen soils. 1 — high stress level, rapid failure; 2 — relatively high stress level, primary creep is rapidly followed by tertiary creep; 3 — medium stress level, failure occurs eventually after critical strain has been reached; 4 — plastic frozen fine-grained soil with consolidation contributing to behavior; 5 — ice-rich soil, secondary creep governs at low stress level; 6 — ice-poor granular soil at low stress level, no long-term settlement.

$$\epsilon = \left(\frac{\sigma}{\sigma_{cuT}} \right)^n \left(\frac{\dot{\epsilon}_c t}{b} \right)^b \tag{4.6}$$

where $\dot{\epsilon}_c$ and σ_{cuT} are the reference values for strain rate and stress, and n and b are material constants (Ladanyi, 1983).

The upper bound of the secondary creep rate of ice-rich soils with little grain-to-grain contact can be estimated based on the creep behavior of ice. Nixon (1978), based on the earlier work by Nixon and McRoberts (1976), has estimated the secondary creep rate $\dot{\epsilon}$ of ice when the uniaxial stress σ is below 100 kPa (15 lb/in²) by the simple equation

$$\dot{\epsilon} = B\sigma^n \tag{4.7}$$

where $B = 5.9 \times 10^{-4}(T + 1)^{-2.93}$
$n = 1.35(T + 1)^{0.2}$
T = temperature below freezing, °C

Equation (4.7) is valid only for units of kilopascals and years. One should keep in mind that salinity accelerates the creep rate of frozen soil and reduces its strength due to the higher contents of unfrozen water. This phenomenon has been studied by Nixon and Lem (1984), among others.

TABLE 4.4 Constants for Equation (4.5)

Frozen soil	m	λ	ω MPa·h$^\lambda$/°C^k	ω [(lb/in^2)·h$^\lambda$/°F^k]	k metric	k (U.S. customary)
Suffield clay	0.42	0.14	0.73	(93)	1.2	(1.0)
Bat-Baioss clay	0.40	0.18	1.25	(130)	0.97	(0.97)
Hanover silt	0.49	0.074	4.58	(570)	0.87	(0.76)
Callovian sandy loam	0.27	0.10	0.88	(90)	0.89	(0.89)
Ottawa sand	0.78	0.35	44.72	(5500)	1.0	(0.97)
Manchester fine sand	0.38	0.24	2.29	(285)	1.0	(0.97)

SOURCE: Adapted from Sayles and Haines (1974). Data also from Vyalov (1962) and Sayles (1968).

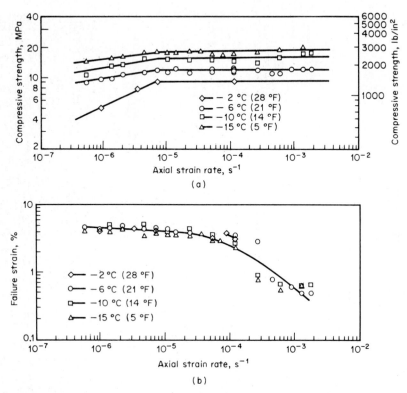

Figure 4.16 (a) Strength and (b) failure strain of frozen sand as a function of strain rate. (Adapted from Bragg and Andersland, 1980.)

It has been found that the strain corresponding to the beginning of tertiary creep is relatively constant in long-term loading for a specific frozen soil and temperature (Fig. 4.16). The long-term strength can also be approximated from the creep formulas after the critical strain has been found.

4.2.3 Thermal Properties

Temperature is one of the basic factors affecting the behavior of soils in cold regions. Thermal properties of soils, that is, thermal conductivity and specific heat or volumetric heat capacity, are fundamental parameters for determining the freeze-thaw boundaries as well as for the more complex computations of the ground thermal regime and its dynamics.

Kersten (1949) has studied extensively the thermal conductivities of frozen and unfrozen soils. Based on his work, some average thermal conductivity values are given in Fig. 4.17. The thermal conductivity of a

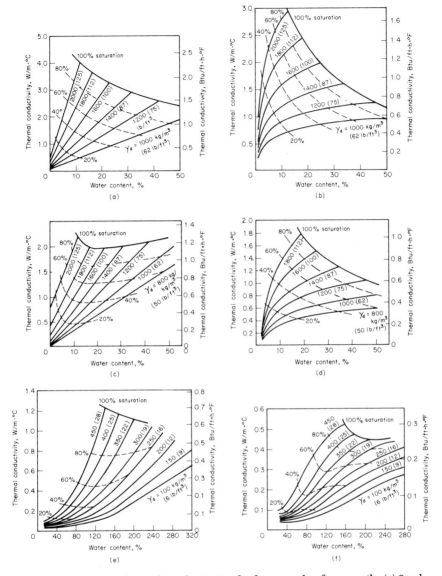

Figure 4.17 Average thermal conductivities for frozen and unfrozen soils. (*a*) Sandy soil, frozen. (*b*) Sandy soil, unfrozen. (*c*) Silt and clay, frozen. (*d*) Silt and clay, unfrozen. (*e*) Peat, frozen. (*f*) Peat, unfrozen. (*Harlan and Nixon, 1978.*)

soil increases with density and saturation but generally also as the soil freezes because the thermal conductivity of water, $0.60\ \text{W}/(\text{m}\cdot{}^\circ\text{C})$ [$0.35\ \text{Btu}/(\text{ft}\cdot\text{h}\cdot{}^\circ\text{F})$], is smaller than that of ice, $2.2\ \text{W}/(\text{m}\cdot{}^\circ\text{C})$ [$1.3\ \text{Btu}/(\text{ft}\cdot\text{h}\cdot{}^\circ\text{F})$].

It is usually assumed that all moisture in soil freezes at $0\,^\circ\text{C}\,(32\,^\circ\text{F})$. The heat capacity of soil is determined by the heat capacities of the soil particles and of water or ice. The volumetric heat capacity of unfrozen soil C_u can be estimated by

$$C_u = \frac{\gamma_d}{\gamma_w}(0.18 + 1.0w)C_w \qquad (4.8)$$

and the volumetric heat capacity for frozen soil C_f by

$$C_f = \frac{\gamma_d}{\gamma_w}(0.18 + 0.5w)C_w \qquad (4.9)$$

where γ_d = dry density of soil
 γ_w = density of water
 w = total water or ice content of dry weight of soil
 $C_w = 4.19\ \text{MJ}/\text{m}^3\cdot{}^\circ\text{C}\ (62.4\ \text{Btu}/\text{ft}^3\cdot{}^\circ\text{F})$

The specific heat capacity or mass heat capacity c can be calculated simply from the volumetric heat capacity by

$$c = \frac{C}{\gamma} = \frac{C}{\gamma_d(1 + w)} \qquad (4.10)$$

The latent heat of moisture $L_w = 334\ \text{J/g}\ (144\ \text{Btu/lb})$ is assumed to be absorbed at $0\,^\circ\text{C}\,(32\,^\circ\text{F})$. The volumetric latent heat of soil is then

$$L = \gamma_d w L_w \qquad (4.11)$$

However, in fine-grained soils a substantial part of the moisture may remain unfrozen at temperatures considerably below $0\,^\circ\text{C}\,(32\,^\circ\text{F})$. In this case it is more accurate to use the apparent specific heat capacity (expressed here per unit mass of mineral solids),

$$c = c_s + c_i(w - w_u) + c_w w_u + \frac{1}{\Delta T}\int_T^{T+\Delta T} L_w \frac{\partial w_u}{\partial T}\, dT \qquad (4.12)$$

where w is the total moisture content, w_u the unfrozen moisture content, and c_s, c_i, and c_w are the specific heat capacities of dry soil ice and water (Anderson et al., 1973). A comprehensive review of the thermal properties of soils is presented in Farouki (1981).

4.3   Ground Thermal Regime

The dynamics of the freeze-thaw boundary in the ground is of fundamental importance in cold region foundation design. It is also often
important to be able to estimate the temperature profile of frozen
ground. In order to analyze the dynamics of the ground thermal regime,
all basic modes of heat transfer, that is, conduction, convection, and
radiation, have to be considered. Significant convection may be associated with the movements of ground and surface waters. Radiation is an
important factor in the energy balance on the ground surface. However,
conduction is generally the dominant factor in the thermal computations.

4.3.1   Steady-State Solutions

The heat flow by conduction in a homogeneous soil is governed by

$$\frac{\partial T}{\partial t} = \alpha \left[\frac{\partial^2 T}{\partial x^2} + \frac{\partial^2 T}{\partial y^2} + \frac{\partial^2 T}{\partial z^2} \right] \qquad (4.13)$$

where the thermal diffusivity $\alpha = k/C$. In some cases the thermal boundary conditions remain practically unchanged for a long period of time. In
such cases it may be sufficient to study the steady-state problem by
setting $\partial T/\partial t = 0$. The thermal field and the freeze-thaw boundary will
eventually approach the solution given by Eq. (4.13).

Brown (1963) has developed some useful steady-state solutions for
frost engineering. The temperature distribution for discontinuous
ground surface temperature is given by

$$T(x, z) = T_1 + \frac{T_2 - T_1}{\pi} \tan^{-1} \left(\frac{z}{x} \right) + Gz \qquad (4.14)$$

where T_1 and T_2 are ground surface temperatures and G is the geothermal gradient. The solution for a long linear heated or cooled strip of
width $2a$ can be written

$$T(x, z) = T_1 + \frac{T_2 - T_1}{\pi} \tan^{-1} \left(\frac{2az}{x^2 + z^2 - a^2} \right) + Gz \qquad (4.15)$$

Examples of computed thermal fields are given in Fig. 4.18. The solutions can be used in estimating the steady-state temperature fields near
shores, beneath rivers, and around heated or cooled structures.

4.3.2   Progress of Freeze-Thaw Boundary

In most engineering applications it is not desirable to ignore the dynamics of the freeze-thaw interface. Assuming that all soil freezes at $0\,^{\circ}C$

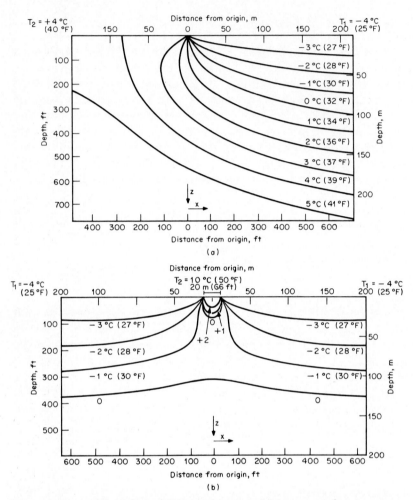

Figure 4.18 Computed ground thermal regimes. (*a*) At junction between warm and cold areas, where ground surface temperature is discontinuous. (*b*) Around a heated strip. The geothermal gradient is assumed to be 3°C/100 m (1.6°F/100 ft).

(32°F), the boundary condition at the interface in the one-dimensional case is

$$k_f \frac{\partial T_f}{\partial z} - k_u \frac{\partial T_u}{\partial z} = L \frac{dZ_f}{dt}\bigg|_{z=Z_f} \tag{4.16}$$

where Z_f = depth of freeze-thaw interface
  k_f, k_u = thermal conductivities of frozen and unfrozen soils
    L = latent heat of fusion of soil moisture

Assume that a soil initially at the uniform temperature of 0°C (32°F) is subjected to a sudden temperature change at its surface. If the temperature distribution within the active layer is assumed to be linear, the progress of freezing or thawing can be integrated readily from Eq. (4.16). If T is the change in surface temperature, the solution is given by

$$Z = \sqrt{\frac{2kT}{L}} \, t = \alpha \sqrt{t} \tag{4.17}$$

where k is the thermal conductivity of the frozen or thawed top soil layer. Since the influence of the latent heat of soil is usually large compared to that of its heat capacity, this solution, known as Stefan's solution, is reasonably accurate.

It is important to note that according to Stefan's solution frost penetrates much faster into coarse-grained relatively dry soils than into fine-grained soils with typically higher moisture content. This is mainly due to the differences in the latent heat of fusion of these soils.

Stefan's solution is also often used in a more accurate modified Berggren form,

$$Z = \lambda \sqrt{\frac{2knI}{L}} \tag{4.18}$$

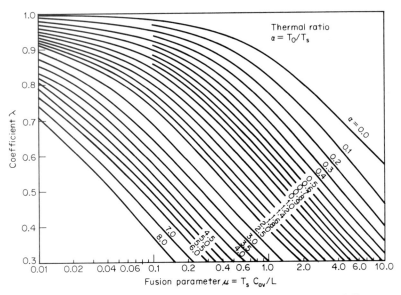

Figure 4.19 Coefficient λ for modified Berggren equation. T_0—difference between mean annual site temperature and 0°C (32°F); T_s—surface freezing or thawing index divided by length of freezing or thawing season; C_{av}—average volumetric heat capacity, $= 0.5(C_u + C_f)$; L—latent heat of fusion of soil. (*Aldrich and Paynter, 1953.*)

TABLE 4.5   Typical Values for n_f and n_t

Surface type	Freezing conditions n_f	Thawing conditions n_t
Snow	0.5	–
Sand and gravel	0.6–1.0	1.4–2.0
Turf	0.5	0.8
Spruce trees, moss over peat soil	0.3 (under snow)	0.4
Trees cleared, moss over peat soil	0.25 (under snow)	0.7
Concrete pavement	0.6–0.9	1.3–2.1
Asphalt pavement	0.4–1.0	1.6–2.3
Shaded surface	0.9	1.0

SOURCE: Lunardini (1978); Linell and Lobacz (1980).

where k is taken as the average thermal conductivity, $k_{av} = \frac{1}{2}(k_u + k_f)$. The coefficient λ, given in Fig. 4.19, takes into account the temperature changes in the soil mass. The coefficient n gives the relation between the air freezing or thawing index I and the surface freezing or thawing index I_s. Some typical values for n are given in Table 4.5.

There are still inaccuracies in the solution. For example, the effects of possible water suction toward the freezing front are ignored, the assumption of a linear temperature distribution is inaccurate, and irregular temperature changes are neglected. However, these drawbacks can be considered secondary compared to the difficulties in estimating the moisture content of soil and the correct value for the factor n.

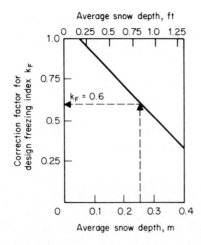

Figure 4.20   Effect of average snow depth on design freezing index. *(Mäkelä, 1982.)*

There are some extensions to the Stefan's solution that are of practical interest. The penetration of frost or thaw for a two-layered system becomes

$$Z = \left[\left(\frac{k_2}{k_1}H\right)^2 + \frac{2k_2T_s(t-t_0)}{L_2}\right]^{1/2} + \left(1 - \frac{k_2}{k_1}\right)H \qquad (4.19)$$

where T_s = absolute value of surface temperature
$\quad H$ = depth of upper layer
$\quad k_1, k_2$ = thermal conductivities of upper and lower layers
$\quad t_0$ = time to freeze or thaw upper layer, = $H^2L_1/2k_1T_s$
$\quad L_1, L_2$ = latent heat of fusion of moisture in upper and lower layers

Instead of Eqs. (4.18) and (4.19), Stefan's solution with a reduced design freezing index can be used to estimate frost penetration into snow-covered ground (Fig. 4.20). However, the average snow depth should be selected carefully. The snow cover is typically thin during the coldest winters.

The progress of thaw or frost around a hot or cold pipe buried in a medium originally at 0°C (32°F) is given by

$$2\bar{R}^2 \ln(\bar{R}) - \bar{R}^2 + 1 = \frac{4kT_s t}{a^2 L} \qquad (4.20)$$

where a = pipe radius
$\quad \bar{R}$ = normalized thaw radius, = R/a
$\quad k$ = thermal conductivity of soil within the radius of phase change

More accurate solutions for insulated pipes, taking into account the average ground temperatures and pipe burial depths, are given in Fig. 4.21.

In foundation design for permafrost areas it is also important to be able to estimate the warmest ground temperatures beneath the active layer. Linell and Lobacz (1980) have estimated the temperature T_z at depth z below the permafrost table by

$$T_z = A_0 \left[1 - \exp\left(-z \sqrt{\frac{\pi}{\alpha p}}\right)\right] \qquad (4.21)$$

where A_0 is the temperature of the permafrost at the depth where no annual variation occurs, $\alpha = k/C$ is the thermal diffusivity, and p the period of the thermal wave (365 days).

Lunardini (1982) has presented more analytical phase-change solutions for insulated geometries, including infinite strips, rectangular buildings, circular storage tanks, and buried pipes. However, in nature the boundary conditions may be complex, and soil conditions may vary

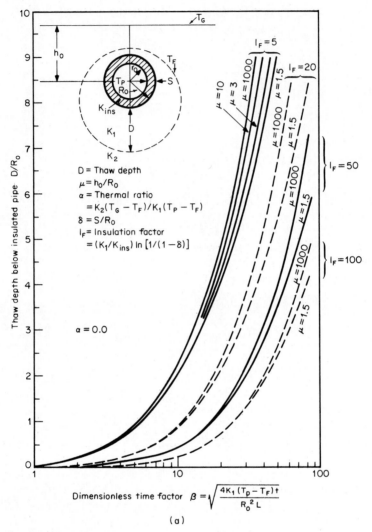

Figure 4.21 Thaw depth versus time for different values of α. (*Hwang et al., 1980.*)

considerably. Although analytical solutions are acceptable for most common and simple cases, there are also a number of more sophisticated construction projects that require the application of numerical methods in thermal analysis in two or three dimensions, including the effects of moisture transport and consolidation. These methods are discussed in Sec. 4.10.

EXAMPLE 4.1 Estimate the depth of thaw and the highest permafrost temperatures for the soil profile shown in Figure 4.22a. The mean ground surface

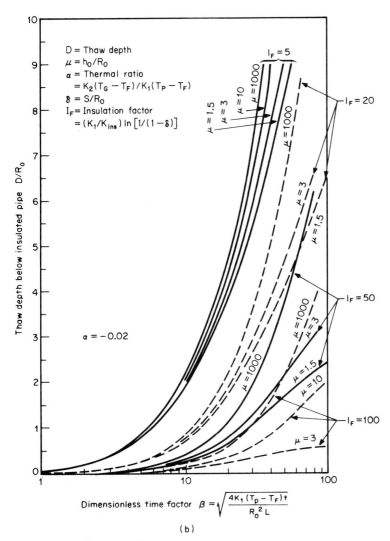

Figure 4.21 *(Continued)*

temperature is $-3°C$ $(26.6°F)$, the design thawing index $I = 1800°C \cdot days$ $(3240°F \cdot days)$, and the design thaw period is 150 days. The area is assumed to be shaded by an elevated building, which gives $n_t = 1.0$.

The time required to thaw the frozen gravel layer can be solved from Eq. (4.18). An average value is used for thermal conductivity, $k_1 = 0.5(1.5 + 1.0) = 1.25$. With thermal ratio $\alpha = 3/(1800/150) = 3/12 = 0.25$, an average volumetric heat capacity $C = 1700/1000[0.18 + 0.75(0.06)](4.19)(10)^6 = 1.6 \times 10^6$ J/(m³ · °C)[23.86 Btu/(ft³ · °F)] and a fusion parameter $\mu = 12(1.6)(10)^6/1700(0.06)(334)(10)^3 = 0.56$, we get $\lambda = 0.83$ (Fig. 4.19). Now t_0 can be solved from

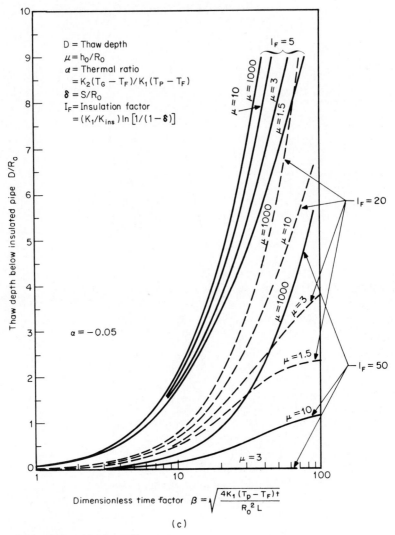

Figure 4.21 *(Continued)*

$$t_0 = \frac{H^2 L_1}{2\lambda^2 k_1 T_s} = \frac{0.9^2(1700)(0.06)(334)(10)^3}{2(0.83)^2(1.25)(12)} = 1.34 \times 10^6 \text{ s}$$
$$\approx 15 \text{ days}$$

Because the moisture content of silt is large, Eq. (4.19) is reasonably accurate in computing the total thaw penetration. We get

$$Z = \left\{ \left[\frac{1.2}{1.5}(0.9)\right]^2 + \frac{2(1.2)(12)(135)(24)(3600)}{1350(0.33)(334)(10)^3} \right\}^{1/2} + \left(1 - \frac{1.2}{1.5}\right)(0.9)$$
$$= 1.85 \text{ m (6.1 ft)}$$

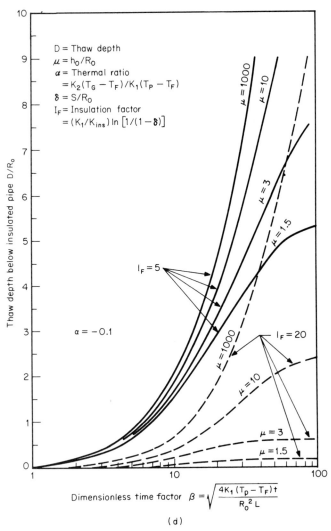

D = Thaw depth
$\mu = h_0/R_0$
α = Thermal ratio
$\quad = K_2(T_G - T_F)/K_1(T_P - T_F)$
$\delta = S/R_0$
I_F = Insulation factor
$\quad = (K_1/K_{ins}) \ln\left[1/(1-\delta)\right]$

$\mu = 1000$ $\mu = 10$

$\mu = 3$

$\mu = 1.5$

$I_F = 5$

$\mu = 1000$

$I_F = 20$

$a = -0.1$

$\mu = 10$

$\mu = 3$

$\mu = 1.5$

Thaw depth below insulated pipe D/R_0

Dimensionless time factor $\beta = \sqrt{\dfrac{4K_1(T_P - T_F)t}{R_0^2 L}}$

(d)

Figure 4.21 *(Continued)*

The temperature at the depth below the influence of annual temperature fluctuations is assumed to be $-2.7°C$ ($27.1°F$). The temperature distribution is given by Eqs. (4.9) and (4.21). At a depth of 5 m (16.4 ft), it is

$$T = -2.7\left[1 - \exp\left(-3.15\sqrt{\pi\left[\frac{2.0}{1.95\,(10)^6}(365)(24)(3600)\right]^{-1}}\right)\right]$$

$$= -1.7°C\ (28.9°F)$$

The result is presented graphically in Fig. 4.22*b*.

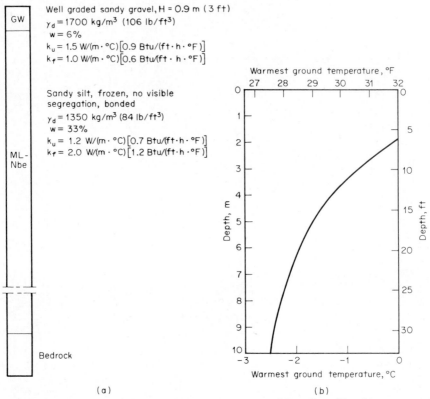

Figure 4.22 Example of temperature computation. (*a*) Soil profile. (*b*) Warmest ground temperature.

4.4 Frost Action

Three important phenomena are associated with frost action: frost heave, thaw settlement, and thaw weakening. Heaving may cause structural distresses and frost jacking. During thaw soil consolidates and may loose a significant part of its prefreezing strength. Not all soils are adversely affected by freezing and thawing. Numerous criteria have been suggested for frost-heaving and thaw-weakening susceptibility, as discussed in Chamberlain (1981). Although frost heaving and thaw weakening are not always associated with one another, engineers commonly speak about frost susceptibility in general. It is often defined based on the particle size distribution. A Canadian criterion is shown as an example in Fig. 4.23.

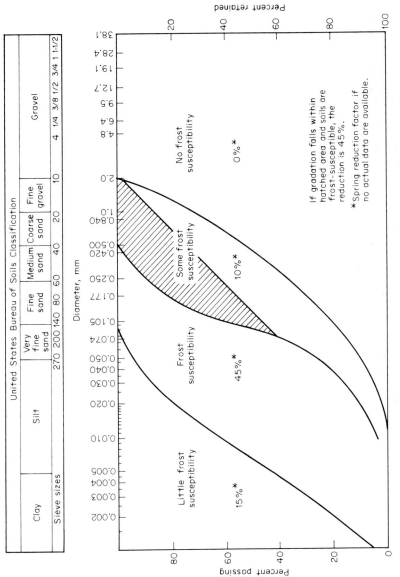

Figure 4.23 Grain distribution limits for frost susceptibility according to Canadian Department of Transport. (*Adapted from Armstrong and Csathy, 1963.*)

4.4.1 Frost Heaving

Frost heave is a result of ice segregation as water drawn to the freezing front forms ice lenses which replace the soil. The "in-place conversion" of soil moisture into ice may also contribute to the heave, but it is generally of minor importance. Three conditions must be met for ice segregation to occur: freezing front, source of water, and frost-susceptible soil.

Casagrande concluded as early as in 1931 that under natural freezing conditions and with sufficient water supply one should expect considerable ice segregation in nonuniform soils containing more than 3% of grains smaller than 0.02 mm (0.0008 in) and in very uniform soils containing more than 10% of grains smaller than 0.02 mm (0.0008 in). This can be explained by the fact that the grain size distribution and especially the size of the small grain factor (D_{10}) are related to factors such as void structure, capillarity, and permeability, which govern the movements of pore water within the soil. Capillary forces that attract moisture to the freezing front and pressures that cause the heave are inversely proportional to the void size. On the other hand, the permeabilities of very-fine-grained soils are low and the water movement to the freezing front is thus limited. Typical values of permeability and capillary rise are given in Table 4.6.

One theory explaining the mechanics of ice segregation and giving a basis for estimating heaving pressures has been presented in Everett and Haynes (1965).

The ice segregation process is quite complex. It depends on the rate of heat removal, the water supply, and other factors. An ice lense continues

TABLE 4.6 Typical Values for Permeability and Capillary Rise

Soil type	Permeability k, m/s (ft/s)	Capillary rise h_c, m (ft)
Coarse sand	$10^{-1}-10^{-2}$ $(3 \times 10^{-1}-3 \times 10^{-2})$	$0.03-0.15\,(0.1-0.5)$
Medium sand	$10^{-2}-10^{-3}$ $(3 \times 10^{-2}-3 \times 10^{-3})$	$0.1-0.5$ $(0.3-1.7)$
Fine sand	$10^{-3}-10^{-4}$ $(3 \times 10^{-3}-3 \times 10^{-4})$	$0.3-2$ $(1-7)$
Coarse silt	$10^{-4}-10^{-5}$ $(3 \times 10^{-4}-3 \times 10^{-5})$	$1-5$ $(3-17)$
Medium silt	$10^{-5}-10^{-7}$ $(3 \times 10^{-5}-3 \times 10^{-7})$	$2-8$ $(7-26)$
Fine silt	$10^{-7}-10^{-8}$ $(3 \times 10^{-7}-3 \times 10^{-8})$	$6-15$ $(20-50)$
Clay	$10^{-8}-10^{-10}\,(3 \times 10^{-8}-3 \times 10^{-10})$	>8 (>26)
Till	$10^{-4}-10^{-9}$ $(3 \times 10^{-4}-3 \times 10^{-9})$	$1-15$ $(3-50)$

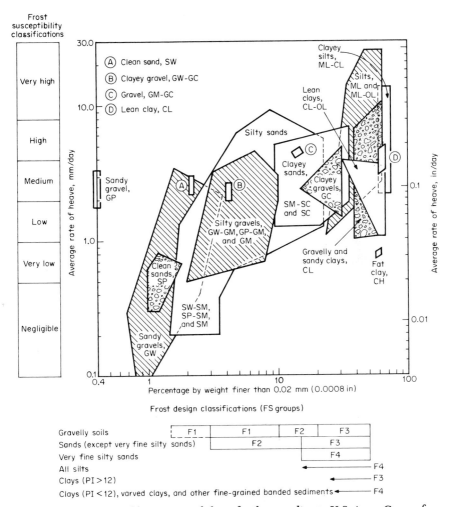

Figure 4.24 Degree of frost susceptibility of soils according to U.S. Army Corps of Engineers freezing tests. *(Adapted from Kaplar, 1974.)*

to grow as long as it can attract moisture from the adjacent soil at the rate of freezing. When the heat removal rate exceeds the moisture supply, the freezing front advances and a new ice lense forms. The total amount of heave is often close to the thickness of segregated ice layers. In permafrost areas the amount of moisture in the active layer is often limited, whereas in seasonal frost areas an unlimited source may be provided from the groundwater table. Thus in principle seasonal frost areas have greater potential for frost heaving than permafrost areas.

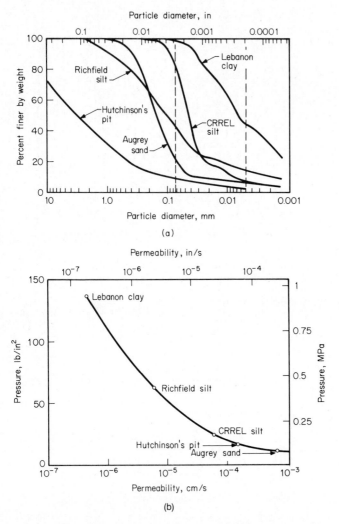

Figure 4.25 (a) Grain-size distribution and (b) heave pressure versus permeability for different soils. (*Hoekstra et al., 1965.*)

Ideal conditions for frost heaving occur when silty soil freezes steadily but at a slow rate. Large ice lenses develop as a continuous flow of water can be maintained to the freezing front. The permeabilities of clays and the capillarity values of coarser soils are too low to produce large heaving. Typical heaving rates for different soils are illustrated in Fig. 4.24. Models simulating ice segregation and frost heaving have been developed by several authors (see, for example, O'Neill and Miller, 1982), but it may still take some time before these models become widely applicable in practice.

The heave itself may cause unacceptable deformations in such structures as pavements or pipelines. In foundation design it is important to know also about the associated pressures that act parallel to the direction of frost penetration. Clays develop the highest pressures, which may exceed 1 MPa (145 lb/in²) (Fig. 4.25). However, these pressures decrease rapidly if some deformation is allowed to occur. An example of the effect of stress on heaving is given in Fig. 4.26.

Freezing ground may adhere to structures. As it heaves, lifting forces are exerted on structures such as piles and cold foundation walls. If the structures are not capable of resisting the tangential forces, they are progressively lifted up.

4.4.2 Thaw Consolidation

Ice segregation and heaving during freezing of frost-susceptible soil is naturally followed by consolidation when the soil thaws. Considerable loss of strength and poor deformation properties may also result because the soil has a loosened structure and a much higher moisture content compared to its prefreezing state. If thawing is rapid and there is insuffi-

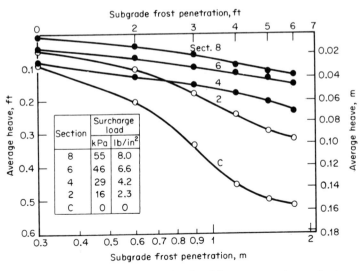

Figure 4.26 Effect of surcharge load and frost penetration on frost heave. (*Adapted from Aitken, 1974.*)

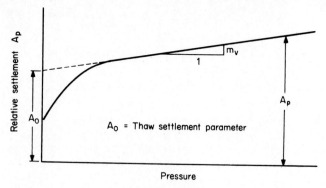

Figure 4.27 Generalized thaw settlement curve. *(Watson et al., 1973.)*

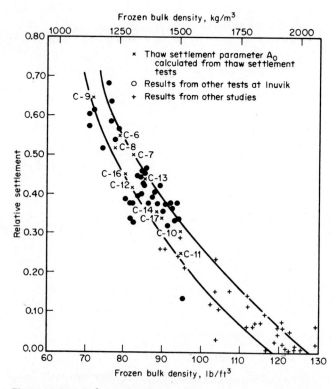

Figure 4.28 Relationship between relative thaw settlement and frozen bulk density. *(Watson et al., 1973.)*

cient drainage, incomplete dissipation of excess pore water pressures may occur, causing further weakening of the soil.

The thaw settlement of a uniform soil layer can be estimated simply by the sum of the settlements caused by the phase change of segregated ice and the subsequent consolidation (Crory, 1973; Watson et al., 1973),

$$s = A_0 Z + m_v \int_0^Z \sigma' \, dz \qquad (4.22)$$

with

$$A_0 = \frac{e_f - e_{th}}{1 + e_f} = 1 - \frac{\gamma_{df}}{\gamma_{dth}} \qquad (4.23)$$

where Z = thickness of originally frozen layer
$\quad m_v$ = coefficient of compressibility
$\quad \sigma'$ = effective stress
$\quad e_f$ = void ratio of frozen soil
$\quad e_{th}$ = void ratio of thawed soil
$\quad \gamma_{df}$ = dry density of frozen soil
$\quad \gamma_{dth}$ = dry density of thawed soil

This is illustrated in Fig. 4.27.

Estimates of the initial thaw settlement can also be made based on the frozen bulk density, as shown in Fig. 4.28, or even based on the thickness of ice lenses in very ice-rich soils.

4.4.3 Thaw Weakening

Thaw consolidation occurs as water is squeezed out from the ground by the weight of the soil itself and the applied load. If the rate of thawing is rapid in fine-grained soils, water may be produced at a rate that exceeds the discharge capacity. Part of the effective stress is transferred to the pore water. The strength of thawed soil, essentially based on friction, is thus reduced.

The situation of thaw weakening is clearly illustrated in Fig. 4.29 by the resilient modulus (total stress divided by recoverable strain) determined from a repeated load-plate bearing test on a pavement silt subgrade system. The resilient modulus, like the strength of frozen soil, is high, but decreases sharply during thaw. The recovery occurs as the possible excess pore water pressures dissipate, followed by ordinary consolidation and desaturation. The conditions for recovery are better in seasonal frost areas than in permafrost areas because downward drainage is not generally blocked by an impervious layer except during the short thawing season.

The development of excess pore water pressures can be estimated based on the one-dimensional thaw consolidation theory proposed by Morgenstern and Nixon (1971), which combines Stefan's solution for the thaw progress and the linear consolidation theory. The key parameter is the thaw consolidation ratio R given by

$$R = \frac{\alpha}{2c_v^{1/2}} \tag{4.24}$$

where α is the constant in Stefan's solution and the coefficient of consolidation c_v is related to the coefficient of compressibility m_v and the permeability k by

$$c_v = \frac{k}{m_v \gamma_w} \tag{4.25}$$

The thaw consolidation ratio R expresses the ratio between the rate of water production at the thaw front and the rate at which water can be discharged upward. The basic solutions for pore pressure buildup in saturated ground corresponding to loading by the soil's own weight and under applied load p_0 are given in Fig. 4.30. It can be seen that for low values of R excess pore pressures are small, but as R becomes greater than

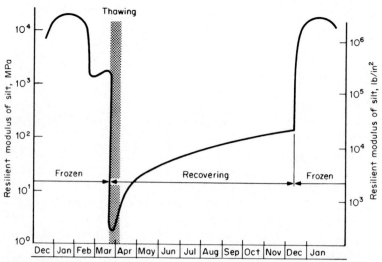

Figure 4.29 Seasonal variations for resilient modulus of a silt subgrade. (*Johnson et al., 1978.*)

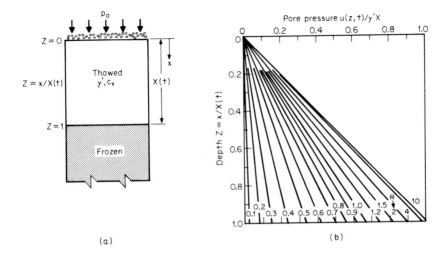

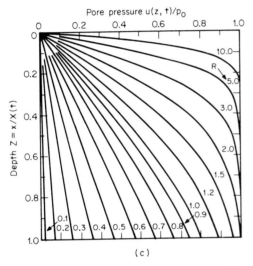

Figure 4.30 One-dimensional thaw consolidation. R — thaw consolidation ratio. (a) Schematic illustration. (b) Excess pore pressures with no applied external load ($p_0 = 0$). (c) Excess pore pressures for weightless material ($\gamma' = 0$). (*Adapted from Morgenstern and Nixon, 1971.*)

1.0, the effective stresses near the thaw line approach zero and instability may occur.

McRoberts (1973) has applied the thaw consolidation theory to the stability analysis of an infinite slope. The expression for the factor of safety F in a saturated slope is given by

$$F = \frac{c'}{\gamma Z \cos \theta \sin \theta} + \frac{\gamma'}{\gamma} \frac{1}{1 + 2R^2} \frac{\tan \phi'}{\tan \theta} \qquad (4.26)$$

where Z = depth of thaw
γ = total unit weight of soil
γ' = effective unit weight of soil (submerged)
c' = effective cohesion
ϕ' = effective angle of internal friction
R = thaw consolidation ratio
θ = slope angle

Sometimes human activities causing changes in drainage patterns may trigger slides.

The thaw consolidation theory has been extended in several ways. Solutions for layered systems and nonlinear consolidation have been presented. Also the role of residual stresses or initial effective stresses that reduce the consolidation and excess pore pressures during the thaw have been pointed out. A review of these extensions is presented in Nixon and Ladanyi (1978).

Thaw settlement and weakening are major considerations in road design. Heavy maintenance efforts or weight restrictions are required in the spring on poorly constructed and improperly drained roads. In permafrost areas thermal disturbances pose another concern as the resulting uneven thaw settlements of the previously frozen soil may damage structures such as buildings, pipelines, and embankments.

4.5 Measurement of Frost-Related Properties of Soils

Extensive laboratory and in situ test programs may be required in addition to the conventional survey of soil conditions in order to control frost action. In seasonal frost areas the frost-related behavior of soil can often be anticipated from the soil profile, groundwater table, gradation curves, densities, permeabilities, moisture and organic contents, and other results of conventional geotechnical investigations. In permafrost areas a number of special in situ and laboratory tests may be carried out to determine the behavior of frozen and thawing soils in addition to sampling and conventional tests.

The frost susceptibility of soil is a property of great economic importance. Yet the gradation criteria commonly used for the evaluation of frost susceptibility cannot be considered fully acceptable. Factors such as mineralogy, soil structure, and the actual field conditions are neglected, and therefore the criteria are generally on the conservative side. Furthermore there is no sharp difference between frost-susceptible and non-frost-susceptible soils. Where fully non-frost-susceptible soils are sparse, marginal soils could be utilized for example in road construction. A comprehensive investigation program is naturally required to confirm that the extent of frost action is within tolerable limits, at least when additional frost control methods are applied. The heaving characteristics can be studied with standardized heaving-rate tests (Fig. 4.31), and the actual field soil moisture and freezing conditions should be simulated as closely as possible. Although heaving and ice segregation are related to thaw weakening, the latter phenomenon can also be investigated separately by thaw California bearing ratio (CBR) tests (Jessberger, 1975).

In permafrost areas the depth of thaw and ground temperatures are key parameters in the design. The progress of thaw can be measured simply by handprobing with steel rods or by frost tubes buried into the ground, which have an indicator solution that changes color upon freezing. Ground temperatures are generally measured with thermal sensors placed in backfilled boreholes. Soil strength and deformation properties can also be studied in situ by pressuremeter, cone penetration, and plate-bearing tests or even by test loading actual foundation structures. These tests are discussed in detail in a review by Ladanyi and Johnston 1978).

In addition to the obvious frozen bulk density and moisture content measurements, thaw consolidation tests are commonly made in a laboratory by using frozen soil samples. The test described in Crory (1973) can be made with a standard consolidation test apparatus. Also unconfined and triaxial compression tests and direct shear tests may be performed with special equipment using frozen samples in order to measure the strength and creep properties of soils. Special care should be taken, however, to avoid sublimation and mechanical or thermal disturbance of the samples on their way from the test pit or core hole to the laboratory.

4.6   Foundation Design

4.6.1   Foundations for Seasonal Frost Areas

Foundations are generally extended below the reach of frost penetration in seasonal frost areas in order to avoid large heaving forces that may

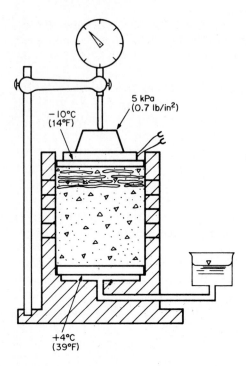

5 kPa
(0.7 lb/in²)

−10°C
(14°F)

+4°C
(39°F)

Figure 4.31 Schematic illustration of typical frost susceptibility test apparatus. *(Chamberlain, 1981.)*

damage the structures. However, the depth of frost penetration around foundations does not necessarily correspond to that in the field. Heat flow from buildings and thermal frost protection may considerably reduce the penetration of frost.

Non-frost-susceptible soils and solid bedrock are not affected by frost action. In this case only ordinary geotechnical criteria govern the foundation design. Frost-susceptible soils are often replaced by non-frost-susceptible ones to eliminate frost effects. The final foundation design with measures to counter frost action is naturally governed by economic considerations.

Heave effects   Some examples of the effects of frost heaving on foundation designs are shown in Fig. 4.32. The vertical heave forces that develop when frost penetrates beneath the foundation represent the most important component, but tangential and lateral heave forces should also be considered. The heave forces are large. In extreme cases pressures up to 1 MPa (150 lb/in²) may develop under footings, and adfreeze strengths of up to 500 kPa (70 lb/in²) have been measured on steel piles (Penner and Goodrich, 1983). Some typical values for tangential heave forces on piles and columns with diameters of 200 to 350 mm (8 to 14 in) are given in Table 4.7. In difficult situations even higher forces may develop. The structure has to be able to resist these forces with an adequate factor of safety.

Sufficient foundation depth and non-frost-susceptible backfill with proper drainage are common approaches to eliminate heave forces. Thermal protection can be used to reduce the foundation depth. Footing enlargements are effective in resisting tangential heave forces.

Piles are generally anchored against frost action, although in principle it is also possible to use low-adhesion coatings, sleeves, non-frost-susceptible backfill, or thermal protection to control uplift. The cohesion or skin friction of unfrozen soil depends principally on loading duration and burial depth. Typical values for the cohesion of clays range from 5 to 25 kPa (0.7 to 3.5 lb/in^2), and the skin friction of silts can be estimated by $T = 4Z$ (kPa), where Z is the depth in meters (Torgersen, 1976). Unloaded piles have to be driven typically to a depth of 7 to 10 m (25 to 35 ft) to resist frost jacking.

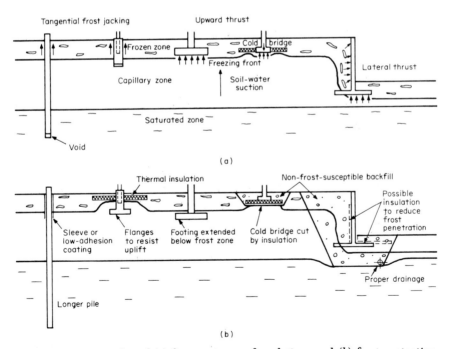

Figure 4.32 Examples of (a) frost action on foundations and (b) frost protection measures.

TABLE 4.7 Typical Uplift Forces [kN (kips)] in Clay on Piles with Diameters of 200 to 350 mm (8 to 14 in)°

Material	Design freezing index, h·°C (°F·days)		
	10,000 (750)	30,000 (2250)	50,000 (3750)
Steel	50 (11)	100 (22)	140 (31)
Wood	30 (8)	50 (11)	60 (13)
Concrete	30 (8)	80 (18)	100 (22)

° The values in silt are about half those listed when the freezing index is 10,000 h·°C (750°F·days) and about one-third when the freezing index is 50,000 h·°C (3750°F·days).
SOURCE: Torgersen (1976).

Foundations for cold structures The foundation depth often has considerable cost implications. That is why the actual thermal conditions in the ground and the design alternatives should be carefully considered. In the case of cold structures, such as unheated garages or outside structures, the foundation depth can be decreased from the value appropriate for full frost penetration by using thermal protection. Some design examples are illustrated in Fig. 4.33.

The use of thermal insulation as frost protection is based on its ability to slow down the penetration of frost. In case of a cold foundation or a ground-supported slab the required frost protection can be obtained from Table 4.8. The decline of insulation properties from normal values in severe conditions (insulation exposed to frost action, moisture, or heavy loads), as discussed in Chap. 5 (see Fig. 5.35), should, however, be taken into account.

Frost protection is most effective in shallow depths, but at least 0.2 to 0.3 m (8 to 12 in) of soil is generally required for cover. A narrow layer of non-frost-susceptible soil placed below the foundation and the insulation is also required.

The frost protection design of cold foundations and floors is discussed in detail in Aho (1977) and Algaard (1976). The width of the frost protection surrounding foundations can be determined from Fig. 4.34. Column footings require greater insulation widths than strip footings. In unheated halls the ground-supported slab has to be insulated also if frost movements are not acceptable. Situations where frost may develop below the foundation because of conduction through the structure should be avoided. This is why rigid insulation is sometimes placed underneath the foundation.

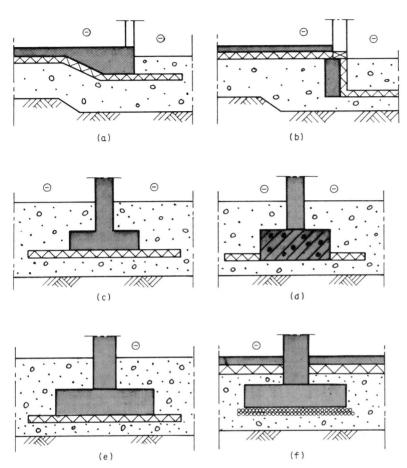

Figure 4.33 Cold foundations. Notice how the cold bridge through the structure to the foundation soil has been cut. (*a*), (*b*) Unheated buildings. (*c*), (*d*) Strip footings. (*e*), (*f*) Individual footings.

EXAMPLE 4.2 Design the frost protection for a cold garage with ground-supported slab and foundation wall (Fig. 4.33*b*). The design freezing index (once in 50 years) is $1500\,°C \cdot$ days ($2700\,°F \cdot$ days), and the mean annual air temperature $+3\,°C$ ($37.4\,°F$).

Assuming that the thickness of the non-frost-susceptible fill under the insulation is 0.6 m (2 ft) the required thermal resistance of the frost protection according to Table 4.7 is 1.5 $(m^2 \cdot C)/W$ [8.5$(ft^2 \cdot h \cdot °F)$/Btu]. If expanded polystyrene with a practical thermal conductance of 0.06 $W/(m \cdot °C)$[0.034 Btu/ $(ft \cdot h \cdot °F)$] is used, the required insulation thickness is 1.5 (0.06) = 0.09 m (3.6 in). The insulation is placed directly under the floor slab and

TABLE 4.8 Requirements for Frost Protection of Cold Structures

Non-frost-susceptible layer below insulation, m (ft)	20,000 (1500) +2 (36)	+3 (37)	≥+4 (≥39)	30,000 (2250) +1 (34)	+2 (36)	+3 (37)	≥+4 (≥39)	40,000 (3000) +1 (34)	+2 (36)	+3−+4 (37−39)	50,000 (3750) +1 (34)	+2 (36)	≥60,000 (4500) +0−+1 (32−34)
						Thermal resistance for frost insulation, $(m^2 \cdot °C)/W$ [$(ft^2 \cdot h \cdot °F)/Btu$]							
0.2 (0.7)	1.6 (9.1)	1.4 (8.0)	1.2 (6.8)	3.2 (18)	2.6 (15)	2.2 (12)	1.8 (10)	4.2† (24)	3.5 (20)	2.8 (16)	°	4.6† (26)	°
0.4 (1.3)	1.4 (8.0)	1.1 (6.2)	0.8 (4.5)	2.6 (15)	2.1 (12)	1.7 (9.7)	1.4 (8.0)	3.5 (20)	2.8 (16)	2.2 (13)	4.6† (26)	3.8 (22)	°
0.6 (2.0)	1.0 (5.7)	0.7 (4.0)	0.5 (2.8)	2.1 (12)	1.7 (9.7)	1.3 (7.4)	1.0 (5.7)	2.8 (16)	2.2 (13)	1.6 (9.1)	3.8 (22)	2.9 (17)	5.0† (28)
0.8 (2.6)	0.6 (3.4)	0.4 (2.3)	0.3 (1.7)	1.7 (9.7)	1.3 (7.4)	1.0 (5.7)	0.7 (4.0)	2.2 (13)	1.6 (9.1)	1.3 (7.4)	2.9 (17)	2.2 (13)	3.8 (22)
1.0 (3.3)	0.4 (2.3)	0.3 (1.7)	0.2 (1.1)	1.3 (7.4)	1.0 (5.7)	0.7 (4.0)	0.5 (2.8)	1.6 (9.1)	1.2 (6.8)	1.0 (5.7)	2.2 (13)	1.7 (9.7)	2.8 (16)
1.5 (4.9)	0	0	0	0.8 (4.5)	0.6 (3.4)	0.4 (2.3)	0.2 (1.1)	1.0 (5.7)	0.7 (4.0)	0.5 (2.8)	1.4 (8.0)	1.0 (5.7)	1.8 (10)

° Foundation depth requires increase.
† Increase of foundation depth preferred.
SOURCE: Mäkelä and Tammirinne (1979).

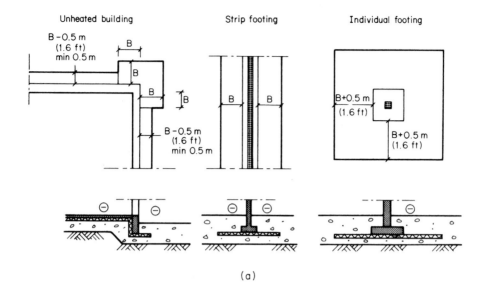

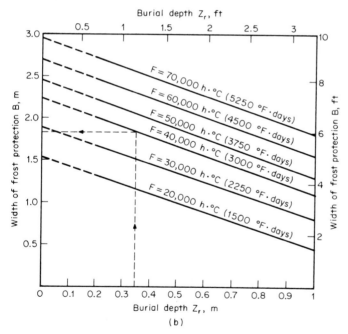

Figure 4.34 Width of frost protection for cold structures. *(Adapted from Aho, 1977.)*

extended with 0.3 m (1 ft) of cover (Fig. 4.34), $1.8 - 0.5 = 1.3$ m (4.4 ft) outside the foundation wall except in corners, where the extension is 1.8 m (6 ft).

The frost-free foundation depth without thermal insulation for compacted gravel with $w = 6\%$, $k_{av} = 1.7$ W/(m·°C) [1.0 Btu/(ft·h·°F)], and $\gamma_d = 1900$ kg/m³ (118 lb/ft³) can be computed from Eq. (4.18). If $n = 0.9$, the length of the freezing season is 180 days,

$$\alpha = \frac{3}{0.9(1500)/180} = 0.4$$

and

$$\mu = 7.5 \frac{(1.9)[0.18 + 0.75(0.06)](4.18)(10)^6}{(1.9)(0.06)(334)(10)^6} = 0.35$$

we get (Fig. 4.19) $\lambda = 0.84$. The maximum depth of frost becomes

$$Z = 0.84 \sqrt{\frac{2(1.7)(0.9)(1500)(24)(3600)}{1.9(0.06)(334)(10)^6}} = 2.7 \text{ m (9 ft)}$$

Although some deformation could be tolerated and the foundation depth could be reduced somewhat, the frost-protected alternative is clearly more appealing than a solution with a gravel pad about 2 m (7 ft) thick or with elevated floor and foundations extended to that depth.

Foundations for heated structures   Thermal losses through the ground are effective in reducing frost penetration around the foundations in heated buildings. The situation is illustrated in Fig. 4.35 for two typical cases: building with a ground-supported floor slab and building with only lightly ventilated creeping space, which remains warm throughout the year. Because the principle of design is to prevent frost penetration into frost-susceptible soil under the foundations, the foundation depth can be reduced from the value of frost penetration in the open field.

Typical frost-free foundation depths for these two cases are given in Table 4.9. The following considerations should be recognized:

1. The results are applicable only to warm buildings with inside temperatures above $+17°C$ (63°F). If the temperatures are between $+17°C$ (63°F) and $+5°C$ (41°F), the foundation depth should be increased by 0.1 to 0.2 m (4 to 8 in).

2. The thermal resistance of the soil can be taken into account when the total thermal resistance of the structure is estimated in the case of the ground-supported slab (energy conservation requirements). This can be done roughly according to Table 4.10 and Fig. 4.36.

3. The foundation depth for parts that are outside the surface of the foundation wall should be increased by their distance from this surface. However, the foundation depth should not be larger than that for cold structures.

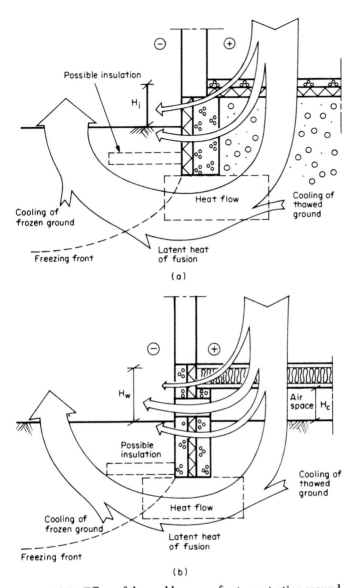

Figure 4.35 Effect of thermal losses on frost penetration around heated buildings in two typical cases. (*a*) Ground-supported slab. (*b*) Elevated slab.

TABLE 4.9   Frost-Free Foundation Depth [m (ft)] for Heated Buildings Wider than 4 m (13 ft) °

Type of foundation	Part of foundation	Design freezing index, $h \cdot °C$ (°F·days)		
		35,000 (2625)	50,000 (3750)	65,000 (4875)
Ground-supported floor slab, total thermal resistance of floor structure 4 $(m^2 \cdot °C)/W$ [22.7 $(ft^2 \cdot h \cdot °F)/Btu$]; thermal insulation at outer surface of foundation wall	Wall Corner	1.0/1.2 (3.3/3.9) 1.3/1.6 (4.3/5.2)	1.3/1.5 (4.3/4.9) 1.6/2.0 (5.2/6.6)	1.6/1.9 (5.2/6.2) 2.0/2.4 (6.6/7.9)
Ground-supported floor slab, total thermal resistance of floor structure 2 $(m^2 \cdot °C)/W$ [11.4 $(ft^2 \cdot h \cdot °F)/Btu$]; thermal insulation at outer surface of foundation wall	Wall Corner	0.8/1.0 (2.6/3.3) 1.1/1.4 (3.6/4.6)	1.1/1.3 (3.6/4.3) 1.5/1.9 (4.9/6.2)	1.5/1.8 (4.9/5.9) 1.9/2.2 (6.2/7.2)
Creeping space, ventilation from outside 0.6 $L/(m^2 \cdot s)$ [7 $ft^3/(ft^2 \cdot h)$]; total thermal resistance of floor slab 4 $(m^2 \cdot °C)/W$ [22.7 $(ft^2 \cdot h \cdot °F)/Btu$]	Wall Corner	1.1/1.4 (3.6/4.6) 1.4/1.8 (3.6/5.9)	1.4/1.8 (4.6/5.9) 1.7/2.2 (5.6/7.2)	1.8/2.2 (5.9/7.2) 2.1/2.6 (6.9/8.5)

° The effects of snow are ignored. The smaller foundation depth is used for fine-grained soils and the larger one for coarse-grained soils and moraines.
SOURCE: Mäkelä and Tammirinne (1979).

4. The thermal resistance of a properly insulated foundation wall should be at least 1 $(m^2 \cdot °C)/W$ [5.7 $(ft^2 \cdot h \cdot °F)/Btu$]. Insulation should be placed on the outer surface of the foundation wall or close to it. It should penetrate at least halfway to the foundation depth. If insulation is placed at the inner surface of the foundation wall, frost may penetrate underneath the foundation. However, if the foundation is constructed of a material such as lightweight concrete with adequate thermal resistance, no additional insulation is required.

5. The distance between the upper surface of the floor insulation and the surrounding ground H_i should not be larger than 0.6 m (2 ft) in case a (Fig. 4.35). The height of the ventilation space H_c should not be larger than 1 m (3.3 ft), and the thermal resistance of the foundation wall should be increased from the minimum value if the distance H_w is larger than 0.6 m (2 ft) in case b.

TABLE 4.10 Typical Conductivities and Resistance Values for Foundation Soils in Different Areas under a Ground-Supported Slab

Soil type	Thermal conductivity, normal conditions, W/(°C·m) [Btu/(ft·h·°F)]	Thermal resistance of foundation soil, (m²·°C)/W [(ft²·h·°F)/Btu]			
		Under building		Outside foundation wall	
		Outer border area	Inner border area	0–1 m (0–3.3 ft) below ground surface	1–2 m (3.3–6.6 ft) below ground surface
Clay, sand, and gravel; good drainage to depth of 1.2 m (4 ft)	1.4 (0.81)	0.80 (4.54)	3.20 (18.17)	0.40 (2.27)	1.60 (9.09)
Silt, sand, and gravel; poor drainage, till	2.3 (1.33)	0.50 (2.84)	2.00 (11.36)	0.25 (1.42)	1.00 (5.68)
Rock	3.5 (2.02)	0.30 (1.70)	1.20 (6.81)	0.15 (0.85)	0.60 (3.41)

SOURCE: Finnish Ministry of the Interior (1978).

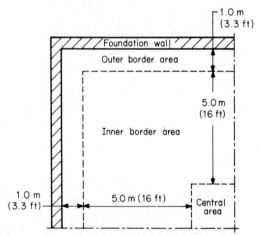

Figure 4.36 Ground-supported slab forming the basis for calculations in Table 4.10. *(Finnish Ministry of the Interior, 1978.)*

Although insulation is beneficial to the energy budget and the physical functioning of a building, unnecessary overinsulation should be avoided. For example, insulation of a ground-supported slab outside the border areas surrounding the foundation wall is not very effective in reducing heat losses. Heavy insulation of the floor slab increases the frost-free foundation depth, as does overefficient ventilation of the space underneath an elevated floor.

If frost protection is used as indicated in Fig. 4.35, the frost-free foundation depth can be further reduced. The frost protection for the two typical cases can be derived from Figs. 4.37 to 4.39. If the thermal resistance of the structure is larger than 5 $(m^2 \cdot °C)/W$ [28 $(ft^2 \cdot h \cdot °F)/$ Btu], the foundation depth should be increased by at least 0.1 m (4 in), and if the temperature of the building is between $+17$ and $+5°C$ (63 and 41°F), the foundation depth should be increased by 0.1 to 0.3 m (4 to 12 in).

The frost protection should extend at least 0.8 m (2.6 ft) from the foundation wall. If the foundation wall is poorly insulated with insulation within or on the inner surface of the foundation wall, the width of the insulation should be increased to a minimum of 1.2 to 1.5 m (4 to 5 ft). At the corners the thickness of insulation should be increased by about 40% and the width by 0.5 m (20 in), or the foundations should be extended to a greater depth.

The foundation depth should also be increased in the case of extensions from the foundation wall, such as footings, by the length of the

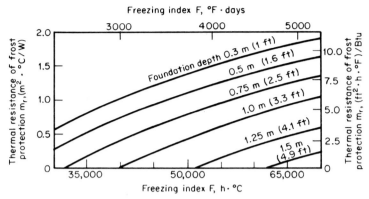

Figure 4.37 Required frost protection for heated building with well-insulated foundation wall. The results are applicable to (1) a ground-supported slab case when thermal resistance of floor is 2 $(m^2 \cdot {}^\circ C)/W$ [11.4 $(ft^2 \cdot h \cdot {}^\circ F)/Btu$]; and (2) an elevated floor when thermal resistance of floor is 2 $(m^2 \cdot {}^\circ C)/W$ [11.4 $(ft^2 \cdot h \cdot {}^\circ F)/Btu$] and ventilation rate of air space beneath building is 0.3 $L/(s \cdot m^2)$ [0.37 $gal/(min \cdot ft^2)$] or less. *(Adapted from Mäkelä and Tammirinne, 1979.)*

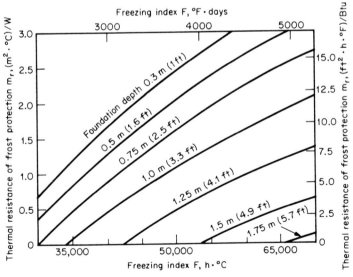

Figure 4.38 Required frost protection for heated building with ground-supported slab. Thermal resistance of floor is 4 $(m^2 \cdot {}^\circ C)/W$ [22.7 $(ft^2 \cdot h \cdot {}^\circ F)/Btu$]; insulation of foundation wall is located at its inner surface. *(Adapted from Mäkelä and Tammirinne, 1979.)*

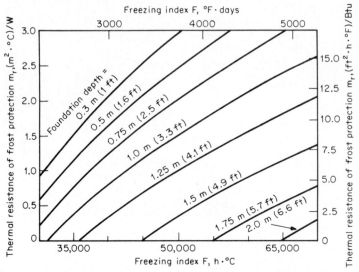

Figure 4.39 Required frost protection for heated building with elevated floor. Thermal resistance of floor 4 (m²·°C)/W [22.7 (ft²·h·°F)/Btu]; ventilation rate of air space beneath building 0.6 L/(s·m²) [0.74 gal/(min·ft²)]; foundation wall is well insulated. *(Adapted from Mäkelä and Tammirinne, 1979.)*

extension. In many cases this can be done conveniently if the soil beneath the foundation is replaced by non-frost-susceptible soil to the required depth.

Junctions between foundations of heated and cold structures and utility line connections to buildings have proved to be especially vulnerable to damage and require special attention in the design.

EXAMPLE 4.3 Determine the foundation depth for a heated structure with a ground-supported floor having a thermal resistance requirement of 4 (m²·°C)/W [23 (ft²·h·°F)/Btu]. Compare the solution with the alternative solution of having outside frost protection. The design freezing index (once in 50 years) is 1500°C·days (2700°F·days).

The thermal resistance of the foundation soil according to Table 4.10 is 0.8 (m²·°C)/W [4.54 (ft²·h·°F)/Btu] for the outer border area and 3.2 (m²·°C)/W [18.17 (ft²·h·°F)/Btu] for the inner border area, assuming good drainage arrangements. The thermal resistance of the floor is hence $R = 4.0 - 0.8 = 3.2$ (m²·°C)/W [18.17 (ft²·h·°F)/Btu] at the outer border area and $R = 4.0 - 3.2 = 0.8$ (m²·°C)/W [4.54 (ft²·h·°F)/Btu] at the inner border area. If polystyrene with an average thermal conductivity of 0.06 W/(m·°C) [0.03 Btu/(ft·h·°F)] is used, about 18 cm (7 in) of polystyrene is needed at the 1-m (3.3-ft)-wide outer border area and 5 cm (2 in) at the

Figure 4.40 This building founded directly on ice-rich permafrost has deformed seriously as a result of ground thawing and subsidence. *(Courtesy of Glenn Johns.)*

5-m (17-ft)-wide inner border area. This is quite heavy for floor insulation. If 2 (m²·°C)/W [11.4 (ft²·h·°F)/Btu] had been selected as the total thermal resistance of the system, 7 cm (2.8 in) of polystyrene at the outer border area only would have sufficed. In this case a cold floor could cause some inconvenience.

The frost-free foundation depth for this case, assuming gravel underfill, according to Table 4.9, is 1.2 m (4 ft) except in the corners, where it is 1.6 m (5.3 ft). A conservative estimate for the alternative frost-protected design is

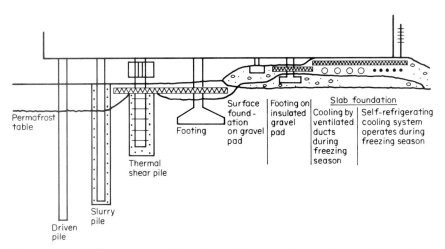

Figure 4.41 Different types of foundations for permafrost conditions. Notice how heat flow from building has been eliminated to prevent thawing of permafrost. *(Eranti and Lee, 1983.)*

(a)

(b)

Figure 4.42   Examples of pile foundations. *[Photo (d) Courtesy of J. P. Zarling.]*

given in Fig. 4.38. If we select a foundation depth of 0.5 m (1.7 ft), the required thermal resistance of frost protection is about 0.9 (m²·°C)/W [5 (ft²·h·°F)/Btu]. Assuming that the foundation wall has insulation at the outer surface, 1-m (3-ft)-wide and 5-cm (2-in)-thick polystyrene board surrounding the foundation wall would be sufficient for frost protection, except in the corners of the building, where the required insulation thickness would be 7 cm (2.76 in) and the width 1.5 m (5 ft). The solutions are on the safe side as far as frost protection is concerned because Table 4.8 and Fig. 4.38 are based on a uniform insulation layer with a thermal resistance of 4 (m²·°C)/W [22.7 (ft²·h·°F)/Btu].

The frost-free foundation depth for gravel in this case, according to Example 4.2, is 2.7 m (8.9 ft). When the heat flow from the building is considered, the foundation depth is reduced to 1.2 m (3.9 ft), and with moderate frost protection a foundation depth of 0.5 m (1.6 ft) is quite feasible. When the design freezing index is larger, the importance of selecting the proper design approach becomes even more obvious.

(c)

(d)

4.6.2   Foundations for Permafrost Areas

In permafrost areas structural loads are generally also transmitted through the active layer to the thermally stable layer. In this case this layer is perennially frozen ground, which may have to be protected from heat flow in order to avoid thaw settlements (Fig. 4.40). There are only a few exceptions to this rule. Sometimes structures can be founded directly on non-frost-susceptible ground or ice-free bedrock. Small rigid structures may be subjected to substantial deformations without loss of serviceability. Finally, in case of temporary structures, harmful deformations may not have time to develop.

Some typical foundation methods for permafrost areas are illustrated in Fig. 4.41. The loads are transmitted through the active layer by piles, columns, and footings or thaw-stable material. Permafrost is prevented from thawing under heated buildings by means of ventilation or artificial cooling. Thermal insulation can be used in this case to slow down the progress of thaw and to limit the depth of the active layer. Gravel pads are often used around structures to improve the conditions for construction and operation, to provide additional thermal protection, and to help snow control and drainage.

The most difficult foundation conditions are often encountered in discontinuous permafrost areas where permafrost is warm and in a very unstable form. The active layer is thick, and layers of talik, that is, unfrozen ground, may exist within the permafrost. Areas free from permafrost are often preferred in site selection.

If structures are founded on permafrost, artificial cooling may be used to add safety. Sometimes permafrost is prethawed and consolidated or replaced with non-frost-susceptible soil.

Pile foundations   Large buildings constructed in permafrost areas are generally founded on piles and elevated to preserve the thermal regime of the ground. Pile material may be steel, timber, or reinforced concrete. Piles are usually placed on drilled or augered holes, which are backfilled with sand or cement slurry. In some cases it is also possible to drive piles into soft and warm permafrost. Some examples of ordinary pile foundations are shown in Fig. 4.42.

Pile design in permafrost is usually based on the adfreeze bond that develops between the pile and the surrounding material. This bond has to support the pile loads and to resist the possible heaving forces. Because the strength of the bond depends on the temperature, artificial cooling may sometimes be applied to increase the capacity. Also thermal insulation may be applied to reduce the penetration of thaw and heaving forces and to increase the adfreeze length. End bearing may contribute significantly to the capacity when piles are short and provided with a bearing plate or when a very strong and nondeformable stratum such as bedrock or a dense granular layer without excessive ice is encountered. The end-bearing capacity can be estimated in the same way as for footings in permafrost.

The way in which the pile loads are transmitted into the ground depends importantly on the duration of loading. This is illustrated in Fig. 4.43. Initially as the pile is loaded, the loads are carried by the upper layer of the soil and only part of the pile is compressed. The loads are then

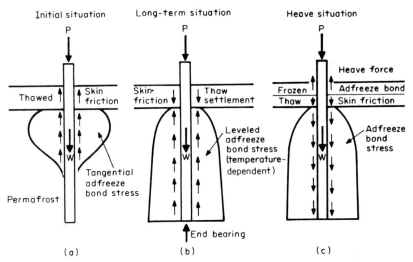

Figure 4.43 Schematic presentation of pile design with applied forces and distribution of loads.

gradually transmitted downward due to the creep of the surrounding permafrost and possible ruptures of the adfreeze bond.

The long-term bearing capacity of a cylindrical pile can be estimated from

$$P = 2\pi r \int_0^z \tau(T) \, dz \qquad (4.27)$$

where τ is the temperature-dependent long-term adfreeze bond strength and T corresponds to the warmest temperature profile. Some suggested values for bond-strength – temperature dependence are given in Fig. 4.44. The short-term bond strength is much higher than the long-term strength. However, especially long piles are not very efficient for short-term loads because of the nature of the shear stress distribution along the pile.

The surface character of the pile and the type of slurry have some effect on the adfreeze strength. Reducing the design value should be considered if the piles are treated with a low-friction coating against rust or rotting. This applies also to cases where the moisture content of the slurry is high and water migrates to the pile during construction, forming a thin ice layer on the surface of the pile. On the other hand, high design loads may be applicable for corrugated or tapered piles. (Wood piles are often installed butt down to resist heaving forces.) Shear piles provided

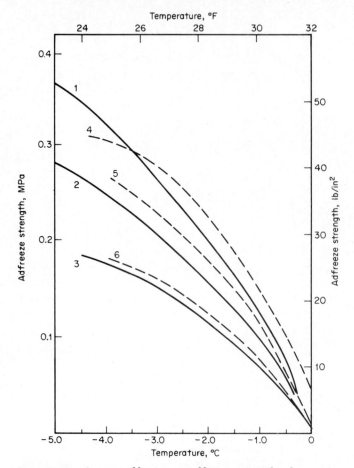

Figure 4.44 Suggested long-term adfreeze strength values. 1 — sandy soil (U.S.S.R. design code *SNiP II-18-76, 1977*); 2 — clayed soil (U.S.S.R. design code *SNiP II-18-76, 1977*); 3 — ice *(Sanger, 1969)*; 4 — saturated fine sand *(Sanger, 1963)*; 5 — average strength at ultimate pile-bearing capacity, silt-water slurried steel pipes (Linell and Lobacz, 1980); 6 — average sustainable adfreeze strength, silt-water slurried steel pipes (Linell and Lobacz, 1980).

with blades have the largest bearing capacity since the long-term shear strength of soil is somewhat higher than the bond strength (Long, 1973).

Piles have a tendency to settle significantly under loads well below the long-term capacity in ice-rich permafrost. Nixon and McRoberts (1976) expressed the steady-state pile settlement rate $\dot{s}$ by

$$\dot{s} = \frac{3^{(n+1)/2}}{n-1}\, Ba\tau_a^n \qquad (4.28)$$

where a is the pile radius and B and n are temperature-dependent creep constants. Their predictions seem to be in reasonable agreement with settlement rates actually measured in the field, as shown in Fig. 4.45.

The settlement rates are considerably lower in granular soils with no excess ice than in ice-rich soils. Weaver and Morgenstern (1981) have developed a method to estimate pile settlements based on the creep parameters of the soil in ice-poor permafrost. In important projects it may be necessary to verify the design assumptions by in situ loading tests, as described in Crory (1968).

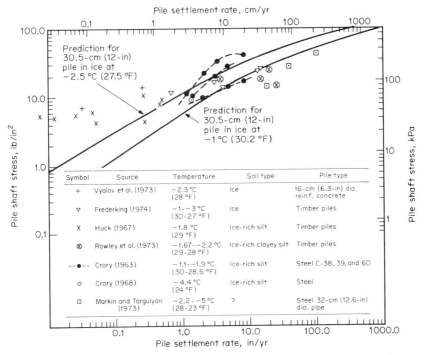

Figure 4.45 Comparison of predicted and measured pile settlement rates. *(Nixon and McRoberts, 1976.)*

Heave forces should also be taken into account in pile design in permafrost areas if the loads are light. An often mentioned rule of thumb is that piles should be embedded into permafrost at least twice the thickness of the active layer. However, at the moment when frost heave forces reach their maximum, permafrost temperatures are at their warmest and the rule is not necessarily on the safe side.

Measures that are sometimes used to avoid frost jacking in permafrost areas are quite similar to those used in seasonal frost areas, that is, replacing the soil in the active layer by a non-frost-susceptible soil around the structure or anchoring the pile against the heave. Thermal insulation can be used around the piles to limit the thickness of the active layer, just as in the areas of seasonal frost. In this case, however, it is the depth of thaw that is reduced. This kind of thaw protection increases the effective length of the pile and decreases the heaving forces. Different sleeve systems can also be used (Fig. 4.46). Concrete piles should naturally have sufficient reinforcement so that heaving forces do not cause excessive cracking.

EXAMPLE 4.4 Determine the design load on a 25-cm (10-in) steel pipe pile installed into a silt-slurry backfilled hole to the depth of 8 m (26 ft). The soil profile and the ground temperatures are given in Fig. 4.22.

The average sustainable pile capacity can be estimated for the steel-pile–silt-slurry system from Fig. 4.44. We get

Depth, m (ft)	Temperature, °C (°F)	Sustainable strength, kPa (lb/in²)	Capacity, kN (kips)
2–3 (7–10)	−0.4 (31.3)	30 (4.4)	24 (5.4)
3–4 (10–13)	−1.0 (30.2)	80 (11.6)	63 (14.1)
4–5 (13–16)	−1.5 (29.3)	110 (16.0)	86 (19.3)
5–6 (16–20)	−1.8 (28.8)	120 (17.4)	94 (21.1)
6–7 (20–23)	−2.0 (28.4)	125 (18.1)	98 (22.0)
7–8 (23–26)	−2.2 (28.0)	130 (18.8)	102 (22.9)
			Total 467 (104.8)

If the factor of safety is selected to be 2, the design load is $P = 467/2 \approx$ 230 kN (52 kips).

Because a properly mixed silt water slurry and the adjacent soil have relatively low moisture contents, the settlement rates are not expected to be critical. However, settlements up to 2 cm (0.8 in) per year could be experienced in very ice-rich silts at this loading level (Fig. 4.45).

The heave forces on the pile depend most importantly on the moisture conditions and on the soil temperatures in the active layer. In difficult conditions the adfreeze bond strength on the upper part of the pile may be as high as

Figure 4.46 Sleeve design to protect piles from heaving forces.

250 kPa (36.3 lb/in²). Assuming that this stress acts over the full depth of seasonal freezing, we get $F = 1.9\pi(0.25)(250) = 370$ kN (83 kips). If the minimum load on the pile is 130 kN (29 kips), the factor of safety against frost jacking is FS = 600/370 = 1.6, which could be considered low in certain cases.

Piles may also be subjected to tensile forces due to structural loads, such as when they are used as anchors. The analysis is quite similar to that for compressed piles and is discussed in Johnston and Ladanyi (1972). Gravity anchors, embedded plate anchors, and screw anchors represent alternatives to pile anchors in the case of power lines, for example (Fig. 4.47).

Piles have good resistance to lateral loads in permafrost. In case of short-term loading the problem can be analyzed by the theory of a beam on an elastic foundation. The behavior of a pile subjected to long-term lateral loads is governed by the creep of frozen soil. This case has been analyzed in Rowley et al. (1975) and Nixon (1984).

Thermal piles are often recommended for warm permafrost conditions [temperatures close to $-1\,°C$ ($30\,°F$) or warmer]. There are two reasons for this. First, the thermal equilibrium of warm permafrost may easily be disturbed as a result of construction (change in the surface conditions or conduction along the pile causing contact thaw). Second, the strength and creep properties of warm permafrost are not very good and difficult to estimate for pile design purposes.

There are active and passive methods to cool the ground around the piles. Active methods such as heat exchangers and forced circulation of refrigerant can be used throughout the year, but they are quite expensive. Passive methods function only when air temperatures are below

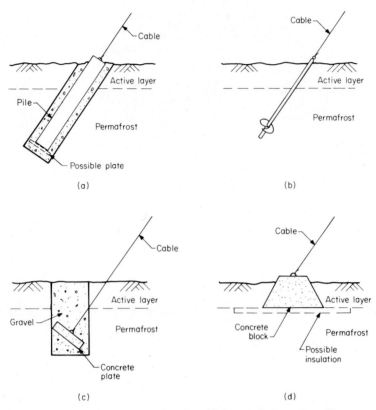

Figure 4.47 Different types of anchor designs. (*a*) Pile anchor. (*b*) Screw anchor. (*c*) Embedded plate anchor. (*d*) Gravity anchor.

freezing. The principle of commercially available two-phase passive refrigeration systems and some applications are shown in Fig. 4.48. A suitable working fluid in the pile or pipe evaporates as it is heated up by the permafrost. The vapor rises to the radiator, where it cools, condenses, and returns back to the evaporator. Typical heat removal rates range from 1 to 10 million kJ (0.95 to 9.5 million Btu) per freezing season, depending on factors such as air temperature, wind, radiator area, soil temperature, and soil conditions (Long, 1978). The thermal design aspects of these piles are discussed for example in Linell and Lobacz (1980).

Passive thermal devices lower the warmest permafrost temperatures only marginally, as shown in Fig. 4.49. However, even a small change can improve the strength and creep properties of warm permafrost significantly. Passive refrigeration systems are also effective in creating permafrost, such as when there are thawed layers within the permafrost. Rapid

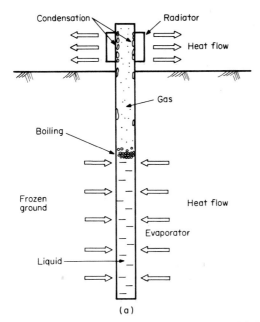

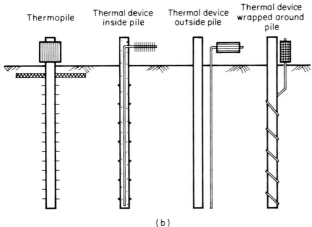

Figure 4.48 (a) Principle of two-phase passive refrigeration system. (b) Examples of its use in pile design.

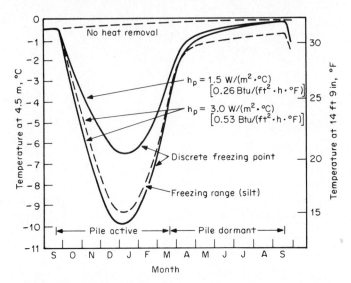

Figure 4.49 Predicted ground temperature adjacent to a thermopile at 4.5-m (15-ft) depth in $-0.5°C$ ($31.1°F$) permafrost. h_p—heat-transfer coefficient (per unit below ground pile surface area). *(Jahns et al., 1973.)*

freezeback of piles can be secured when temperatures are below freezing. Finally these thermopiles begin to operate early in the fall, cooling the ground and creating an expanding frost bulb in the thawed layer (no heaving forces because freezing occurs radially). This improves the anchorage of the pile and reduces the maximum heaving forces.

Deep footings   Deep footings extended to the permafrost represent alternatives for pile foundations. They are not as commonly used as piles due to difficulties connected with construction. If a strong stratum can be found at shallow depth, end-bearing piles, which can be treated as a special type of deep footing, may be applicable.

Both long-term bearing capacity and settlements should be considered in the geotechnical design of footings in permafrost. The bearing capacity is generally determined based on the classical Terzaghi bearing capacity theories using long-term strength values for frozen soils. The settlements are estimated by substituting the creep properties of frozen soil into the stress field under the foundation. The cavity expansion theory presented in Ladanyi and Johnston (1974) and Ladanyi (1976) offers an alternative approach to predict creep settlements and the time-dependent bearing capacity for frozen ground.

A general formula for the bearing capacity of a footing (Hansen, 1961) is given by

$$q_{\text{ult}} = \frac{P_{\text{ult}}}{A'} = \tfrac{1}{2}\,\gamma' B N_\gamma s_\gamma d_\gamma i_\gamma + c N_c s_c d_c i_c + \bar{q} N_q s_q d_q i_q \qquad (4.29)$$

where P_{ult} = bearing capacity
A' = effective footing area, $= (B - 2e_B)(L - 2e_L)$ in case of a rectangular footing
B = width of foundation
L = length of foundation
γ' = effective unit weight of soil
c = cohesion
q = effective surcharge pressure
N_γ, N_c, N_q = bearing capacity factors, which depend on effective angle of friction
s_γ, s_c, s_q = foundation shape factors
d_γ, d_c, d_q = foundation depth factors
i_γ, i_c, i_q = load inclination factors

Values for bearing capacity factors, shape factors, depth factors, and inclination factors are given in basic geotechnical literature.

The bearing capacity of centrally loaded footings buried in ice-rich permafrost (friction neglected), calculated by the simplified version of Eq. (4.29), is

$$q_{\text{ult}} = c N_c s_c + \gamma D N + \gamma D N_q s_q \qquad (4.30)$$

where D is the foundation depth.

For a frictionless soil, $N_q = 1.0$ and $N_c = 5.2$. The values for s_q and s_c are equal to 1.0 for a strip footing. For a rectangular footing they are $s_q = 1.0$ and $s_c = 1 + 0.2B/L$. The long-term cohesion can be determined by uniaxial compression tests using Eq. (4.3) and the relationship $c = \tfrac{1}{2}\sigma_f$. Some nominal bearing capacity values used in the U.S.S.R. are shown in Fig. 4.50. The allowable foundation load is determined by dividing the capacity with a factor of safety.

In principle, settlements should be computed based on the deviator stress distribution under the footing in a nonlinear viscous medium (Nixon, 1978). In practice simple solutions based on linear elasticity generally provide reasonable accuracy. The vertical stress at any point under a rectangular footing can be estimated from Fig. 4.51 using superposition. (For example, the stress under the center of a rectangular footing is four times the value calculated for one quarter.)

The creep values can be estimated based on empirical data (Fig. 4.52) or on laboratory measurements. Especially large footings in ice-rich permafrost have proved to be vulnerable to intolerable settlements.

EXAMPLE 4.5 Determine the design load on a 1.2- by 1.2-m² (4- by 4-ft²) footing buried to the depth of 2.4 m (8 ft) and underlain by 0.3 m (1 ft) of

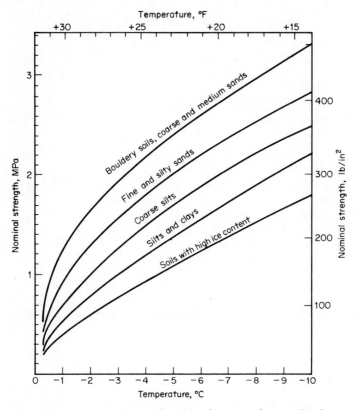

Figure 4.50 Nominal strength values for permafrost under footings according to U.S.S.R. design code SNiP II-18-76 (1977). The bearing capacity is obtained by multiplying the nominal strength with a circumstance factor, which has a typical value of 1.3 and depends on live load and soil temperature.

tamped gravel. The soil profile and the ground temperatures are given in Fig. 4.22 and the creep constants in Table 4.4.

The temperature of permafrost at the bottom of the gravel layer is about $-0.6°C$ (31°F). If the friction of frozen soil is neglected, the long-term strength of frozen soil could be determined from Eq. (4.3) using uniaxial test data, and the long-term cohesion is given by $c = \frac{1}{2}\sigma_f$. In this case we use Table 4.3 and estimate the long-term cohesion to be about 100 kPa (14.5 lb/in²) at $-0.6°C$ (31°F). The bearing capacity of the soil is given by Eq. (4.30). With $N_c = 5.2$, $s_c = 1.2$, $\gamma = 18$ kN/m³ (115 lb/in³), $D = 2.7$ m, $N_q = 1.0$, and $s_q = 1.0$, we get

$$q_{ult} = 100(5.2)(1.2) + 18(2.7)(1)(1) = 670 \text{ kPa (97.2 lb/in²)}$$

If the factor of safety is selected to be 2.0, the design load is

$$P = \frac{670}{2}(1.44) = 480 \text{ kN (110 kips)}$$

This is a conservative design value: the moisture content of the frozen silt ($w = 33\%$) is not very high and the effect of friction may be significant.

The settlement of the footing has to be determined in order to confirm the serviceability. Because the primary creep is dominant for silt of low moisture content, Eq. (4.5) is used. With a service life $t = 20$ years, creep constant $\theta_c = 1\,°C$, $m = 0.49$, $\lambda = 0.074$, $k = 0.87$, $\omega = 4.58$ MPa·h$^\lambda$/°C^k (Table 4.4), and contact stress under the foundation of 340 kPa (49 lb/in²), we compute the settlement for the vertical stress distribution given by Fig. 4.51 (see page 264).

Because the settlement estimate was made based on the warmest temperature distribution, with much lower temperatures experienced during winter, the estimate is on the safe side. The settlement $s = 0.7$ cm (0.3 in) is small for this low-ice-content silt. However, the settlement rate for very ice-rich silt or

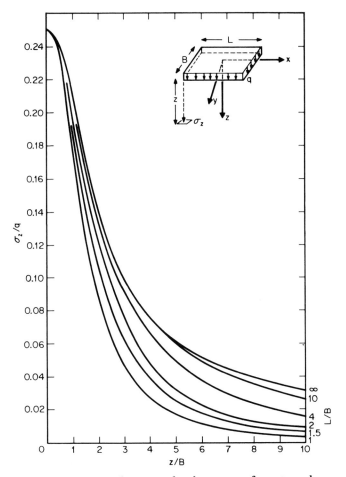

Figure 4.51   Vertical stress under the corner of a rectangular footing.

Computation of Settlement in Example 4.5

Layer, m (ft)	Average maximum temperature, °C (°F)	Average vertical stress, MPa (lb/in²)	Strain	Settlement s, cm (in)
2.7–3.2 (8.9–10.5)	−0.7 (30.7)	0.32 (46)	$\left[\dfrac{0.32\,[(20)(365)(24)]^{0.074}}{4.58\,(1+0.7)^{0.87}}\right]^{1/0.49} = 0.011$	0.011(50) = 0.53 (0.21)
3.2–4.0 (10.5–13.1)	−1.1 (30.0)	0.18 (26)	$\left[\dfrac{0.18\,(2.44)}{4.58\,(1+1.1)^{0.87}}\right]^{2.04} = 0.0022$	0.0022(80) = 0.18 (0.07)
4.0–5.0 (13.1–16.4)	−1.5 (29.3)	0.065 (9.4)	$\left[\dfrac{0.065\,(2.44)}{4.58\,(1+1.5)^{0.87}}\right]^{2.04} = 0.00021$	0.00021(100) = 0.02 (0.01)
5.0–8.0 (16.4–26.2)	−2.0 (28.4)	0.014 (2.0)	$\left[\dfrac{0.014\,(2.44)}{4.58\,(1+2)^{0.87}}\right]^{2.04} = 0.00001$	0.00001(300) ≈ 0.00 (0.00)
				Total 0.73 (0.29)

Long-term strength, kPa (lb/in²)	Temperature, °C (°F)	Time, years
357 (52)	−3 (26.6)	0.5
250 (36)	−1 (30.2)	0.5
220 (32)	−3 (26.6)	100
171 (26)	−1 (30.2)	100

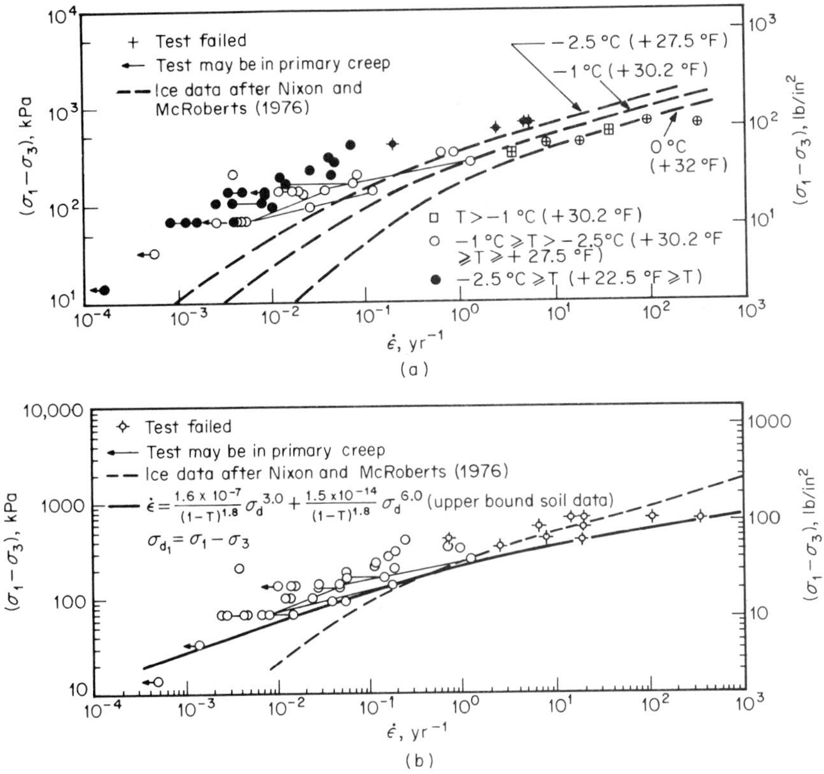

Figure 4.52 Creep rate data for ice-rich silt. (a) Creep rates for Norman Wells silt. (b) Creep rates adjusted to −1°C (30.2°F). (McRoberts et al., 1978.)

ice, when estimated from Fig. 4.52, may in this case exceed 0.1 m (4 in) per year, and failure of the foundation is obvious.

Shallow foundations Foundations can also be constructed directly on a thaw-stable gravel pad in permafrost areas. In this case the pad should be designed to prevent the penetration of thaw into the frost-susceptible soil. In cold structures such an arrangement causes no special concern. In heated structures the heat flow to the ground must either be eliminated by open ventilation or countered by cooling ducts or artificial refrigera-

Figure 4.53 Wooden pad foundations in Dawson City, Yukon Territory.

tion. The design of foundations on a gravel pad is based on ordinary geotechnical criteria. However, in cases of heavy loads and warm permafrost, the bearing capacity and settlements of the subsoil should be checked in the same manner as for deep footings.

Small structures with light loads are often founded on sills, pads, or shallow footings (Figs. 4.53 and 4.54). The ventilation space is generally on the order of 0.5 to 1 m (2 to 3 ft), but may be larger under wide heated structures to prevent degradation of the permafrost. The pad thickness is usually determined by normal thermal computations. However, preventive measures are sometimes required to eliminate changes in the ground thermal regime caused by improper drainage or blocked ventilation due to snow drifts.

Figure 4.54 Shallow concrete footings. Notice erosion effects endangering the corner foundation.

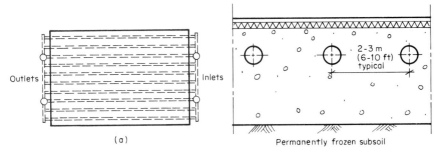

Figure 4.55 Principle of ventilated pad design. (*a*) Plan view. (*b*) Cross section.

A lifted floor system with pile or footing foundations becomes quite expensive for heated heavily loaded industrial buildings, large garages, aircraft hangars, oil tanks, and other such structures. An insulated ground-supported slab provided with artificial cooling is an alternative and often preferable solution. The traditional method is to use ventilation through cooling ducts placed in the pad (Fig. 4.55).

The principle of duct ventilation design is to remove sufficient heat from the soil in the winter so that the heat flow from the structure is compensated and the soil that has thawed during the summer is refrozen and chilled during the winter with a reasonable margin of safety. The total heat flux F for a unit area achieved by ventilation is given by

$$F = \frac{Q C_a \Delta T}{ls} \tag{4.31}$$

where Q = airflow rate
s = duct spacing
l = duct length
C_a = volumetric heat capacity of air, = 1330 J/(m³·°C)
    [0.02 Btu/(ft³·°F)]
ΔT = temperature rise of ventilated air in ducts

Typical ventilation requirements are on the order of 0.01 m³/(m²·s) [2 ft³/(ft²·min)]. It may be necessary to use a limited value for the temperature rise of ventilated air, for example, $\Delta T = 3\,°C$ (5 °F), in order to maintain effective cooling also near the outlets of the ducts.

The required ventilation for structures with restricted slab area may be obtained by normal air circulation through straight open ducts (Fig. 4.56). Centralized air intakes operated by electric fans or stack effect are required to deliver sufficient amounts of air to the duct network in case of larger structures. The air intakes are generally closed in the spring when air temperatures approach the thawing limit and opened again after the freezing season has begun.

Figure 4.56 Open-duct ventilation system for oil tank. Notice alternative solution with heavy piling.

The required pad thickness is determined based on the depth of thaw under the slab during the summer [Eq. (4.19) with $t_0 = 0$]. A thorough discussion of ventilated pad design is presented in Nixon (1978).

Active cooling systems provide a more efficient alternative for cooling ducts in very large heated structures, especially in warm permafrost areas. However, two-phase thermopipes that operate according to the same passive principles as thermopiles generally offer a more practical solution (Fig. 4.57). The heat-transfer rate of a thermopipe strongly depends on factors such as air temperature, wind speed, radiator area, ground thermal regime, and soil thermal properties. A typical value might be on the order of 50 $(W/°C) \cdot \Delta T$ [90 $(Btu/(h \cdot °F)) \cdot \Delta T$], where ΔT is the difference between the air and soil temperatures adjacent to the pipe. Examples of thermopipe applications are discussed for example in Yarmak and Long (1983).

Numerous failures have been experienced when the protection of permafrost under heated structures was based purely on thermal insulation. The thaw bulb eventually penetrated into the permafrost, causing uneven thaw settlements. This kind of approach is now commonly regarded as unfeasible due to numerous setbacks. However, when permafrost is cold and the lateral dimensions of the heated structure are small, the thaw bulb can be restrained within a reasonable insulation layer (Fig. 4.58). For example, a 10-m (33-ft)-wide heated building with a floor surface temperature of 20°C (68°F) built on permafrost with a mean temperature of −5°C (23°F) requires about 20 cm (8 in) of polystyrene insulation. The solution with a gravel pad, a ground-supported floor, and a small addition to the floor insulation thickness may well prove to be quite attractive when the costs of piles with an elevated floor or of an artificial cooling system are considered.

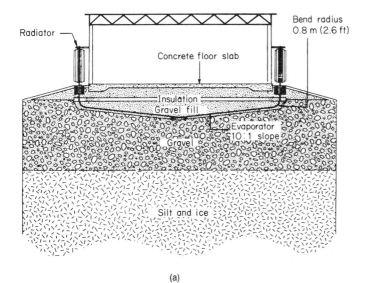

(a)

(b)

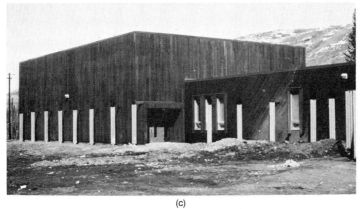

(c)

Figure 4.57 (a) Foundation schematic and (b), (c) application of thermopipes at Ross River School. Spacing is 3 m (10 ft). *(Hayley, 1981.)*

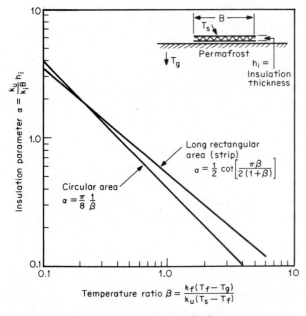

Figure 4.58 Required insulation thickness to prevent thawing of permafrost beneath heated building. k_i, k_f, k_u—conductivities of insulation, frozen soil, and thawed soil; T_f—freezing point of soil. *(Nixon, 1983.)*

4.7 Frost Action on Roads

Frost is an essential condition to be considered in the design of roads for cold regions. The problems are in principle quite similar in seasonal frost and permafrost areas. Factors such as uneven frost heave, thaw settlement, and thermal contraction may greatly reduce the serviceability of the pavement. Thaw weakening, on the other hand, should be taken into account when embankment strength is in question. In permafrost areas degradation of frozen ground and thermoerosion may also be subjects of concern.

This section concentrates on methods to eliminate the adverse effects of frost action on roads. Adequate non-frost-susceptible fill and proper drainage are key factors. The general design principles for roads in cold regions are also widely applicable in the design of other similar structures such as railways, runways, and pads.

4.7.1 Pavement Failures

Failures of roadway pavements are often caused by the combined effect of frost, moisture, and traffic. Frost action, thermal contraction, and traffic initiate cracks in the pavement. Water penetrating through these cracks may form ice within the base and cause some additional heave. On the other hand, water or deicing solutions may create uneven support conditions for the pavement and cause additional distress during thawing seasons. Occasionally traffic loads actually pump some fine material out from the base through cracks, and a raveling type of failure will form. Frost action and dynamic traffic loads may also pump fine material from the subgrade into the base, making the pavement eventually frost-susceptible. Separation phenomena associated with construction often contribute to these failures.

Moisture is released at the thawing front as the thaw progresses. Additional moisture may enter the base and the subbase through cracks and shoulders. In many cases thaw progresses rapidly under the dark pavement while shoulders protected by snow are still frozen, which blocks the drainage. The loss of bearing capacity due to pore water buildup and reduced soil density causes fatigue damage and unevenness under heavy loads and leads eventually to cracking and potholing.

Some typical examples of pavement failures on improperly designed, constructed, or maintained roads are shown in Fig. 4.59. The majority of potholes are created in the spring or winter. The primary mechanisms involved are improper base support, creating fatigue failure of the pavement and raveling failure due to the combined effects of water and mechanical action. In thick pavements a crack network may be created by thaw softening before any potholes occur.

Transverse cracking caused by thermal contraction may penetrate through the pavement and the base deep into the subgrade. It can be controlled by reducing the moisture content of the soil. Sometimes the cracking is limited only to the bituminous pavement layer, and design of the paving mixture should provide solutions to the problem.

Longitudinal cracking is often created by frost action in nonuniformly shaped transverse cross sections. Another common reason for longitudinal cracking is uneven frost penetration into frost-susceptible subgrade, when the center area of the road is exposed to freezing temperatures but the shoulders are insulated by snowdrifts. In permafrost areas it is often caused by changes in the ground thermal regime under the side slopes of the embankment.

4.7.2 Design against Frost Action

The general design principles for pavement structure are well established and described by many authors, such as in Yoder and Witczak (1975). These principles should also be followed in areas of seasonal frost and permafrost. Furthermore, deformations resulting from frost action should be restricted to acceptable limits, and the bearing capacity should be maintained at acceptable levels during the spring thaw also.

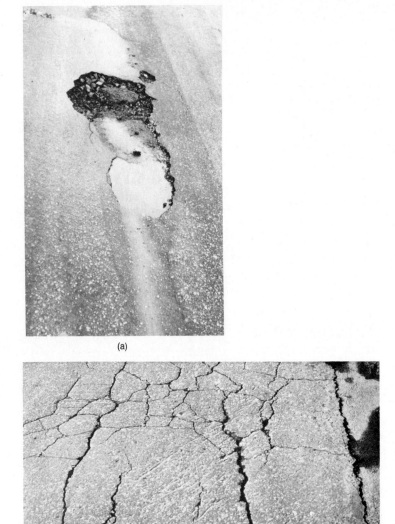

(a)

(b)

Figure 4.59 Typical pavement failures. (*a*) Classic combination of moisture and pothole. (*b*) Crack network created by thaw softening. (*c*) Transverse crack. (*d*) Longitudinal cracking. *(Courtesy of Finnish Roads and Waterways Administration.)*

(c)

General design principles   Three design methods are used by the U.S. Army Corps of Engineers to control frost action in pavement structure:

1. Complete protection
2. Limited subgrade frost penetration
3. Reduced subgrade strength

Naturally factors such as proper drainage and permafrost degradation should also be considered. The final design is governed by serviceability requirements and by investment and maintenance costs.

When complete protection is desired, the thickness of the pavement and the non-frost-susceptible base course is determined so that the active layer does not penetrate into the frost-susceptible subgrade. In many cases this method is overly conservative and expensive. It is used primarily in areas of shallow seasonal frost or continuous permafrost or for roads with the highest quality requirements. The use of this method may be justified also when the subgrade conditions are highly frost-susceptible or variable.

The aim of the limited subgrade frost penetration method is to provide sufficient combined thickness of pavement and non-frost-susceptible base course so that surface deformations remain acceptable [total heave less than or about 10 cm (4 in)]. A design chart is given in Fig. 4.60. However, the reduced bearing capacity during the spring thaw should be taken into account. Design curves for various subgrade conditions are given in Fig. 4.61. The methods can be used together, but if a special approach is adopted to eliminate uneven frost action locally, the reduced bearing capacity method may provide an acceptable result. In any case, it is good practice to use a 100- to 200-mm (4- to 8-in) filter layer or filter fabric between the base and the subgrade.

Insulation   An undesirably high embankment may be required to limit frost action in areas of deep seasonal frost and permafrost. The rate of frost or thaw penetration may be reduced somewhat by using materials with high moisture content and latent heat of fusion. However, the use of thermal insulation may prove to be the most economical alternative. This is illustrated in Fig. 4.62.

Theoretically thermal insulation is most effective close to the surface of the pavement. The insulation must, however, have sufficient strength to resist the traffic loads. Pavements have also been reported to experience severe surface icing conditions, which are hazardous to traveling, when in paved roadbeds insulation has been placed too close to the surface.

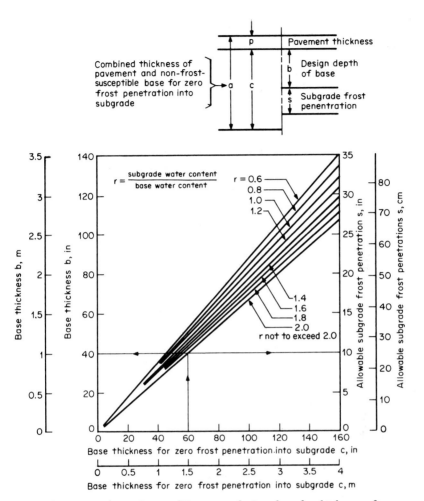

Figure 4.60 U.S. Army Corps of Engineers design chart for thickness of non-frost-susceptible base for limited subgrade frost penetration. *(Adapted from Lobacz et al., 1973.)*

The depth of frost penetration in an insulated embankment can be computed following the principles given in Sec. 4.3.2. If the moisture content and the latent heat of fusion of the soil are small, the analytical solution given in Lachenbruch (1959) is often applied. An example of the thermal design of the insulated road section is given in Fig. 4.63. If the filter layer below the insulation (frost-free layer) is thicker than h_{min}, the thickness of insulation can be reduced as shown in Table 4.11.

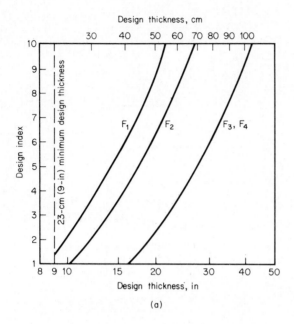

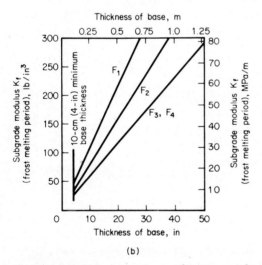

Figure 4.61 U.S. Army Corps of Engineers design curves for reduced subgrade strength. (*a*) Flexible highway pavements. Design thickness includes both pavement and non-frost-susceptible base. (*b*) Rigid highway pavements. Subgrade modulus value is used in normal condition design. (*Adapted from Lobacz et al., 1973.*)

Group	Subgrade Description
F_1	Gravelly soils containing grains between 3 and 10% finer than 0.02 mm (0.0008 in) by weight
F_2	a. Gravelly soils containing grains between 10 and 20% finer than 0.02 mm (0.0008 in) by weight b. Sands containing grains between 3 and 15% finer than 0.02 (0.0008 in) by weight
F_3	a. Gravelly soils containing grains more than 20% finer than 0.02 mm (0.0008 in) by weight b. Sands, except very fine silty sands, containing more than 15% of grains finer than 0.02 mm (0.0008 in) by weight c. Clays with plasticity index greater than 12
F_4	a. All silts b. Very fine silty sands containing more than 15% of grains finer than 0.02 mm (0.0008 in) by weight c. Clays with plasticity index less than 12 d. Varved clays and other fine-grained banded sediments

Figure 4.61 *(Continued)*

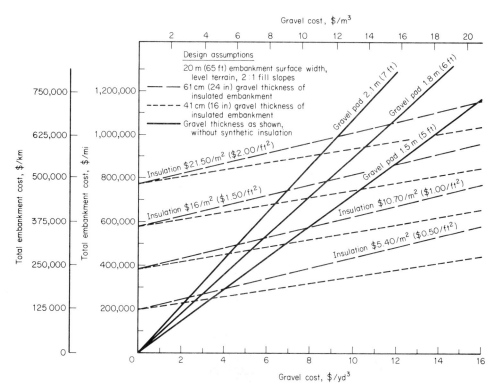

Figure 4.62 Effect of thermal insulation on embankment design and economics. (*Wellman et al., 1977.*)

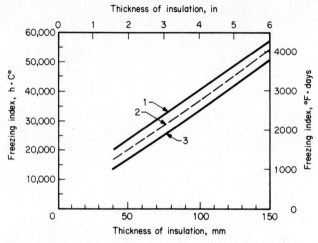

Figure 4.63 Required thickness of polystyrene insulation when protective granular cover is at least 0.7 m (28 in) and filter layer beneath insulation is 100 to 200 mm (4 to 8 in). 1—extruded polystyrene, dry conditions (embankments); 2— extruded polystyrene, wet conditions (cuts), and expanded polystyrene, dry conditions; 3—expanded polystyrene, wet conditions.

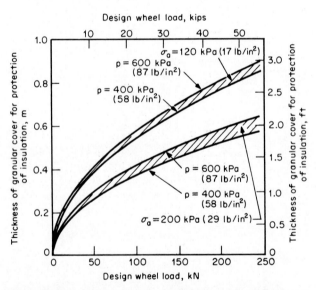

Figure 4.64 Granular cover requirements for different wheel loads. p—tire pressure; σ_a—allowable stress on insulation, depending on insulation type and number of vehicle passes. *(Nixon, 1979.)*

TABLE 4.11   Corrected Thickness of Insulation for Frost-Free Layers Thicker than h_{min}, [mm (in)]

Thickness of insulation according to Fig. 4.63, mm (in)	Thickness of frost-free layer below insulation			
	$h_{min} +$ 200 mm (8 in)	$h_{min} +$ 400 mm (16 in)	$h_{min} +$ 600 mm (24 in)	$h_{min} +$ 800 mm (32 in)
40 (1.6)	30 (1.2)	30 (1.2)	30 (1.2)	30 (1.2)
50 (2.0)	35 (1.4)	30 (1.2)	30 (1.2)	30 (1.2)
60 (2.4)	40 (1.6)	30 (1.2)	30 (1.2)	30 (1.2)
70 (2.8)	50 (2.0)	35 (1.4)	30 (1.2)	30 (1.2)
80 (3.2)	60 (2.4)	40 (1.6)	30 (1.2)	30 (1.2)
90 (3.6)	70 (2.8)	50 (2.0)	35 (1.4)	30 (1.2)
100 (4.0)	80 (3.2)	60 (2.4)	40 (1.6)	30 (1.2)
110 (4.3)	90 (3.6)	70 (2.8)	50 (2.0)	35 (1.4)
120 (4.7)	100 (4.0)	80 (3.2)	60 (2.4)	40 (1.6)
130 (5.1)	110 (4.3)	90 (3.6)	70 (2.8)	50 (2.0)
140 (5.5)	120 (4.7)	100 (4.0)	80 (3.2)	60 (2.4)
150 (5.9)	130 (5.1)	110 (4.3)	90 (3.6)	70 (2.8)

The allowable stress on insulation depends on the insulation type and the number of vehicle passes. For heavier polystyrene and polyurethane boards [35 to 40 kg/m³ (2.2 to 2.5 lb/ft³)] the long-term strength corresponding to 2% deformation ranges from 50 to 200 kPa (7 to 29 lb/in²). The required protective granular cover can be estimated by the Boussinesq theory (McDougall, 1977),

$$Z = \frac{W/\pi p}{(1 - \sigma_a/p)^{2/3} - 1} \qquad (4.32)$$

where Z = required depth
σ_a = allowable horizontal stress
p = tire inflation pressure
W = design wheel load

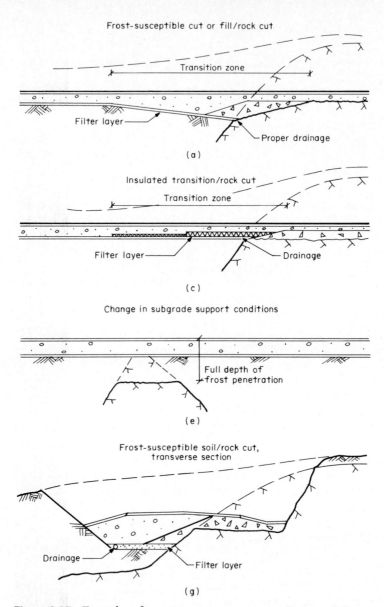

Frost-susceptible cut or fill/rock cut

Transition zone

Filter layer

Proper drainage

(a)

Insulated transition/rock cut

Transition zone

Filter layer

Drainage

(c)

Change in subgrade support conditions

Full depth of frost penetration

(e)

Frost-susceptible soil/rock cut, transverse section

Drainage

Filter layer

(g)

Figure 4.65 Examples of transition arrangements to eliminate differential frost action.

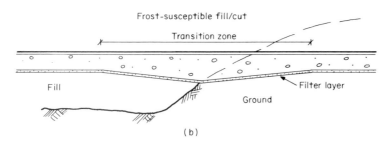

Frost-susceptible fill/cut

Transition zone

Fill

Filter layer

Ground

(b)

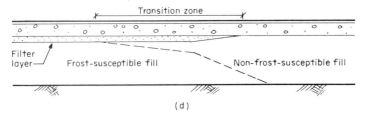

Frost-susceptible fill/non-frost-susceptible fill

Transition zone

Filter layer

Frost-susceptible fill

Non-frost-susceptible fill

(d)

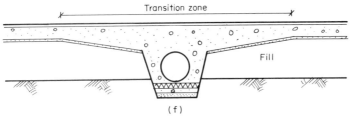

Culvert transition

Transition zone

Fill

(f)

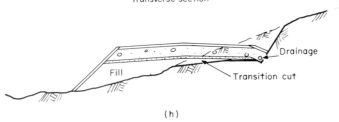

Frost-susceptible cut/fill,
transverse section

Drainage

Fill

Transition cut

(h)

An ordinary granular cover of 0.5 to 0.8 m (20 to 32 in) is quite sufficient in a typical design with wheel loads on the order of 50 kN (5 tons)
and tire inflation pressures of up to 800 kPa (116 lb/in²) (Fig. 4.64).

Transitions, side slopes, and cuts   Abrupt differential heaving and thaw
settlement often occur at discontinuities of subgrade conditions, especially when the groundwater table is high. Pavement failures typically
appear at cut and fill intersections, transition zones in subsoil conditions,
areas where boulders or uneven rock profile exist at shallow depths, and
locations of culverts with diameters on the order of 50 cm (20 in) or
larger. Sections where water penetrates under the pavement from the
side, such as side cuts, are also very prone to frost damage. Frost action
can be controlled with tapered transition fills that may reach full depth of
frost penetration and have slopes of 1 : 20 to 1 : 40 depending on the frost
conditions and quality requirements for the pavement. An alternative
approach is to use insulated sections that are stepped up in less than
25-mm (1-in) lifts. Examples are shown in Fig. 4.65.

In the areas of discontinuous permafrost the transitions between
thawed ground and permafrost are often especially troublesome, because the constructed embankment changes the thermal equilibrium of
the ground. Progressive permafrost degradation and sagging due to thaw
settlements may occur in the boundary areas. Special measures such as
artificial cooling or removal of permafrost may be required to achieve
acceptable stability (Fig. 4.66).

Figure 4.66 Thermopipes used at intersection of permafrost and
thawed ground to stop thaw and settlement of ice-rich peat bog beneath pavement.

Figure 4.67 Shoulder cracking along roadway. *(Courtesy of Glenn Johns.)*

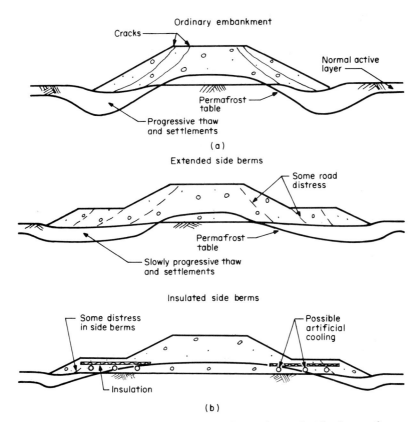

Ordinary embankment

Cracks

Normal active layer

Permafrost table

Progressive thaw and settlements

(a)

Extended side berms

Some road distress

Permafrost table

Slowly progressive thaw and settlements

Insulated side berms

Some distress in side berms

Possible artificial cooling

Insulation

(b)

Figure 4.68 (*a*) Roadway cracking due to thaw beneath side slopes of embankment. (*b*) Control of problem with side berms.

The problem with side slopes is their vulnerability to loss of stability due to frost action. Sloughings and erosion generally occur in the spring when a thawed slope is saturated and weakened. Runoff of surface water and seepage from the thawing soil may further worsen the situation. The stability of the slope may be improved by a layer of coarse non-frost-susceptible material on the surface or by a built-up stabilizing cover.

In warm permafrost areas longitudinal cracks often appear in roadway embankments as a result of long-term thaw-related settlements under the side slopes (Fig. 4.67). Soils under the side slopes do not necessarily refreeze each winter because of thick snow insulation, water ponding, and other changes in surface conditions. Extension of the side berms and application of insulation can be used to slow down the thaw progress and to eliminate the problem, as shown in Fig. 4.68. In some cases artificial cooling of the side berms may also be necessary (Esch, 1983).

Cuts made in ice-rich permafrost represent another potential source of thaw-related problems. Extensive mud slides and thermoerosion occur as a result of the thawing of the slope. The situation eventually stabilizes within a few years, but in the meanwhile drainage problems are difficult to manage (Fig. 4.69). The stability of the slopes in ice-rich soils may be preserved by a protective layer of granular fill and insulation designed for complete protection of permafrost (Fig. 4.70).

Marginal conditions   Sometimes it is not economically feasible to acquire good-quality non-frost-susceptible fill in amounts normally required. The performance of marginal soils may prove to be quite acceptable in the road base if special attention is given to measures such as:

Improving the drainage

Lowering the groundwater table

Making the subgrade conditions even more uniform than normally (removal of boulders, bedrock peaks, and silt pockets, blending the top of the subgrade to avoid stratification, use of tapered transition fills, etc.)

Careful soil compacting to optimum density

Increasing the height of the grade line

Making sure the base conditions are uniform

Experience gained from unpaved low-volume roads constructed of frost-susceptible gravels and processed tills has been quite favorable. The roads are strong and firm during the summer because the material can hold moisture much longer than non-frost-susceptible gravels. Consequently maintenance efforts are reduced. The degree of frost-suscepti-

Figure 4.69 Thermoerosion and stabilization of road cut in Alaska. Note how trees have been cut, but the organic layer has been preserved above the cut to provide thermal protection for ice-rich soil. *(Courtesy of Glenn Johns.)*

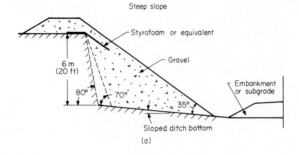

Steep slope

Styrofoam or equivalent

Gravel

6 m
(20 ft)

80°   70°

35°

Embankment
or subgrade

Sloped ditch bottom

(a)

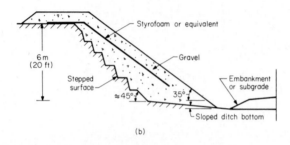

Moderately steep slope

Styrofoam or equivalent

Gravel

6 m
(20 ft)

Stepped
surface

≈45°   35°

Embankment
or subgrade

Sloped ditch bottom

(b)

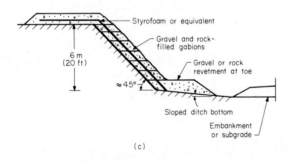

Moderately steep slope

Styrofoam or equivalent

Gravel and rock-
filled gabions

Gravel or rock
revetment at toe

6 m
(20 ft)

≈45°

Sloped ditch bottom

Embankment
or subgrade

(c)

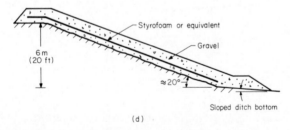

Low-angle slope

Styrofoam or equivalent

Gravel

6 m
(20 ft)

≈20°

Sloped ditch bottom

(d)

Figure 4.70 Methods for thermal protection of road
cut in permafrost. *(Pufahl and Morgenstern, 1980.)*

Figure 4.71 Severe disruption of roadway due to thawing of massive ice lenses in permafrost. *(Courtesy of Glenn Johns.)*

bility of the soil material that can be tolerated without causing excessive loss of bearing capacity during the spring thaw requires careful evaluation in each case.

The frost-related properties of soils may be improved by washing out fine grains or by adding coarse material to reduce the portion of fine grains. There are also some special methods to eliminate the adverse effects of frost action in frost-susceptible soils. Moisture barriers have been used to some extent to prevent the migration of water to the base. The capillary cutoff can be arranged with a coarse granular layer above the bottom of the ditches, but clay or an impermeable membrane can also be used for this purpose. Methods to encapsulate frost-susceptible soils with impermeable membranes to prevent frost action and increase base strength are also under development. Finally there are different methods to stabilize the frost-susceptible soil either by cementing action or by modifying its frost-related properties. These special methods are discussed in detail, for example, in Johnson et al. (1975).

The thermal equilibrium of the ground is often extremely delicate in areas of discontinuous permafrost. It may be uneconomical or even unfeasible to avoid the slow degradation of permafrost under the embankment. This is the reason why ice-rich subgrade conditions should be avoided and estimates should be made on the magnitude and unevenness of the thaw settlement (Fig. 4.71). The serviceability is in this case secured by additional maintenance.

(a)

(b)

Figure 4.72   Drainage problems. (*a*) Thaw softening caused by improper drainage. (*b*) Attempt to avoid water ponding on road by removing snow and ice slush from ditch. (*Courtesy of Finnish Roads and Waterways Administration.*)

4.7.3 Drainage

Moisture is an essential component in different frost-related pavement distresses such as frost heave, thaw softening, rutting, fatigue cracking, and potholing (Fig. 4.72). Proper surface and subsurface drainage is among the most effective measures to control frost action.

Water may enter the pavement structure from the groundwater table, through the pavement, permeable surface, or shoulders, or by seepage from higher ground. Surface drainage can be handled effectively by adequate slopes and grade inclinations. Shoulders are often paved to block surface water penetration. Interceptive subdrains and base-course drainage layers are also used for subsurface drainage. Where the groundwater table is high, deepened ditches provide an alternative to raised grade for partial frost control.

Figure 4.73 Culverts blocked by ice. *(Courtesy of Glenn Johns.)*

The problem with culverts is that they are often blocked by ice during spring runoff when the need for drainage is the greatest (Fig. 4.73). Ice formation may be a result of ordinary icing or great heat losses and accumulation of frozen slush at the ends of the culvert. Freezing may also occur in the middle of the culvert, especially in permafrost areas. Some solutions to these problems are discussed in Carey (1977). Obviously the use of larger ditches and culverts or small bridges improves the situation. Small steam pipes or electric heating cables can be installed into the culvert. A pivot channel is easily opened into the culvert and is rapidly enlarged as water runs through. Fuel heaters are also used to open blocked culverts.

Sometimes the ends of culverts are closed up at the beginning of the winter and the closures are removed in the spring, thus permitting runoff through a clear culvert. Another similar approach is the use of staggered culverts, that is, two or more culverts at different levels of the fill. The higher culvert is above the design icing level and carries the initial spring runoff. Later in the spring the lower culvert becomes cleared of ice and carries the rest of the flow.

Ditching should be avoided in areas of ice-rich permafrost in order to protect the permafrost. On the other hand ponding of water may also cause thermal degradation of permafrost. Thus special attention should be given to drainage arrangements in permafrost areas. Culverts are often used, but narrow spacing may be required to avoid ponding and to minimize any disturbance to the natural drainage patterns. If interceptive drains are constructed, they should be placed away from the embankment and have small gradients and coarse granular lining.

Heat transfer through the culvert may cause frost heave and in permafrost areas also thaw settlements under the culvert if it is not properly surrounded with thaw-stable material or insulation. In extreme cases the culvert and the embankment may experience rapid collapse due to the

(a)

(b)

(c)

Figure 4.74 Drainage failures in permafrost areas. (*a*) Ponding of water has caused a thermokarst depression. Notice steam pipes installed in failed culvert. (*b*) Excessive flow and rapid thermoerosion breaking entire embankment. (*c*) Road culvert has concentrated the flow and created a gully type erosion channel into ice-rich permafrost. [*Photo (c) Courtesy of H. Mäkelä.*]

thawing of ice-rich permafrost and erosion. Concentrated drainage may also create gully-type erosion into ice-rich permafrost (Fig. 4.74).

Icing   Icing or aufeis, as it is also called, means in this context a sheetlike mass of ice formed as water has penetrated onto the surface and frozen in subsequent layers. Three different types of icing are distinguished, depending on the source of water: river icing, spring icing, and ground icing. River icing may be related to ice jamming in the winter, whereas spring icing and ground icing relate to ordinary cold region drainage problems. Icing poses the most potential hazard to roads and railways, but its possible occurrence should be evaluated also in the selection of routes for pipelines and utility lines and in the site selection for dwellings, communities, and so on. A good review of different types of icing is presented in Carey (1973).

The most extensive icing occurs typically but not necessarily in wide and shallow rivers and streams with braided channels in a very cold climate. These channels may become blocked with frazil ice or freeze solid to the bottom in midwinter. The subriverbed groundwater flow may also be cut by permafrost, bedrock, or an impervious layer (Fig. 4.75). Water in this partly or totally blocked channel must penetrate to the surface through cracks and holes in the ice cover and riverbanks. It spreads out at the surface and freezes in thin subsequent layers. Icing may continue until the end of the winter season, covering large areas.

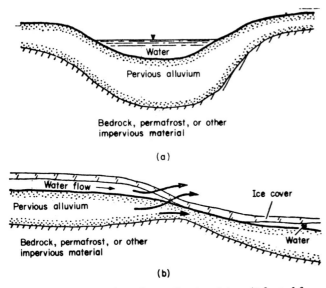

Figure 4.75 Typical conditions for river icing. *(Adapted from Carey, 1973.)*

(a)

(b)

Figure 4.76 (a) Roadway and bridge inundated due to river icing are being cleared by bulldozer. (b) Same location in summer after ice has melted. *(Carey, 1973.)*

The process of river icing is discussed in detail, for example, in Kane (1981).

Icing may block roads, cause water damage on buildings, and endanger structures because of erosion during the thawing period (Fig. 4.76). The maximum possible extent of an icing can be estimated based on its winter discharge. River icings of varying magnitudes often occur in the same area year after year, providing an indication to avoid the site or to prepare counteractive measures. Sometimes, however, river icings occur as a result of engineering activities. For example, when a bridge is built across a river, the hydraulic gradient may be altered abruptly. The bridge deck prevents the formation of an insulative snow cover and hence increases the heat loss. Potential for man-made icing is created by altering the environmental circumstances.

It may sometimes be economical to prevent river icing by modifying the channel. Cross sections with small width-to-depth ratios do not favor icing. However, more often earth embankments are used as a permanent protection (see Veldman and Yaremko, 1978, for design alternatives). To protect the embankments from erosion during spring thaw, heavy riprap should be used. Some tolerance should be maintained so that the embankments can be raised without affecting the riprap, since uncertainties may often exist concerning icing and flooding levels.

Spring and ground icing also causes considerable problems in the maintenance of transportation networks, although it is not as extensive as river icing. Spring icing can be separated from ground icing by noting that the source of water is well-defined and the water discharge is continuous during the winter. Ground icing is caused by trapping or restricting the groundwater flow between the freezing front and an impervious layer, such as rock, clay, or permafrost, typically in sloping terrain. Icing may also develop in rock cuts and tunnels due to gravity flow of water in the cracks and seams. The severity of ground icing varies annually, depending on the groundwater situation and the climatological characteristics of the winter. These factors also govern the rate of icing during the winter.

Spring and ground icing are common features in cold regions except for the arctic areas where perennial springs are rare and the active layer above permafrost is thin. Ground icing is often triggered by human activities. The construction of cuts and fills, removal of snow and vegetation, as well as drainage arrangements may abruptly change the thermal and hydrological regime of the ground.

If the accumulation of seasonal icing can be estimated beforehand, raising the grade or providing storage area may be an acceptable permanent solution. An insulated subsurface drainage network may be applicable especially in seasonal frost areas, but also in permafrost areas, as shown in Fig. 4.77. In permafrost areas frost belts can be created for temporary protection by removing the vegetation cover and accelerating frost penetration. The belt blocks the groundwater flow and triggers icing at a safe distance from the protected objects. Temporary ice fences can be used in a marginal situation or in cases of emergency. These fences are often quite lightly constructed wooden fences covered by polyethylene, reinforced paper, or other impervious materials. Wind may be the governing factor in the design of fences, because the pressures caused by ice and thin water layers are small (Fig. 4.78).

Tunnels and cuts may also occasionally experience icing problems in seasonal frost areas. Drainage facilities become blocked and high mainte-

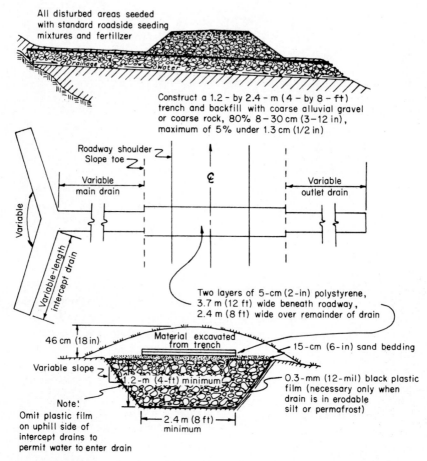

All disturbed areas seeded
with standard roadside seeding
mixtures and fertilizer

Construct a 1.2 - by 2.4 - m (4 - by 8 - ft)
trench and backfill with coarse alluvial gravel
or coarse rock, 80% 8 - 30 cm (3–12 in),
maximum of 5% under 1.3 cm (1/2 in)

Roadway shoulder
Slope toe

Variable
main drain

Variable
outlet drain

Variable

Variable-length
intercept drain

Two layers of 5-cm (2-in) polystyrene,
3.7 m (12 ft) wide beneath roadway,
2.4 m (8 ft) wide over remainder of drain

46 cm (18 in)

Material excavated
from trench

15-cm (6-in) sand bedding

Variable slope

1.2-m (4-ft) minimum

0.3-mm (12-mil) black plastic
film (necessary only when
drain is in erodable
silt or permafrost)

Note!
Omit plastic film
on uphill side of
intercept drains to
permit water to enter drain

2.4 m (8 ft)
minimum

Figure 4.77 Principle of insulated subdrain designed to carry springwater down-slope from road and to avoid icing problems. *(Livingston and Johnson, 1978.)*

nance efforts are required to keep traffic lanes open. Ice formation and melting sometimes loosen pieces of fractured rock. Ice may also damage heat, water, and electricity installations in tunnels.

Pedersen (1977) has described the use of injection, insulated aluminum arches, and waterproof concrete linings for protection against icing. Injection provides a fairly good protection against small leakages in a good-quality rock. More often, however, insulated aluminum sheets are used along the entire length of the frost zone. Membrane-waterproofed concrete linings are expensive, but they are the only acceptable solution in poor-quality rocks with major water leaks.

Figure 4.78 Attempt to block ground icing with light ice fence. *(Courtesy of Glenn Johns.)*

4.8 Utility Lines

The basic concerns in the design of utility lines for cold regions are (1) to prevent the lines from freezing and (2) to protect the lines from the adverse effects of frost action. Several alternative methods have been developed to eliminate these problems in both seasonal frost and permafrost conditions. The final design depends on the costs of construction and maintenance.

Freezing of water and sewer lines in seasonal frost areas can be conveniently prevented by burying them below the active layer. However, this is not always practical. On the other hand, especially heating lines and sewer lines and, to a lesser degree, waterlines with significant circulation have thermal reserves. Additional heat may be available if the heat pipes are located near other utility lines. If the lines are protected with sufficient insulation, it is often possible to prevent freezing in frozen ground or even in arctic atmospheric temperatures. Heat may be added to water for further protection. The lines may also be provided with internal or external heating cables. The power requirements for insulated buried lines are small, typically on the order of 2 to 5 W/m [2 to 5 Btu/(ft·h)]. When the lines are provided with thermostats, the heating period is quite limited.

The rate of heat removal Q from an insulated pipe can be estimated by

$$Q \approx \frac{T_W - T_E}{R_I} = \frac{2\pi k_I (T_W - T_E)}{\ln(r_i/r_p)} \tag{4.33}$$

where T_W = water temperature
T_E = temperature of exterior ground or air
R_I = thermal resistance of cylindrical insulation

k_I = thermal conductivity of insulation
r_i = exterior diameter of insulation layer
r_p = exterior diameter of pipe

When, in addition, the water circulation rate, the allowable water temperature drop, and the pipe length are known, the required insulation thickness or additional heat requirement can be calculated. Equation (4.33) also provides the basis for estimating the allowable periods of interruptions in water circulation or heating to prevent the beginning of freeze up. Special attention should be given to the thermal protection of sewer wells, valves, and similar details.

Methods to eliminate frost action include burying below the active layer, reducing the depth of the active layer using insulation, and backfilling with non-frost-susceptible soil. In permafrost areas, utility lines are often elevated and supported by piles. If they are buried, possible thawing and erosion effects should be carefully evaluated.

4.8.1 Utility Lines in Seasonal Frost

Utility lines are generally installed sufficiently deep into the ground, out of the reach of frost penetration in seasonal frost areas. In areas of deep seasonal frost, excavation, blasting, and backfilling generally represent over 50% of the total construction costs. That is why the installation depth has a significant influence on the construction costs. In addition to the sharply increasing amount of work and time, difficulties may be encountered with groundwater and with the stability of the walls in deep excavations. Access to the pipes is also more difficult in case of service or repair needs. The depth of frost penetration can be influenced somewhat by the selection of the backfilling material, but large reductions can be obtained only by using insulation materials and by using heat.

Some typical arrangements for frost protection of utility lines are illustrated in Fig. 4.79. Cases 1, 2, and 3 represent the traditional construction methods where the utility lines are extended below the design frost penetration depth. In cases 4, 5, and 6 the effect of heat losses and possible heating is recognized. This phenomenon has long been accepted in practice by noting that sewers are less prone to freezing than waterlines at equal depth. Empirical reductions to the depth of full frost penetration have been applied for the burial depths of sewers.

Insulation materials and heat can be used effectively together to eliminate freezing of pipes and frost action in areas of deep seasonal frost, as shown in cases 7, 8, 10, and 11. All heat sources from the pipe group should be utilized and electric cables could be installed in the same excavation. The situation has been analyzed extensively in Gundersen (1976). Design charts are provided in Figs. 4.80 to 4.83.

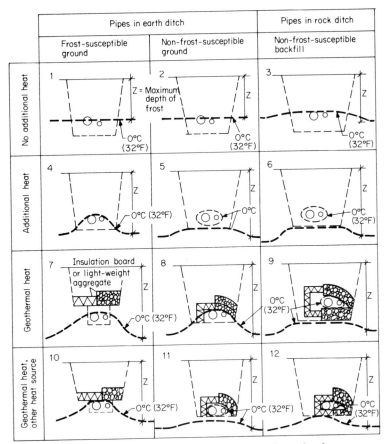

Figure 4.79 Typical frost protection arrangements for utility lines in seasonal frost areas.

Frost protection of utility lines in rock ditches (cases 9 and 12 in Fig. 4.79) is a special case in many respects. The thermal conductivity of rock is good and its water content low. The freezing front penetrates deeply. On the other hand, frost movements are not significant in properly filled rock ditches. It is sufficient to prevent the freezing of the utility lines. This is generally done with the help of insulation and heating cables if the thermal reserves of the circulated fluid are not adequate.

Rock ditches may be separated into broad and narrow ones when thermal analysis is considered. The thermal behavior of broad rock ditches resembles somewhat that of ordinary soil excavations. For example, silty soils can be used as a fill material in nontraffic areas to provide protection against frost penetration and thermal losses. According to a

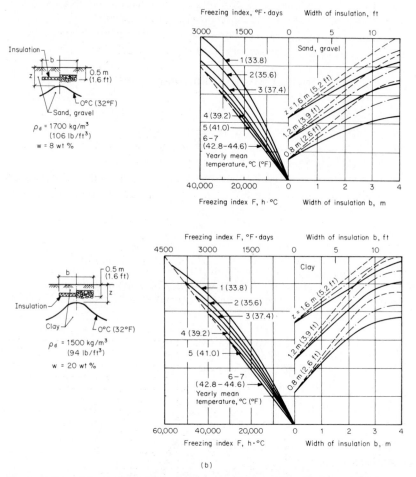

Figure 4.80 Required thermal resistance and lateral insulation width for a given frost penetration depth in (*a*) gravel or sand and (*b*) clay. Thermal resistance for insulation *R*: —1.4 (m²·°C)/W [7.9 (ft²·h·°F)/Btu]; —·— 2.1 (m²·°C)/W [11.9 (ft²·h·°F)/Btu]; ———— 2.8 (m²·°C)/W [15.9 (ft²·h·°F)/Btu]. (*Gundersen, 1976.*)

rough rule of thumb, the width of a broad rock ditch at the surface is at least twice its depth.

Design charts for utility lines in rock ditches are given in Figs. 4.84 and 4.85. However, special care is required when utility lines are installed at shallow depth in narrow rock ditches and the snow cover is thin. In this case the heat requirement should be determined by the lowest surface temperatures rather than by the design freezing index.

Insulation boards or lightweight aggregate are generally used for the thermal protection of utility lines (Fig. 4.86). A well-compacted and

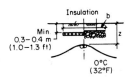

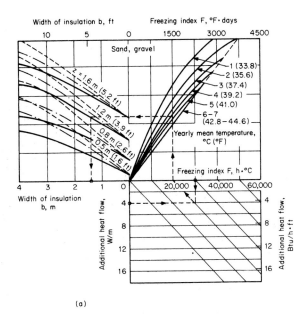

(a)

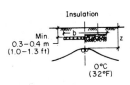

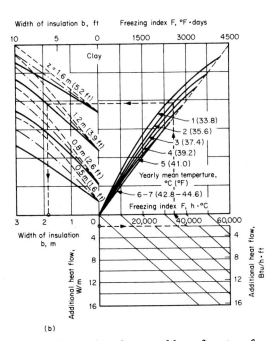

(b)

Figure 4.81 Required thermal resistance and lateral insulation width as a function of heat output to soil for a given frost penetration depth in (*a*) sand or gravel and (*b*) clay. Thermal resistance for insulation R: —— 1.4 $(m^2 \cdot °C)/W$ [7.9 $(ft^2 \cdot h \cdot °F)/Btu$]; —·— 2.1 $(m^2 \cdot °C)/W$ [11.9 $(ft^2 \cdot h \cdot °F)/Btu$]; ———— 2.8 $(m^2 \cdot °C)/W$ [15.9 $(ft^2 \cdot h \cdot °F)/Btu$]. (*Gundersen, 1976.*)

299

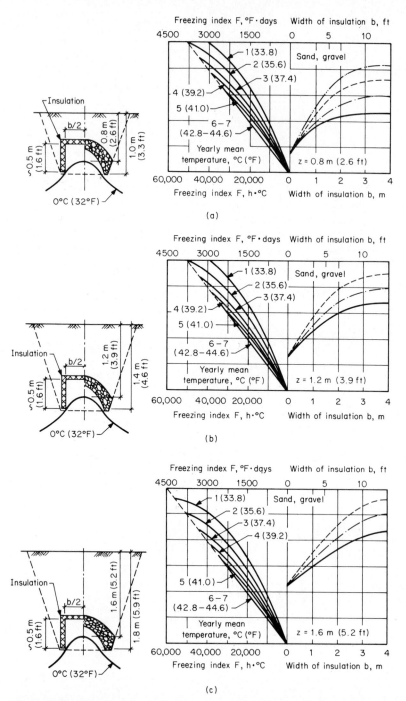

Figure 4.82 Required thermal resistance and U-shaped insulation width in sand or gravel. Allowed frost penetration depth z is (a) 0.8 m (2 ft 7 in). (b) 1.2 m (4 ft), and (c) 1.6 m (5 ft 3 in.) Thermal resistance for insulation R: —1.4 $(m^2 \cdot °C)/W$ [7.9 $(ft^2 \cdot h \cdot °F)/Btu$]; —·— 2.1 $(m^2 \cdot °C)/W$ [11.9 $(ft^2 \cdot h \cdot °F)/Btu$]; ——— 2.8 $(m^2 \cdot °C)/W$ [15.9 $(ft^2 \cdot h \cdot °F)/Btu$]; —··— 4.2 $(m^2 \cdot °C)/W$ [23.8 $(ft^2 \cdot h \cdot °F)/Btu$]. *(Gundersen, 1976.)*

leveled base is required for insulation boards in order to prevent break-age. New insulation units provided with holes for piping have recently been introduced for the thermal protection of pipes (Fig. 4.87). These units are applicable for utility lines at shallow depths in rock ditches and in non-frost-susceptible ground, such as in roadway embankments.

An otherwise well-protected utility line may freeze or heave if the thermal protection of details such as valves or sewer wells is not properly provided. Furthermore the sewer wells have to be designed so that cold air cannot circulate in the sewer lines. Some examples of thermal protection arrangements are given in Fig. 4.88.

4.8.2 Utility Lines in Permafrost Areas

In permafrost areas elevated utility lines represent quite an appealing alternative to buried lines. Some typical design alternatives are illustrated in Fig. 4.89. Because there are no possibilities to avoid frozen ground or freezing temperatures, greater attention is required to prevent the freezing of pipes than in seasonal frost areas.

Problems connected with the burial of utility lines include frost action, thawing of permafrost, changes in drainage patterns, thermoerosion, and seepage-induced erosion. Furthermore excavation works are generally laborious and troublesome during both summer and winter. Well-insulated utility lines, possibly provided with heating cables, can be buried at the bottom of the active layer where the thermal conditions are relatively stable. It is also possible to place the lines into roadway shoulders or directly on insulated ground and to use gravel for protection.

Main utility lines are often placed into underground utilidors in major arctic communities. Problems with traffic arrangements and service and repair access can be conveniently avoided when this solution is applied. Permafrost can be protected from thermal degradation by circulating cold air through the utilidor system in the winter. Proper insulation should be used in utilidors, which have pipes with large heat losses. Precast concrete elements, large-diameter metal pipes, and wood have been used as construction materials for underground utilidors. Drainage arrangements may also be required in utilidors, for example in case of a surface water leakage.

The extent of the changes in the permafrost regime and the frost-related movements of the buried utility lines can be predicted following the guidelines given in Secs. 4.3, 4.4, and 4.10 (see also Zirjacs and Hwang, 1983). Simple solutions for numerous practical cases are provided in Velli et al. (1977).

Piles or light surface foundations are generally used as foundations for elevated utility lines. Problems associated with elevated lines include traffic arrangements, building elevation arrangements to secure re-

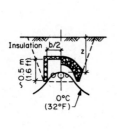

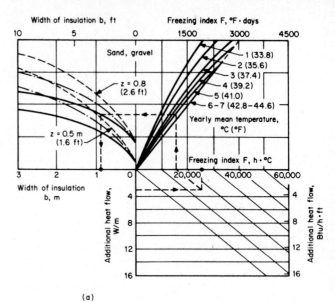

(a)

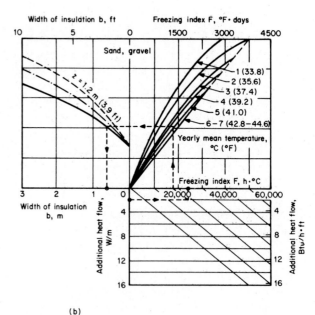

(b)

302

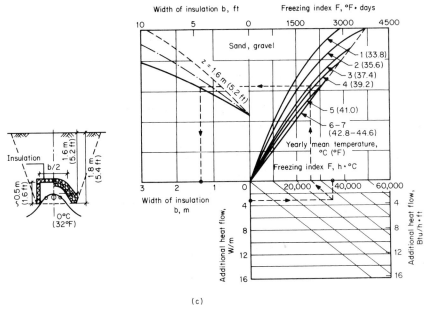

Figure 4.83 Required thermal resistance and U-shaped insulation width as a function of heat output to soil (sand or gravel). Allowed frost penetration depth z is (a) 0.5 or 0.8 m (1 ft 8 in or 2 ft 7 in), (b) 1.2 m (4 ft), and (c) 1.6 m (5 ft 3 in). Thermal resistance for insulation R: —1.4 $(m^2 \cdot {}^\circ C)/W$ [7.9 $(ft^2 \cdot h \cdot {}^\circ F)/Btu$]; —·— 2.1 $(m^2 \cdot {}^\circ C)/W$ [11.9 $(ft^2 \cdot h \cdot {}^\circ F)/Btu$]; ——— 2.8 $(m^2 \cdot {}^\circ C)/W$ [15.9 $(ft^2 \cdot h \cdot {}^\circ F)/Btu$]. *(Gundersen, 1976.)*

quired slopes for sewers, heat losses, control of large thermal movements of pipes, and prevention of freezing as the lines are exposed to arctic temperatures. On the other hand installation and repair works can be executed in a simple and effective manner. Some examples of elevated utility lines and corridors are shown in Fig. 4.90.

4.9 Special Constructions in Permafrost Areas

As the industrial development has spread to subarctic and arctic areas, the need for special structures such as pipelines, tunnels, dams, bridges, piers, and offshore oil platforms has risen. The existence of permafrost may greatly complicate the design and construction of such structures. Problems connected with the behavior of permafrost, especially those dealing with its possible degradation, have to be recognized and solved. On the other hand, innovative solutions have been developed so that subfreezing temperatures actually work to the benefit of the structure.

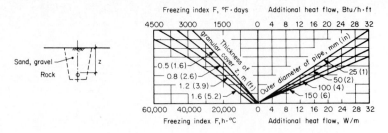

Narrow rock ditch, unsulated pipe

(a)

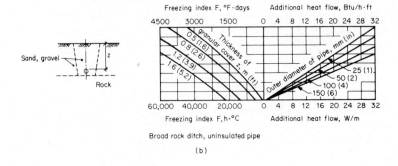

Broad rock ditch, uninsulated pipe

(b)

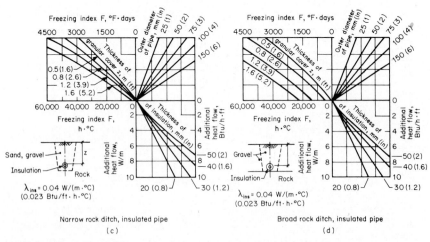

Narrow rock ditch, insulated pipe

(c)

Broad rock ditch, insulated pipe

(d)

Figure 4.84 Required heat input for (a), (b) uninsulated and (c), (d) insulated pipes in narrow and broad rock ditches. *(Adapted from Gundersen, 1976.)*

4.9.1 Pipelines

The use of conventional buried pipeline designs in permafrost areas would cause extensive thermal degradation of frozen soil, loss of bearing capacity, differential settlements, thermokarst, and eventually a total failure of the pipeline. These difficulties can be avoided by lifting the pipeline above the ground. If a pipeline is buried into ice-rich permafrost, a careful control of energy losses has to be practiced, possibly coupled with artificial cooling.

Experience with the elevated parts of the Trans-Alaska Pipeline System (TAPS) (Figs. 4.91 and 4.92) has in general been positive. No significant movements of the pipeline supported by the artificially cooled piles have been reported. The observed environmental disturbances have been induced by the pipeline workpad rather than the pipeline itself. On the other hand, short sections of the pipeline were also buried in permafrost areas. Artificial cooling is used in areas where the ground is not expected to be thaw-stable. In some cases misjudgments were made concerning the occurrence of permafrost or the thaw stability of the ground. Local problems connected with thermoerosion and uneven settlements have been experienced (Krzewinski et al., 1981).

Pipelines are capable of sustaining relatively large differential movements. If the amount of energy input to the ground is limited, the extent of thaw and the magnitude of settlements can be kept within acceptable limits. The design of the Norman Wells low-energy pipeline is an example of this kind of philosophy (Nixon et al., 1984). The maximum extent of thaw settlement is estimated and the situation is analyzed based on the theory of an infinite beam resting on an elastic foundation (Fig. 4.93). In this case special attention should be paid to the possible construction- or seepage-induced thermoerosion effects. Seepage-induced thermoerosion may occur in slopes when surface waters gather and flow along the buried pipeline. Ditch plugs or breakers can be used to prevent the erosion (Vita and Rooney, 1978). Other design considerations include buoyancy forces on submerged unactivated pipelines and the strength of thawed soil.

The approach of limiting differential thaw settlements to acceptable values is applicable also for the design of arctic offshore pipelines in the presence of permafrost (Heuer et al., 1983; Walker et al., 1983). In this case the effects of salt migration on permafrost degradation should also be addressed.

If gas is chilled below $0\,^\circ$C ($32\,^\circ$F) in a buried pipeline, permafrost can be protected conveniently, and at the same time operational power requirements can be reduced. In this case as well as in the case of low-energy oil lines passing through long stretches of permafrost, the major

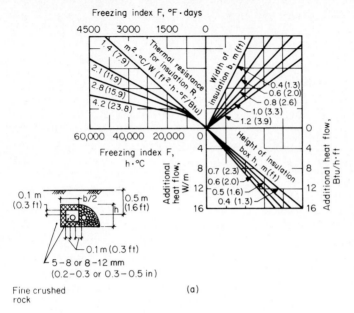

(a)

Fine crushed rock

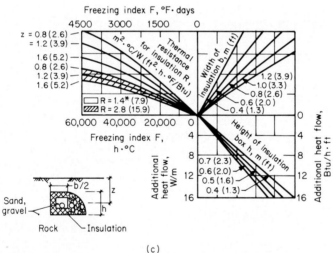

(c)

Figure 4.85 Required thermal resistance and width of box-type insulation for narrow and broad rock ditches. (*a*), (*b*) Pipe depth $z = 0.5$ m (1 ft 8 in). (*c*), (*d*) Pipe depth $z = 0.8$ m (2 ft 7 in), 1.2 m (4 ft), and 1.6 m (5 ft 3 in). (*Adapted from Gundersen, 1976.*)

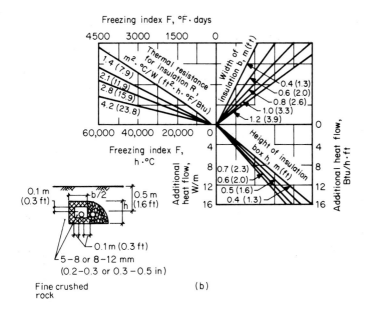

Freezing index F, °F·days

Fine crushed rock

(b)

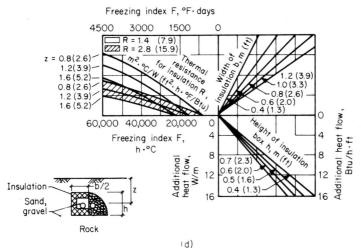

Freezing index F, °F·days

Rock

(d)

Figure 4.86 Installation of U-type thermal protection for utility lines. *(Courtesy of H. Mäkelä.)*

Figure 4.87 Installation details of prefabricated insulation units. *(Courtesy of Alaska Instrument, Inc.)*

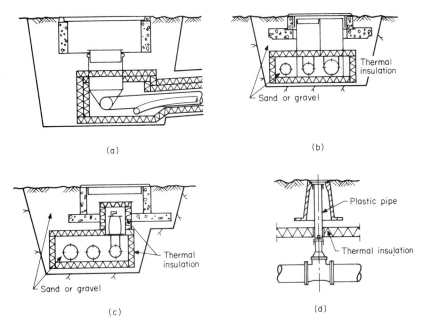

Figure 4.88 Thermal protection arrangements. (*a*) Shallow sewer well. (*b*) Combined surface water and sewer well. (*c*) Shallow fire cock. (*d*) Shallow valve. (*Mäkelä, 1982.*)

problem is differential frost heave when thawed ground is encountered. The problem has been discussed and analyzed, for example, in Jahns and Heuer (1983) and in Konrad and Morgenstern (1984). Insulation, adequate non-frost-susceptible backfill, heating the ground locally around the pipe, and freezing the thawed soil with vertical thermopipes represent potential solutions to the problem.

4.9.2 Tunnels

The construction of tunnels in permafrost is quite similar to that in rock. Conventional excavation methods, including mechanical excavation, drilling, and blasting, can be used. Thermal disturbances and viscoelastic behavior of permafrost represent special features in permafrost tunnel design.

Disturbances in the thermal regime around tunnels may be caused by heat released through ventilation, machinery, lights, people, or other sources. As a result, the creep of permafrost increases. If thawing begins, soil layers may lose their stability because the cementing effect of the ice is lost. The thawing of even permanently frozen rock may be hazardous.

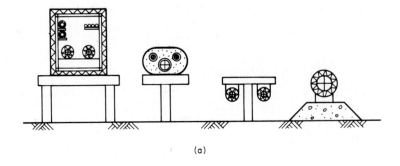

(a)

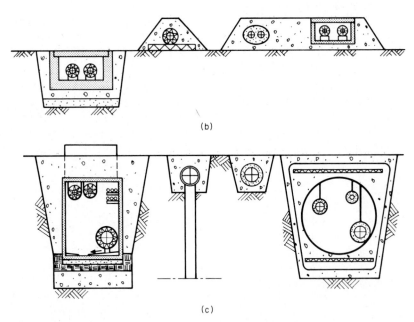

(b)

(c)

Figure 4.89 Utility line arrangements for permafrost areas. (*a*) Above ground installations. (*b*) Surface installations. (*c*) Buried installations.

The stress field is altered by the presence of the tunnel, and the thawing of the ice in the fractures may cause local failures in the rock.

Finally, sublimation of the ice at low relative humidity may cause some problems in permafrost tunnels. One problem is the increasing layer of dry ice-free material. It may be firm but very friable and therefore cause the tunnel air to be very dusty. Spray painting, lacquering, or coating the walls with a nondrying petroleum may be applied to control sublimation.

Tunnels in warm ice-rich permafrost or in ice tend to have high rates of closure (Figs. 4.94 and 4.95). On the other hand, deformations in dense ice-poor permafrost may be negligible. The long-term behavior of tunnels can be predicted with numerical analysis that recognizes the temperature-dependent viscoelastic properties of frozen soil (Thompson and Sayles, 1972).

The failure of a tunnel in permafrost is usually ductile and occurs when the ratio of loading exceeds the ability of the ice or permafrost to adjust to plastic flow. Brittle behavior even under high-velocity shocks seems to be very limited. The problems associated with the maintenance of tunnels in warm or discontinuous permafrost can be helped by circulating cold air through the tunnel during the winter. The effect can be further improved by using refrigeration systems, if necessary.

Not very much is known about the behavior of underground structures such as hydropower tunnels and underground powerhouses in permafrost. So far the general design criterion is that they should be located in hard massive rock (Gevirtz and Mostkov, 1978). The thermal balance of the rock is disturbed by flowing water and air. The thawed zone increases slowly, and it may take decades until a new thermal balance is reached. Special difficulties arise when seamy rock is crossed by tectonic fractures or zones. In addition to conventional reinforced concrete lining, construction with combined reinforced concrete anchors and shotcrete may be applied to secure the stability of the tunnel. Because the temperature of rock is low, heating and chemical admixtures are usually needed to secure the quality of sand-cement mortar, shotcrete, and reinforced concrete.

4.9.3 Dams

Dams are usually built to provide hydroelectric power, to build up a water reservoir or a sewage lagoon, or to control floods. Embankment dams built of local materials are often preferred in permafrost areas. There are two fundamental types of embankment dams (Fig. 4.96). In frozen dams, permafrost may be used as a strong, impermeable material. In thawed dams, the soil will remain thawed during construction and service. The need for an impermeable core in such a dam is obvious. Short reviews of the mostly Soviet experiences on dam construction on permafrost are presented in Tsytovich (1975) and Johnson and Sayles (1980).

Dams of the frozen type are relatively low, usually under 15 m (50 ft). They can remain in a frozen state without artificial freezing only in very cold permafrost areas (Fig. 4.97). Methods of cooling include surface protection from solar radiation, circulation of cold air or cooling solution,

Figure 4.90   Examples of elevated utility line arrangements.

and passive refrigeration by thermopipes. A frozen dam can be constructed of relatively pervious embankment materials because the through seepage is completely sealed off by freezing of the embankment, as long as the frozen state is preserved. At the construction stage, layer by layer systems or artificial freezing are used to create the frozen state.

A very careful thermal analysis is needed in the case of a frozen dam. Special attention should be given to seepage because it causes thermal convection and progressive thawing. This may eventually lead to the loss of tightness and stability of the structure. This is why even frozen dams may be provided with a watertight core of clay or ice. According to Soviet experiences, the areas near the floodgate or the spillway of the dam are

Figure 4.90   *(Continued)*

very vulnerable, and they need special consideration in the analysis and design.

Thermal analysis is the basis for predicting deformations. There will be thaw settlements under the water reservoir and also in the base of the dam. Consolidation of the base and the fill must be estimated, and the necessary overlapping must be provided. It is possible to experience a differential thaw settlement of more than 1 m (3 ft) without failure by employing materials such as sand which can deform easily. However, if the thaw settlements are estimated to be intolerable, preliminary thaw-

(a)

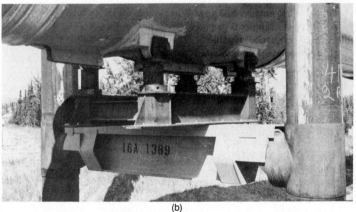

(b)

Figure 4.91   Details of Trans-Alaska Pipeline System (TAPS). (*a*) Transition from buried to elevated mode and zigzag configuration that converts thermal movements of pipe into lateral movements. (*b*) Pipeline support. (*c*) Pipeline crossing a road. (*d*) Gathering line crossing a river at Prudhoe Bay, Alaska.

ing and preconsolidation or removal of ice-rich soil is necessary at the base.

Large dams and dams in areas of warm permafrost are generally designed to be of the thawed type. These dams are usually built on strong bedrock or thaw-stable granular soil. In this case it is important to prevent excessive freezing of frost-susceptible soil during and after the construction. One possibility to cut seepage is to use grout curtains that

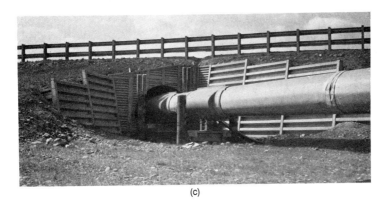

(c)

(d)

reach the permafrost or solid rock in addition to the impermeable core
(Biyanov, 1973).

4.9.4 Coastal and Offshore Structures

The shoreline typically marks a boundary between continuous shallow
permafrost and thawed ground in the Arctic. The thermal equilibrium at
this boundary is quite delicate and will certainly be disturbed by construction. These disturbances have to be predicted, thaw settlements or
heave deformations estimated, and possible counteractive measures designed.

Rivers and streams are often beaded, and the thermal regime in the
ground undergoes constant changes as a result of varying surface conditions, erosion, and sedimentation in the Arctic. If the river flow is restricted with embankments, further erosion and thermal disturbances
can be expected. Heavy erosion protection, possibly combined with
artificial cooling of the ground, for example around bridge foundations,
is recommended in such cases (Fig. 4.98).

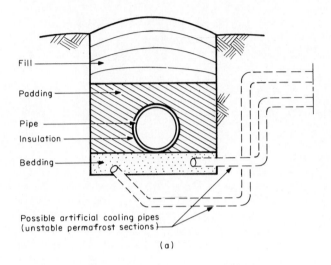

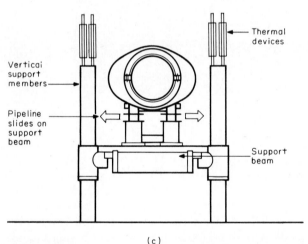

Figure 4.92 Pipeline construction modes for warm Trans-Alaska Pipeline System. (*a*) Buried sections, seasonal frost and permafrost areas. (*b*) Pipe ditch plug to cut seepage and control erosion. (*c*) Elevated sections, permafrost areas. (*d*) Anchor support.

Permanent wharves in the Arctic are commonly constructed of sheet piles backfilled with soil. The thermal regime of soil may change as a result of such construction. Significant frost heave may result depending on the soil conditions. This may also cause additional strains to the structure. On the other hand, changes in the thermal regime can also be used to the advantage of the designer. In areas where good-quality granular material is sparse, ice can be used as the fill material (Fig. 4.99). Freeze-

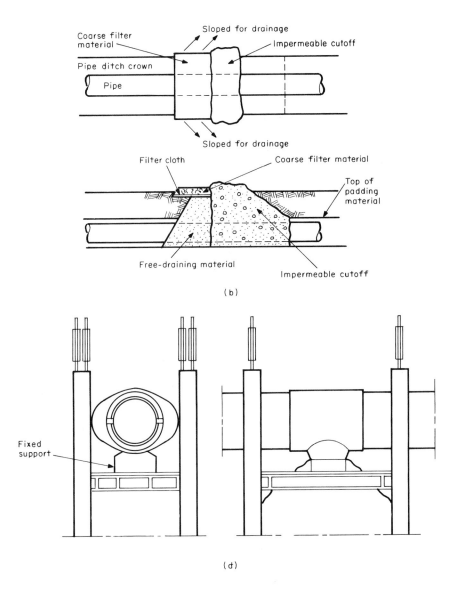

(b)

(d)

back and thermal stability behind sheet piles can be secured with insulation and thermo pipes. A very strong wharf structure can be created in such a manner.

The occurrence of offshore permafrost may also create problems for construction. Structures such as artificial oil production islands may create changes in the permafrost conditions by themselves. However, their actual function, oil production, is the reason for greater concern.

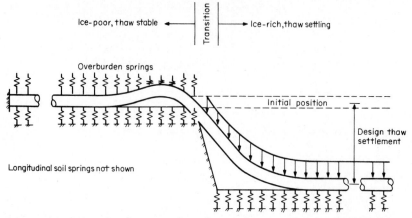

Ice-poor, thaw stable ◀ ─── ───▶ Ice-rich, thaw settling

Overburden springs

Initial position

Design thaw settlement

Longitudinal soil springs not shown

Figure 4.93 Principle of numerical pipeline analysis in a thaw settlement case. *(Nixon et al., 1984, reprinted by permission of American Society of Mechanical Engineers.)*

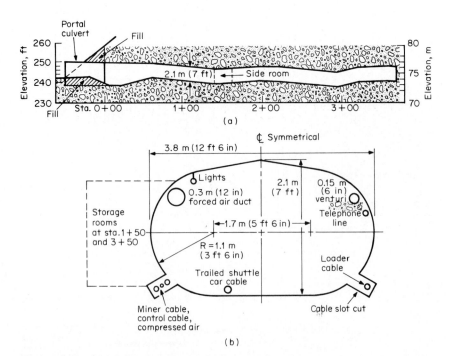

Portal culvert

Fill

Fill

Elevation, ft

260
250
240
230

80
75
70

Elevation, m

2.1 m (7 ft) ◀── Side room

Sta. 0+00   1+00   2+00   3+00

(a)

₵ Symmetrical

3.8 m (12 ft 6 in)

Lights

0.3 m (12 in) forced air duct

2.1 m (7 ft)

0.15 m (6 in) venturi

Telephone line

Storage rooms at sta.1+50 and 3+50

◀─1.7 m (5 ft 6 in)─▶

R = 1.1 m (3 ft 6 in)

Trailed shuttle car cable

Loader cable

Miner cable, control cable, compressed air

Cable slot cut

(b)

Figure 4.94 *(a)* Profile and *(b)* cross section of permafrost tunnel in Alaska. *(Swinzow, 1970.)*

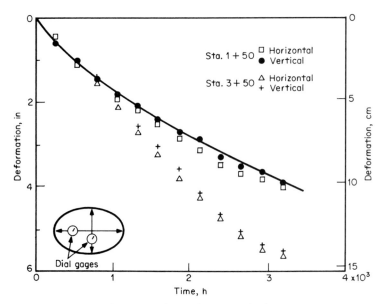

Figure 4.95 Initial closure of tunnel in warm high-ice-content permafrost. *(Swinzow, 1970.)*

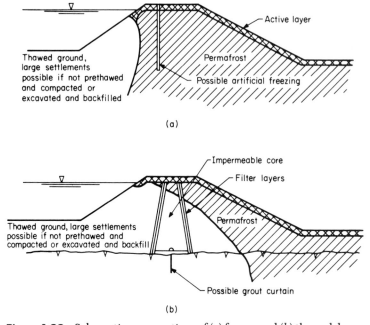

Figure 4.96 Schematic cross sections of (a) frozen and (b) thawed dams.

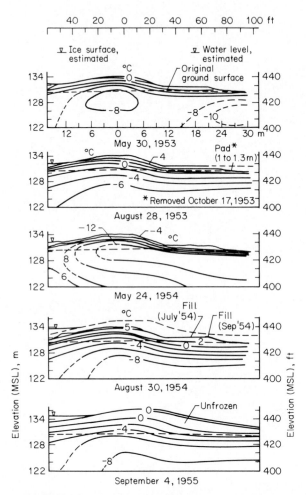

Figure 4.97 Temperature profiles during typical year in a frozen dam in Greenland. *(Fullwider, 1973.)*

The flow of warm oil thaws subsea permafrost at a considerable radius. The resulting thaw weakening and settlement may cause the collapse of the well casing, but it will also affect the stability and serviceability of the structures at the surface. These problems are illustrated in Fig. 4.100.

4.10 Numerical Method for Frost Problems

The purpose of this section is to introduce certain mathematical models in terms of various physical and environmental parameters and to de-

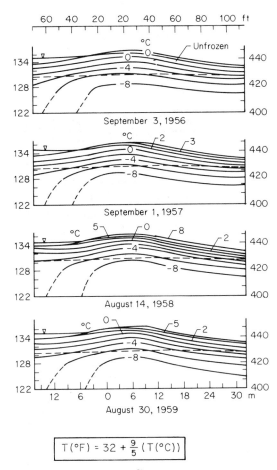

Figure 4.97 *(Continued)*

velop finite element formulations for frost heaves and thaw consolidation. Since the introduction of Galerkin's method to finite element formulation, problems in heat transfer, fluid flow, and other fields have been solved by the finite element approach. For frost heave and thaw consolidation problems, including moisture transport, heat transport, and phase-change conditions, the finite element method appears to be the best analytical tool.

Various attempts have been made to describe frost heaving problems quantitatively. For example, Guyman et al. (1984) have developed a two-dimensional model of coupled heat and moisture flow in frost-heaving soils and obtained a numerical solution by the integrated finite difference and the Galerkin finite element methods. In this section we

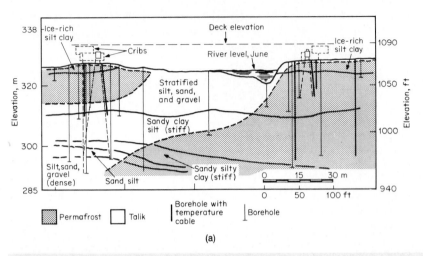

(b)

Figure 4.98 (*a*) Eagle River bridge, Yukon Territory, with permafrost conditions around it as measured by Johnston (1980) and with heavy erosion protection. (*b*) Some indications of thermoerosion can be observed around constructed riverbank.

introduce the subject following the approach used by Berg et al. (1980). This is a one-dimensional model describing simultaneously heat and moisture fluxes in the frozen soil-water system.

4.10.1 Equation of Simultaneous Heat and Moisture Flux

In developing mathematical models of frost heaving and thaw consolidation, the complex coupling relationship of heat and moisture flux in soil-water systems must be considered. In this section a one-dimensional soil column will be analyzed for its moisture transport, heat transport, phase change, and coupling effects.

Moisture transport   In the following mathematical model these assumptions are used:

1. Darcy's law applies to moisture movement in both saturated and unsaturated conditions.

2. The porous media are nondeformable as far as moisture flux is concerned.

3. All processes are single-valued, that is, hysteresis is not present in relationships of the characteristic curve.

4. The effect of dissolved salts in unfrozen water is neglected. Furthermore, vapor flux is neglected, that is, water flux is primarily treated as liquid.

In a one-dimensional soil pore, we assume that the major components of flux are those due to liquid driven by hydraulic gradients, because vapor transfer driven by thermal and hydraulic gradients and liquid flux driven by thermal gradients are generally several orders of magnitude lower than liquid moisture flux due to hydraulic mechanisms.

A one-dimensional moisture flux relationship as described by Darcy's law is given by

$$v = -K \frac{\partial \phi}{\partial z} \qquad (4.34)$$

where z = coordinate axis, positive downward
$\quad v$ = fluid flux in the z direction, positive downward
$\quad \phi$ = total hydraulic head, $= \psi - z$, where ψ is the pore pressure in length units ($\psi < 0$ for unsaturated flow, and $\psi > 0$ for saturated flow)
$\quad K$ = hydraulic conductivity, $= K(\psi, T)$
$\quad T$ = temperature

Substituting Eq. (4.34) into the continuity equation applicable to freezing soils,

$$\frac{\partial v}{\partial z} = -\frac{\partial W}{\partial t} - Q \qquad (4.35)$$

we have

$$\frac{\partial}{\partial z}\left(K \frac{\partial \psi}{\partial z}\right) - \frac{\partial K}{\partial z} = \frac{\partial W}{\partial t} + Q \qquad (4.36)$$

(a)

(b)

Figure 4.99 Construction of ice berm with gravel cover to protect pipeline shore entry at Melville Island, Canada. (a) Ice blocks used for construction. (b) Flooding of ice blocks. (c) Placing gravel on flooded ice blocks. (d) Completed berm. (Courtesy of Fenco Consultants Ltd.)

where W = volumetric liquid water content, = $W(\psi, T, t)$
$\qquad Q$ = rate at which liquid water is converted to ice
$\qquad t$ = time

Let us assume that K and W can be approximated by relationships proposed in Gardner (1958),

$$K = \begin{cases} K_0 & \psi \geq 0 \\ \dfrac{K_0}{-A_K \psi^3 + 1} & \psi < 0 \end{cases} \qquad (4.37)$$

(c)

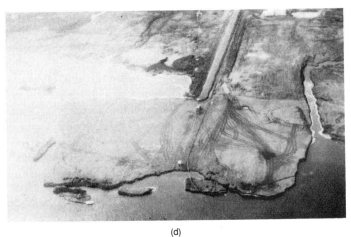

(d)

$$W = \begin{cases} W_0 & \psi \geqslant 0 \\ \dfrac{W_0}{-A_W \psi^3 + 1} & \psi < 0 \end{cases} \qquad (4.38)$$

where K_0 = saturated hydraulic conductivity
$\quad\ W_0$ = saturated volumetric moisture content (porosity)
A_K, A_W = parameters depending on soil

Taking the partial derivative of Eq. (4.37) with respect to z and the partial derivative of Eq. (4.38) with respect to t, we get

$$\frac{\partial K}{\partial z} = \frac{3 A_K K_0 \psi^2}{(-A_K \psi^3 + 1)^2} \frac{\partial \psi}{\partial z} \qquad \psi < 0 \qquad (4.39)$$

$$\frac{\partial W}{\partial t} = \frac{3 A_W W_0 \psi^2}{(-A_W \psi^3 + 1)^2} \frac{\partial \psi}{\partial z} \qquad \psi < 0 \qquad (4.40)$$

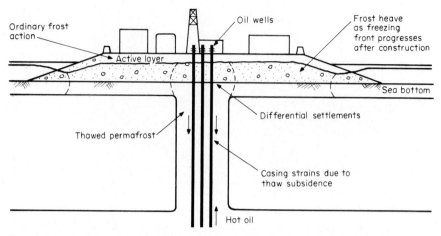

Figure 4.100 Permafrost problems in arctic offshore oil production islands.

Substituting Eqs. (4.39) and (4.40) into Eq. (4.36), the result is

$$\frac{\partial}{\partial z}\left(K\frac{\partial \psi}{\partial z}\right) - \overline{K}\frac{\partial \psi}{\partial z} = \overline{W}\frac{\partial \psi}{\partial t} + Q \qquad (4.41)$$

where

$$\overline{K} = \frac{3A_K K_0 \psi^2}{(-A_K \psi^3 + 1)^2}$$

$$\overline{W} = \frac{3A_W W_0 \psi^2}{(-A_W \psi^3 + 1)^2}$$

$\overline{K}$ and $\overline{W}$ being functions of ψ and parameters of the soil.

Equation (4.41) is nonlinear; ψ is the primary state variable, and the parameters K, $\overline{K}$, and $\overline{W}$ are functions of ψ. The parameter Q can be specified a priori or computed from suitable empirical relationships. Equation (4.41) is strictly applicable to moisture flux in unsaturated zones or partially frozen soil.

Considering two types of boundary conditions, specified boundary conditions and moisture flux boundary conditions, we have for the upper boundary,

$$\psi = \psi_u \qquad z = 0 \qquad t > 0$$

$$\frac{\partial \psi}{\partial z} = F_u \qquad z = 0 \qquad t > 0$$

where ψ_u is a specified pressure (it may be a function of time), and F_u is a known function of time. Similarly, for the lower boundary,

$$\psi = \psi_L \qquad z = L \qquad t > 0$$

$$\frac{\partial \psi}{\partial z} = F_L \qquad z = L \qquad t > 0$$

The auxiliary conditions are initial conditions of the form

$$\psi = \psi_0 \qquad 0 \leqslant z \leqslant L \qquad t = 0$$

where $\psi_0 = \psi_0(z)$ and is known a priori.

Heat transport   The one-dimensional heat-transport model is based on the following assumptions:

1. Freezing or thawing is an isothermal process, and latent heat generation can be accounted for by a systematic recording technique.

2. Radiation heat transfer is considered only at the interface between the soil and the atmosphere. It can be accounted for in the boundary conditions of the soil surface.

3. Convected heat is primarily in the fluid form. Vapor-convected heat is negligible.

The energy balance equation is given by

$$\dot{E}_c + \dot{E}_v = \dot{E}_t \tag{4.42}$$

where $\dot{E}_c$ = rate of heat conduction into elemental volume
$\dot{E}_v$ = net rate of heat convection into elemental volume
$\dot{E}_t$ = total rate of heat energy stored in elemental volume

From Eq. (4.42) we can get

$$\frac{\partial}{\partial z}\left(K_t \frac{\partial T}{\partial z}\right) - C_w v \frac{\partial T}{\partial z} = C_a \frac{\partial T}{\partial t} \tag{4.43}$$

where C_w = volumetric heat capacity of fluid
v = fluid flux in the z direction
K_t = thermal conductivity of entire mass
C_a = bulk volumetric heat capacity

Because water storage and movement and thermal states are assumed to be coupled through the entire soil column, it is necessary to specify boundary conditions at the same boundaries as those used for the moisture transport equation:

$$T = T_u \qquad z = 0 \qquad t > 0$$

$$\frac{\partial T}{\partial z} = H_u \qquad z = 0 \qquad t > 0$$

and

$$T = T_L \qquad z = L \qquad t > 0$$

$$\frac{\partial T}{\partial z} = H_L \qquad z = L \qquad t > 0$$

The upper or lower specified temperatures may be a function of time, and H indicates a specified heat flux function that may be time-dependent. Equation (4.42) and certain of its auxiliary equations require initial conditions of the following form:

$$T = T_0 \qquad 0 \leqslant z \leqslant L \qquad t = 0$$
$$W = W_0 \qquad 0 \leqslant z \leqslant L \qquad t = 0$$

Phase change   Phase change is the major heat-transport phenomenon associated with the freezing and thawing of soils. In order to describe phase change, a systematic recording process to ascertain the location of the freezing front is required.

The amount of heat that must be removed to freeze an element is determined by

$$L \times V$$

where L is the volumetric latent heat of fusion of the soil water within an element, and V is the volume of the element.

The amount of heat required to change the average temperature of an element by $1\,°C$ is given by

$$C_a \times V$$

Let

$$E = \frac{L \times V}{C_a \times V}$$

where E is the number of "excess degrees" required to freeze the element.

Consider a common node A. After each time increment, the temperature of point A is scanned to determine whether or not it has dropped below the freezing point depression T_d. If node A has not been entirely frozen and the computed temperature has dropped below T, then the quantity of excess degrees evolved during the previously computed time step becomes

$$E_t = T_d - \text{computed temperature}$$

where E_t represents the excess number of degrees removed from node A during the previous time step. If the required amount of latent heat has been extracted, node A and the volume of soil water are considered to be frozen.

The thawing process is handled in a similar manner.

Coupling effects of moisture and heat transport   Equation (4.41) for fluid flux and Eq. (4.43) for heat flux may be coupled together to describe the freezing and thawing characteristics of the soil because of the nature of the parameters used in the two equations.

Parameters K, $\overline{K}$, $\overline{W}$, and Q in Eq. (4.41) are dependent on temperature. At the same time, parameters K_t, v, and C_a in Eq. (4.43) are dependent on water content. The equations can be solved individually and alternatively by keeping their parameters constant in a prescribed time period. Parameters can be updated in the computation at any arbitrary frequency.

Frost heave   The frost heave problem consists of two parts:

1. The actual frost heave during the previous time increment. The porosity of each element is provided in the input data. If the volume of water available for freezing is greater than 90% of the porosity, frost heave occurs. The magnitude of the incremental amount of frost heaving is determined by multiplying the amount of newly frozen water in excess of 90% of the porosity by 1.09.

2. The pore water pressure at the freezing front. At the ice-water interface a pressure difference occurs at the boundary because of energy imbalance. The pressure on the ice may be assumed equal to the overburden stress and surcharge loading.

The pore water tension at the freezing front is given by

$$P_w = \frac{1}{2}\left[\left(S + \sum_{j=1}^{n} \gamma_j H_j\right) + \frac{\sigma_{iw} P_c}{\sigma_{aw}}\right] \qquad (4.44)$$

where P_w = pore water pressure
$\quad\quad S$ = surcharge stress
$\quad\quad \gamma_j$ = bulk density of soil layer j
$\quad\quad H_j$ = thickness of soil layer j
$\quad\quad \sigma_{iw}$ = ice-water surface tension
$\quad\quad \sigma_{aw}$ = air-water surface tension
$\quad\quad P_c$ = critical soil water pressure from moisture characteristic curve

The bulk density of a soil layer changes with time as water flows into or out of the layer. It also changes as ice accumulates in the layer.

Equation (4.44) is used to compute the soil water suction at the freezing front. This pressure drives the water movement, and it can be considered as a moving boundary condition. After each time step, soil water tension can be computed. As water flows into the freezing element from

below, the degree of saturation of the soil voids increases. When saturation exceeds 90%, heaving commences within the element.

4.10.2 Finite Element Formulation

Development of finite element equations   The moisture flux equation and
the heat-transport equation have the general form for one-dimensional
problems:

$$K_1 \frac{\partial^2 u}{\partial z^2} - K_2 \frac{\partial u}{\partial z} = K_3 \frac{\partial u}{\partial t} - K_4 \tag{4.45}$$

where K_1, K_2, K_3, and K_4 are the moisture pressure state parameters or
the heat-transport equation parameters, and u is the state variable, either
the pore water pressure or the temperature. Applying the Galerkin
method, we can get

$$\int_0^L \mathcal{N}^T \left(K_1 \frac{\partial^2 u}{\partial z^2} - K_2 \frac{\partial u}{\partial z} - K_3 \frac{\partial u}{\partial t} + K_4 \right) dz = 0 \tag{4.46}$$

where $\mathcal{N}$ is the shape function in the form of a row matrix and L the length
of the finite element.

Considering natural boundary condition $\partial u/\partial z = 0$, and substituting
the shape function $u = \mathcal{N}\{u\}$, Eq. (4.46) yields

$$\sum_{m=1}^{M} \int_0^L \left(K_1 \frac{\partial \mathcal{N}^T}{\partial z} \frac{\partial \mathcal{N}}{\partial z} \{u\} + K_2 \mathcal{N}^T \frac{\partial \mathcal{N}}{\partial z} \{u\} + K_3 \mathcal{N}^T \mathcal{N} \{\dot{u}\} - K_4 \mathcal{N}^T \right) dz = 0 \tag{4.47}$$

where $\{u\}$ is a column matrix of the nodal state-variable values and $\{\dot{u}\}$ is
the differentation of $\{u\}$ with respect to time t.

Integrating every term in Eq. (4.47), the resulting equation in matrix
form is given by

$$[s]\{u\} + [p]\{\dot{u}\} = \{f\}$$

At the total system level we may write

$$[S]\{U\} + [P]\{\dot{U}\} = \{F\} \tag{4.48}$$

where $[S] =$ square nonsymmetrical matrix, determined by the geometry of the discretization in space and the conductivity parameter

$[P] =$ determined by spatial discretization and the capacitance
parameter

$\{F\} =$ column matrix, determined by the ice sink term and the
boundary conditions

Equation (4.48) must be solved with respect to time.

Dynamic problem—time domain solution   According to the Crank-Nicholson method, using the derivative of the state variable at the beginning and the end of each time step, we can move the solution ahead in time. Letting

$$\{U\}^{i+1} = \{U\}^i + \frac{\Delta t}{2}(\{\dot{U}\}^i + \{\dot{U}\}^{i+1})$$

where i denotes the time step, and premultiplying by the $[P]$ matrix, this equation becomes

$$[P]\{U\}^{i+1} = [P]\{U\}^i + \frac{\Delta t}{2}([P]\{\dot{U}\}^i + [P]\{\dot{U}\}^{i+1})$$

Substituting

$$[P]\{\dot{U}\} = \{F\} - [S]\{U\}$$

into the above equation, we get

$$\left([S] + \frac{2}{\Delta t}[P]\right)\{U\}^{i+1} = \left(\frac{2}{\Delta t}[P] - [S]\right)\{U\}^i + 2\{F\}$$

Because $\{U\}^i$ represents the initial conditions or previously computed values of the state variable, the above equation can be written in the form

$$[W]\{U\}^{i+1} = \{G\} \tag{4.49}$$

From Eq. (4.49), $\{U\}^{i+1}$ can be computed in a step-by-step fashion.

It should be noted again that the above description only refers to the one-dimensional model. There are authors who presented two- or three-dimensional models. For example, Guymon and Hromadka (1977) presented a two-dimensional and a three-dimensional heat-transport model with phase change for freezing soils and volume change induced by freezing and thawing of the soil-water-ice system. In their work, the partial differential equation for transient heat conduction is solved by a finite element analog, using a quadratic weighting function for the discretized spatial domain. The transient problem is solved by the Crank-Nicholson approximation. Phase change is approximated as an isothermal process.

4.10.3   Examples of Numerical Solutions

Problem of heated pipeline in permafrost   The finite element method can be applied to problems of heated oil pipelines in permafrost. Using a finite element model, it is possible to evaluate the spatial and time distributions of pore pressure, temperature, and displacement for the filtration and consolidation of thawed permafrost around a buried heated pipeline.

Based on work by Sykes (1973), the finite element method was used to determine the significance of variable geometry and forced convection as well as the effects of settlement. Thaw boundary conditions incorporated within the model (Sykes et al., 1974a) include the consolidation of the latent heat as a moving source in the energy balance equation, the movement of the heat source and other surfaces due to consolidation, and the use of film coefficients to model radiation at the surface of the active layer and insulation of the heat sources. By using this model for the consolidation of thawed permafrost, Gurtin's method (1964) is used to reduce the thaw and consolidation field equations to boundary-value problems for which functionals are derived by the Galerkin technique.

For reducing the field equations to boundary-value problems, an initially undeformed system was assumed for consolidation. The initial-value problem can be reduced to a boundary-value problem by taking the Laplace transform, dividing by the transform domain variable, and then taking the inverse Laplace transform.

Using the discretization procedure of the finite element method, the generalized coordinates are selected as the pressures, displacements, and temperatures at the nodal points of the finite element idealization.

The reduced stress vector in the solid phase is

$$\{\sigma^m\} = [D^m]\{e^m\} + \{\sigma_0^m\} \tag{4.50}$$

where $[D^m]$ = plane-strain elasticity matrix
$\{e^m\}$ = reduced strain vector
$\{\sigma_0^m\}$ = vector of initial stress

The pore pressure, displacement, and temperature at any point within an element may be expressed in terms of the nodal point pressures, displacements, and temperature matrix form.

Using the Galerkin procedure, the finite element equation for all elements governing heat transfer in permafrost is given in matrix notation involving the convolution product as

$$[K_3]*\{T\} + [C_2]\{T\} - [U]*\{T\} - [P]*\{T\} + g'*[C_s]\{T\} \\ = g'*[L] + [C_2]\{T_0\} + g'*[C_s]\{\bar{T}\} + g'*[F] \tag{4.51}$$

where $[K_3]*\{T\}$ = heat energy conducted out of nodal regions
$[C_2]\{T\}$ = heat energy in nodal regions
$g'*[L]$ = latent heat term
$g'*[F]$ = nodal heat input on heat-flux surface
$-g'*[C_s]\{T\}, g'*[C_s]\{\bar{T}\}$ = nodal heat inputs on heat-transfer surfaces
$[U]*\{T\}, [P]*\{T\}$ = forced convection heat energy of nodes

Similarly, we can get the consolidation finite element equations;

$$[K_1]\{U(t)\} + [C]\{P(t)\} = -\{M_1\} + \{M_2\} + \{\bar{F}\} \qquad (4.52)$$
$$[C]^T\{U(t)\} - g'*[K_2]\{P(t)\} = g'*\{M_3\} - g'*\{Q\} \qquad (4.53)$$

where $[K_1]\{U(t)\}$ = nodal force vector due to straining of soil skeleton

$[C]\{P(t)\}$ = nodal force vector resulting from pore pressure gradient

$\{M_1\}$ = initial stress vector

$\{M_2\}$ = body force vector

$\{\bar{F}\}$ = specified boundary traction vector

$[C]^T\{U(t)\}$ = volumetric strain of soil skeleton

$g'*[K_2]\{P(t)\}$ = fluid inflow due to pore pressure

$g'*\{M_3\}$ = fluid gravity forces

$g'*\{Q\}$ = specified boundary flow

To obtain a solution for a specified time, a step-by-step integration technique is used. To ensure continuity of the nodal variables in the time domain, an interpolation function is used with the end conditions of the time interval as the generalized coordinates. Thus the convolution products in Eqs. (4.51) to (4.53) can be evaluated.

Finally the finite element equations for the equilibrium condition, the continuity condition, and the energy or heat equation can be expanded as

$$[K_1]\{U\}_n + [C]\{P\}_n = -\{M_1\}_n + \{M_2\}_n + \{\bar{F}\}_n$$

$$[C]^T\{U\}_n - \frac{\Delta t}{4}[K_2]\{P\}_n = [C]^T\{U\}_{n-1} + \frac{\Delta t}{4}[K_2]\{P\}_{n-1}$$

$$+ \frac{\Delta t}{4}[\{M_3\}_{n-1} + \{M_3\}_n] - \frac{\Delta t}{2}[\{Q\}_{n-1} + \{Q\}_n]$$

$$\left[\frac{\Delta t}{2}([K_3] - [U] - [P] + [C_s]) + [C_2] + j[L_1]\right]\{T\}_n$$

$$= \left[-\frac{\Delta t}{2}([K_3] - [U] - [P] + [C_s]) + [C_2] + j[L_1]\right]\{T\}_{n-1} \qquad (4.54)$$

Sykes et al. (1974b) analyzed a two-dimensional heated pipeline problem. The authors considered a 1.2-m (4-ft)-diameter uninsulated heated oil pipeline at a temperature of 80°C (176°F) and centered at a depth of 2.4 m (8 ft) in permafrost at an initial temperature of −8.9°C (16°F). Using 181 triangular elements with a total of 109 nodes, the continuum was discretized as shown in Fig. 4.101. The surface air temperature is assumed to be at a yearly average of −2.2°C (28°F).

Insufficient soil property data restrict this analysis to the case of isotropy, resulting in the permeability matrix K_{ij} having values in the prin-

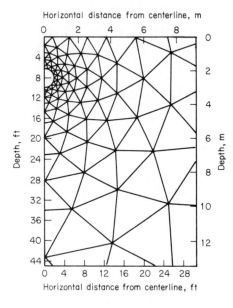

Horizontal distance from centerline, m

Figure 4.101 Discretized continuum for heated oil pipeline. (*Sykes et al., 1974b.*)

cipal coordinate directions only. The same is true for the thermal conductivity matrix. For the elasticity matrix it was evaluated for the isotropic plane strain condition.

In Fig. 4.102 the finite element results and Lachenbruch's (1970) results are plotted. The position of the $0°C$ ($32°F$) phase boundary along the pipe centerline and the pipe settlement are plotted versus the square root of time in Fig. 4.103. Figure 4.104 gives the settlement of the pipe and the air-soil surface for various times. Figure 4.105 shows the pore pressure equipotential lines at time $t = 10$ years from which the flow lines can be generated. The plot of pipe displacement versus the square root of time is given in Fig. 4.106 for three values of C_s.

Problem of warm building foundation   In cold regions under warm building foundations, the process of heat transfer is unique because the soil body consists of a combination of components, such as solid, water, ice, and air. The freezing and thawing of soils are complex phenomena. Such processes require the extraction or addition of the latent heat of fusion.

Let a partially frozen soil body be subjected to a heat flux varying with time. The differential equation expressing heat balance for an infinitesimal element in the body can be written.

Using the variational principle for field problems of heat conduction, the functional of a temperature field introduced over the volume V and boundary S of a continuum can be obtained.

Mohan (1975) devised two-dimensional triangular finite elements which are adequate to solve the problem of temperature distribution in the soil body. For a typical triangular element m, the temperature field can be written as

$$T_m(\bar{x}, t) = \lfloor b_m(\bar{x}) \rfloor \{T(t)\} \qquad (4.55)$$

where $\{T(t)\}$ = column vector of nodal point temperatures
$\lfloor b_m(\bar{x}) \rfloor$ = spatial dependence matrix, which is a constant

Differentiation of Eq. (4.55) with respect to spatial coordinates yields a vector of the temperature gradient,

$$\left\{ \frac{\partial T(\bar{x}, t)}{\partial x_i} \right\} = [a_m(\bar{x})]\{T(t)\} \qquad (4.56)$$

where $[a_m(\bar{x})]$ is the spatial dependence coefficient matrix.

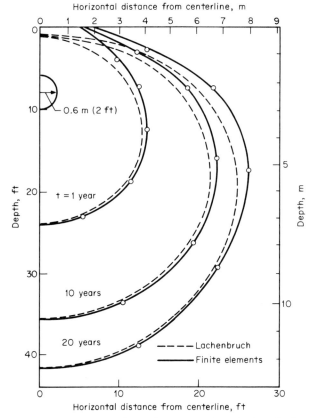

Figure 4.102 Heated pipeline; 0°C (32°F) thaw bulbs. (*Sykes et al., 1974b.*)

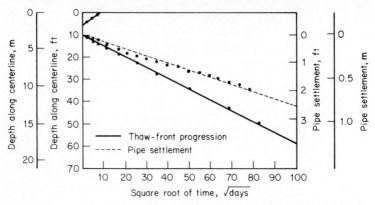

Figure 4.103 Position of 0°C (32°F) phase boundary and pipe settlement versus square root of time along centerline of heated oil pipeline. (Sykes et al., 1974b.)

For the surface temperature and heat flux along a typical boundary element, T_m is expressed as

$$T_m(\bar{x}, t) = \lfloor d_m(\bar{x}) \rfloor \begin{Bmatrix} T_i(t) \\ T_j(t) \end{Bmatrix} \tag{4.57}$$

where $T_i(t)$ = temperature at one end i of boundary element side ij
$T_j(t)$ = temperature at other end j
$\lfloor d_m(\bar{x}) \rfloor$ = spatial dependence coefficient matrix

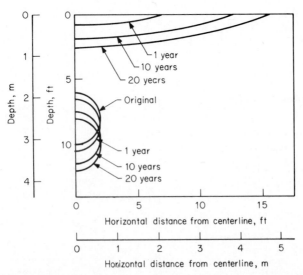

Figure 4.104 Surface and oil-pipe settlement. (Sykes et al., 1974b.)

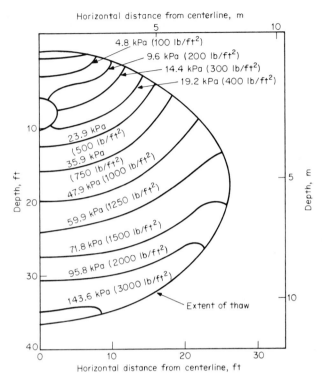

Figure 4.105 Pore-pressure equipotential lines at $t = 10$ years. *(Sykes et al., 1974b.)*

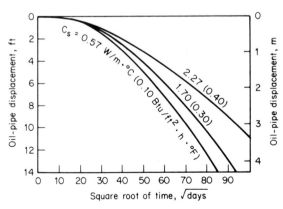

Figure 4.106 Oil-pipe settlement versus square root of time for different surface film coefficients C_s. *(Sykes et al., 1974b.)*

Then the generating functional for the finite element system may be written in summation form over all elements of the system,

$$
\begin{aligned}
\Omega_t(T_m) = \sum_{m=1}^{m} \int_{v_m} \frac{1}{2} \Big\{ & \rho_m C_m * T_m * T_m + \frac{\partial T_m}{\partial x_i} * k_{ij} * \frac{\partial T_m}{\partial x_j} \\
& + 2\rho_m * h_{gm} * T_m + 2\rho_m * h_{Lm} * T_m \\
& - 2\rho_m C_m T_{cm} * T_m \Big\} (\bar{x}, t)\, dv \\
& - \int_{s2} \{ h_{im} S_{nm} * T_m \} (\bar{x}, t)\, ds \\
& - \int_{s3} \Big\{ H_m * \Big(T_e - \frac{T_m}{2} \Big) * T_m \Big\} (\bar{x}, t)\, ds
\end{aligned}
\tag{4.58}
$$

Substituting Eqs. (4.55) to (4.57) into Eq. (4.58) we have

$$
\begin{aligned}
\Omega_t(T) = & \frac{1}{2} \{T(t)\}^T * [K] * \{T(t)\} + \frac{1}{2} \{T(t)\}^T [C] * \{T(t)\} \\
& - \{T(t)\}^T * [N]\{T(0)\} + \{T(t)\}^T * [G] + \{T(t)\}^T * [L] \\
& - \{T(t)\}^T * [F] - \{T(t)\}^T * [M] + \frac{1}{2} \{T(t)\}^T * [Q] * \{T(t)\}
\end{aligned}
$$

Here
$$
\begin{aligned}
[K] &= \sum_{m=1}^{m} \int_{v_m} [a_m(\bar{x})]^T [k] [a_m(\bar{x})]\, dv \\
[C] &= \sum_{m=1}^{m} \int_{v_m} \rho_m(\bar{x})\, C_m(\bar{x}) \lfloor b_m(\bar{x}) \rfloor^T \lfloor b_m(\bar{x}) \rfloor\, dv \\
[N] &= \sum_{m=1}^{m} \int_{v_m} \rho_m(\bar{x}) C_m(\bar{x}) \{T_{Cm}(0)\} \lfloor b_m(\bar{x}) \rfloor^T \lfloor b_m(\bar{x}) \rfloor\, dv \\
[G] &= \sum_{m=1}^{m} \int_{v_m} \rho_m(\bar{x}) \lfloor b_m(\bar{x}) \rfloor^T h_{Gm}(\bar{x}, t)\, dv \\
[L] &= \sum_{m=1}^{m} \int_{v_m} \rho_m(\bar{x}) \lfloor b_m(\bar{x}) \rfloor^T h_{Lm}(\bar{x}, t)\, dv \\
[F] &= \sum_{m=1}^{m} \int_{s_{2m}} h_{im}(\bar{x}, t) S_{nm} \lfloor b_m(\bar{x}) \rfloor^T\, ds \\
[M] &= \sum_{m=1}^{m} \int_{s_{3m}} T_e H_m \{d_m(\bar{x})\}\, ds \\
[Q] &= \sum_{m=1}^{m} \int_{s_{3m}} H_m \{d_m(\bar{x})\}^T \{d_m(\bar{x})\}\, ds
\end{aligned}
\tag{4.59}
$$

where $[K]$ = conductivity matrix
   $[C]$ = heat capacity matrix
   $[N]$ = modal heat energy matrix due to initial conditions
   $[G]$ = heat generation matrix
   $[L]$ = latent heat matrix
   $[F]$ = heat source matrix or sink input from surface temperature
   $[M]$ = modal input matrix or output from the environment
   $[Q]$ = heat transfer matrix due to the environment
and C = specific heat capacity
   ρ = mass density
   h_G = rate of heat generation or absorption
   h_L = rate of latent heat generation or absorption
   k_{ij} = thermal conductivity tensor
   S_h = unit normal to a prescribed heat flux surface
   h_i = prescribed heat flux at a surface
   H = coefficient of surface heat transfer

Letting the functional achieve a stationary value at any time, the following condition may be written:

$$[K]*\{T(t)\} + [C]\{T(t)\} = [N] + [G] + [L] - [F] - [M] + [Q]*\{T(t)\} \quad (4.60)$$

Equation (4.60) is a set of linear equations to be solved for nodal point temperatures of the finite element representation as a function of time.

Using prescribed temperature boundary conditions, Eq. (4.60) can be rearranged in partitional form in such a manner that equations involving unknown temperature $T_a(t)$ and specific temperature $T_b(t)$ can be separated into two groups:

$$[K_{aa}]*\{T_a(t)\} + [C_{aa}]\{T_a(t)\} = \{P_a(t)\} + [C_{aa}]\{T_a(0)\} \quad (4.61)$$

where

$$\{P_a(t)\} = [C_{ab}]\{T_b(0)\} + [G_a] + [L_a] - [F_a] - [M_a]$$
$$- [K_{ab}]*\{T_b(t)\} - [C_{ab}]\{T_b(t)\} \quad (4.62)$$

The effective thermal force matrix $\{P_a(t)\}$ is completely defined, and the solution of the matrix in Eq. (4.61) for unknown temperatures $\{T_a(t)\}$ can be accomplished by a step-by-step time integration scheme. At first Eq. (4.61) is rewritten to reflect the solution at time t in terms of a pseudoinitial value at time $t - \Delta t$,

$$[C]\{T(t)\} + [K]*\{T(t)\} = [C]\{T(t - \Delta t)\} + \{P(t)\} \quad (4.63)$$

Equation (4.63) can be evaluated by assuming a form of interpolation relationship within a time interval Δt.

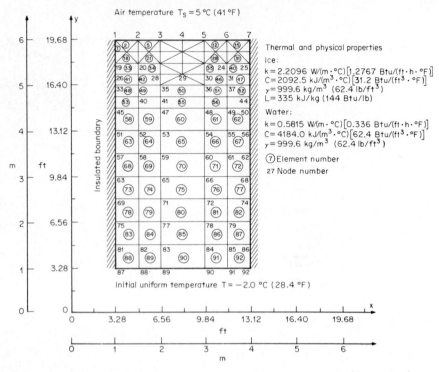

Figure 4.107 Finite element representation of Neumann's problem. *(Mohan, 1975.)*

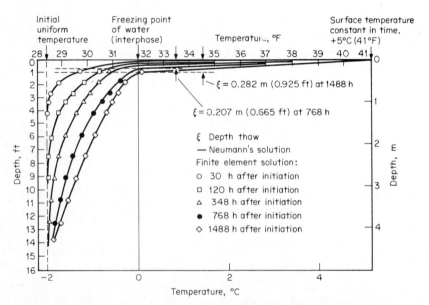

Figure 4.108 Comparison of Neumann and finite element solutions of melt-ice problem. *(Mohan, 1975.)*

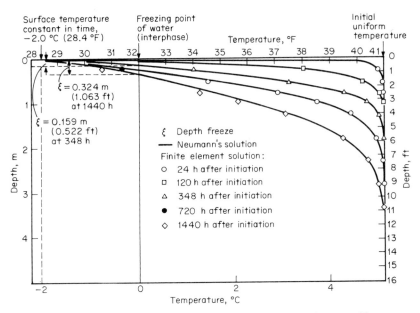

Figure 4.109 Comparison of Neumann and finite element solutions of freezing-water problem. *(Mohan, 1975.)*

The latent heat matrix $[L]$, given in Eq. (4.62), appears in the thermal force matrix $\{P\}$, and is required for the solution of the problem of heat flow in a system involving freezing and melting.

Assuming that the ground is initially at a uniform temperature below freezing $[-2\,°C\ (28.4\,°F)]$, this ice region is superposed by a constant temperature of $5\,°C\ (41.0\,°F)$ on the upper boundary because of the warm building on the ground. In the solution by finite elements it is necessary to limit the real infinity continuum to a finite continuum. The finite region for this problem is shown in Fig. 4.107. The lateral boundaries of the plane region of unit thickness are assumed to be perfectly insulated. The resulting temperature variations with depth at various intervals of time are shown in Fig. 4.108.

If the temperature in the building is lowered, a freezing problem arises. Assume a uniform initial temperature of $5\,°C\ (41.0\,°F)$ and that the process of freezing is started by applying a constant temperature of $-2\,°C\ (28.4\,°F)$ at the top surface. The resulting temperatures as a function of depth and time are obtained as shown in Fig. 4.109.

By using the approach described for the problem of a heated pipeline in permafrost, we can solve the problem of filtration and consolidation of the thawed permafrost under the warm building.

CHAPTER

5

CONSTRUCTION
MATERIALS

The arctic and subarctic environment with wide temperature fluctuations and freezing and thawing cycles in often moist conditions sets stringent requirements for material selection. Many criteria have been established to measure material applicability for cold environments. Only some of the general features and the most important properties of selected construction materials are discussed herein.

The first test for the construction materials may come during construction. Many materials do not respond in a normal manner in cold weather conditions (for example, deform, harden, or adhere). The high initial moisture content of the material may cause accelerated deterioration because of frost action, failures of coatings due to building physical action, or harmful strains due to changes in the moisture content.

In the service stage, exposed structures are subjected to frost action, often in moist conditions and sometimes even in the presence of massive ice. The mechanical properties of many materials at very low temperatures may differ significantly from those observed in more temperate climates. In extreme cases the structure may be strained heavily due to a combination of external loads, thermal loads, and other deformation-induced loads when the temperature is so low that ductility, which guarantees serviceability, is lost. The evaluation of material properties at the service stage is thus important.

5.1 Steel

5.1.1 Fracture Toughness and Other Considerations

The cold arctic environment has many special effects on the engineering properties of steel. It may affect the steel weldability or be a partial factor in the corrosion or fatigue behavior of steel. The low temperature has a significant effect on the mechanical properties of steel. An example of the temperature and loading-rate effect on the yield strength of steel is given in Fig. 5.1. However, by far the most important special consideration in cold region steel construction is the danger of brittle fracture.

It is generally expected that structural steel members deliver ductility in particular at stress concentration points in the case of impacts, normal restrained structural movements, or overloading. This is taken into consideration in normal design specifications. However, at very low temperatures, loss of ductility has to be considered. Typical curves describing the fracture toughness or ductility of various types of steel as a function of temperature are given in Fig. 5.2. It is seen that if fully ductile behavior is required, the very expensive austenitic or Ni steel has to be used at cold region service temperatures. Fortunately, the requirement of sufficient ductility for most structural applications is only partial. A common structural steel grade can be found even for arctic applications if proper methods are used in manufacturing and construction.

One should recognize that the fracture toughness or ductility is not at all a material constant at a given temperature. The nature of loading and the material thickness are very important factors, as well as initial stresses, stress concentrations, and flaws, especially in welds. It is necessary to have a good understanding of the factors that influence the fracture toughness properties and requirements in order to arrive at an

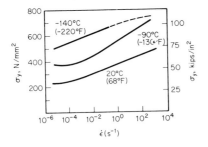

Figure 5.1 Effect of loading rate and temperature on the yield strength of steel. (*Vosikovsky, 1968.*)

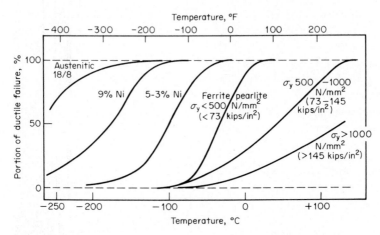

Figure 5.2 Typical curves illustrating ductility of specimen failure as a function of temperature for different steel types. *(Räsänen, 1980.)*

economic and safe structural solution and material selection for critical low-temperature applications.

5.1.2 Methods to Measure Fracture Toughness

There are several test methods available to study the characteristics of steel failure at different temperatures. Unfortunately all the methods have limitations and none of them has been widely accepted as a basis for design. Some most commonly used test methods are described in the following.

Standard specimens are struck at a wide range of temperatures by a falling pendulum in the Charpy V-notch test. The energy required to fracture the specimen is measured in each test. The portions of brittle

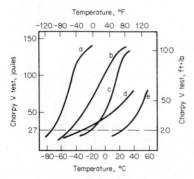

Figure 5.3 Typical Charpy V test curves. *a*—normalized steel for arctic applications; *b* — structural steel with improved notch toughness and weldability properties; *c*— ordinary structural steel; *d*— weld of steel *b*; *e*—weld of steel *c*. *(Courtesy of Rautaruukki Ltd.)*

and ductile failures at the failure surface can also be estimated. Some results from Charpy V tests for typical structural steel grades are shown in Fig. 5.3. At low temperatures the failure is brittle and the energy level low. In the transition zone the fracture starts to show some ductility, and the energy begins to increase rapidly. Finally, fracture becomes fully ductile and the energy reaches a constant value.

The transition temperature in which the fracture begins to show significant ductility typically corresponds to energy levels of 20 to 50 J (15 to 37 ft·lb). However, the dynamic Charpy V-notch test is very effective in causing the fracture. This is why the Charpy V transition temperature is often higher than the lowest service temperature of a structure that is typically subjected to nearly static loads and has only small flaws.

In the NDT (nil-ductility transition temperature) measurement, a weight is dropped on a specimen with a crack starter. Deformation is limited by a stop. The NDT is the highest temperature at which the crack penetrates from the tension surface through the entire specimen. The fracture behavior of brittle material is usually measured by the static and dynamic fracture toughness values K_{IC} and K_{ID} (American Society for Testing and Materials, 1978). These values describe the resistance of fracture initiation in a brittle manner (and also in a partially ductile manner). The CAT (crack arrest temperature) curve, which gives the temperature at which progressive cracking stops at a given dynamic stress level, provides additional information for the design engineer.

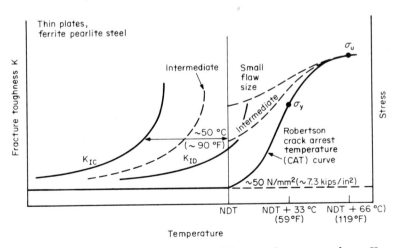

Figure 5.4 Typical example for static and dynamic fracture toughness K_{IC} and K_{ID}, nil-ductility temperature (NDT), and crack arrest temperature (CAT) curve. Note also the effects of loading rate on fracture toughness and of flaw size on crack arrest temperature.

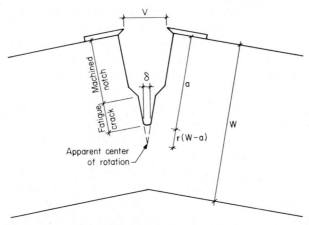

Figure 5.5   COD test setup.

A typical example of the crack toughness values K_{IC} and K_{ID}, the NDT point, and the CAT curve for an ordinary ferrite-pearlite steel is given in Fig. 5.4. The difference between static and dynamic transition temperatures is about 50°C (90°F), and the crack arrest temperature corresponding to the yield strength of steel is about 33°C (60°F) above NDT. However, if the size of the flaws or cracks is limited, a smaller temperature difference is required in order to prevent crack progression (Pellini, 1971). It is again noted that the strength of the material and the plate thickness in particular have some effect on these general features.

Finally, a brief description is necessary of the crack opening displacement (COD) test (British Standard 5762, 1977). The test arrangement is shown in Fig. 5.5. The critical opening of the crack δ_c is computed

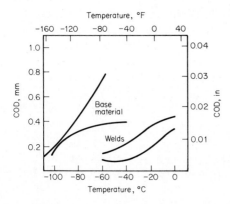

Figure 5.6 Typical COD curves of structural steel grades for cold region applications. *(Courtesy of Rautaruukki Ltd.)*

according to the standard from the plastic component of the opening V_p and the maximum force value P_{max}. Some typical COD curves are given in Fig. 5.6. This method has recently gained popularity as a tool in fracture mechanics design for material behavior in the elastic-plastic range.

5.1.3 Manufacturing Effects

Several factors have an effect on the toughness properties of a steel structure, including the chemical composition and grain size of steel, the type, size, and position of flaws, and the character of the initial stresses. The manufacturing processes of the material, the structural members, and the structure itself provide numerous tools for toughness control.

The strength of steel can be increased by increasing its carbon content, but this also increases its transition temperature. The negative effect of carbon on the toughness properties can be partly eliminated by adding magnesium. However, small amounts of alloying elements are also needed to maintain good welding properties. The steel grades used for cold region applications are usually fully killed, that is, deoxidized before pouring.

The fracture toughness of steel improves with decreasing grain size. Normalizing, that is, heating the steel to the austenitic range and cooling it in a controlled process, is a common method to obtain a small uniform grain size. A temperature-controlled rolling process down to values considerably below the normal lowest rolling temperatures is an alternative, especially for thin plates.

Welding reduces the fracture toughness of steel structures. The main reasons for this are the changes in the microstructure of steel in the heat-affected zone, localized yielding of the material, and the welding stresses. Welds are known to be very common sources of structural failures in normal applications; in cold temperature conditions the role of welds in structural safety considerations cannot be overemphasized.

The control of material composition, welding energy, and temperature are among the important considerations for successful welding. As the welding area cools down rapidly from temperatures above and near the melting point, new hard microstructures tend to form. They increase the risk of weld cracking and the resistance to fracture initiation. When the cooling rates are low, that is, in welding with high welding energy, the tendency of grains to grow decreases the notch toughness values.

A convenient criterion for general weldability and risk of hardening and cracking can be based on the chemical composition of the material. The carbon equivalent C_e of the base metal is often mentioned. It is given by

$$C_e = C + \frac{Mn}{6} + \frac{Cr + Mo + V}{5} + \frac{Ni + Cu}{15} \tag{5.1}$$

where the amounts of the composition elements are given in percentages. A low value of carbon equivalent means good weldability, but at values above 0.41 proper welding becomes increasingly difficult. The welding energy and the need for preheating can be estimated based on the carbon equivalent and the material thickness so that the effect of metallurgical changes in the welding area is under control (Cotton and Macaulay, 1976).

The weld and the heat-affected zone undergo permanent plastic deformations during the welding operation. First the temperature of steel rises locally and plastic compression is produced in the vicinity of the joint. When the welded area cools down, permanent tensile deformations occur. As a result, residual stresses reaching the yield strength of steel develop and the fracture toughness decreases, not only because of metallurgical changes but also because of inelastic deformations, which may have occurred in several cycles. The thick plates with three-dimensional stress states and a tendency to lamellar tearing are especially critical in this respect. Also the fracture toughness in cold deformed sections may be reduced significantly, especially if the cold deformed area is affected by welding heat.

There are no perfect welded connections, and all welds contain some kinds of flaws. The failure of a welded structure commonly occurs as a crack, initiated at a flaw, expands over the weld due to residual stresses and stress fluctuations. This kind of fatigue fracture is especially prone to occur if the steel does not deliver sufficient ductility.

It is desirable, but not usually possible, to avoid welds at severely stressed locations of the structure. However, some measures can be taken to minimize the possibility of fracture initiation at the weld. The composition of the welding material should be selected so that the strength, toughness, and corrosion resistance properties of the weld correspond to those of the base steel grade. Preheating, control of welding conditions and welding energy, and subsequent mechanical treatment of the weld are among measures that may reduce the danger of brittle fracture.

Residual stresses in welds can be reduced by preloading or heat treatment. The effect of preloading is shown in Fig. 5.7. It should be done at a sufficiently high temperature $[T \geqslant \text{NDT} + 30°\text{C} \ (54°\text{F})]$ in order to avoid internal material damage. Heat treatment should be pursued in a process that recognizes the dangers connected with changes in the microstructure of steel, insufficient support, uneven heat distribution, or too rapid and uneven cooling, and so on. The effect of heat treatment on a

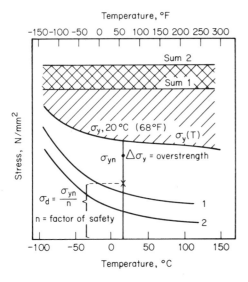

Figure 5.7 Estimation of residual stresses in a welded structure manufactured at 20°C (68°F) after preloading at different temperatures. 1 — preloading with design load; 2 — overloading (90% of design yield strength). Note the effect of the difference between design yield strength σ_{yn} and real yield strength σ_y. *(Räsänen, 1982.)*

structural component is shown in Fig. 5.8. If a steel structure welded at 20°C (68°F) is heated to 500°C (932°F) and then cooled, the residual stresses will be reduced to about 50% of the actual yield strength of the steel. If relieving has occurred at 600°C (1112°F), residual stresses are only about 10% of the yield strength. Heat treatment can also be concentrated on a small portion of a structure, but careful control of the process is required. It should be noted that postheat treatment not only relieves

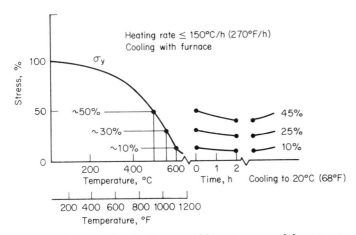

Figure 5.8 Principle of reducing welding stresses with heat treatment. *(Räsänen, 1982.)*

residual stresses but may also improve notch toughness and is thus very beneficial for structures with thick walls.

5.1.4 Design Based on Fracture Mechanics

When the nature of a structural failure begins to show considerable brittleness, the design methods based on traditional yielding criteria begin to lose their reliability. An approach based on fracture mechanics has to be used. The general stress state, stress concentrations or flaws, and the fracture toughness of the structure are used as basic elements in the process.

Figure 5.9 gives an overview of the design criteria and their domains of applicability. The principal domain of limit-state design and plastic instability corresponds to the fully ductile behavior of steel (field 3). When the fracture of steel shows no significant ductile behavior, linear elastic fracture mechanics should be applied (field 1). Field 2 in between is the domain of elastic-plastic fracture mechanics, although limit-state design may also provide acceptable results in a large portion of the area.

Linear elastic fracture design   In linear elastic fracture design the measured value for crack toughness corresponding to the actual loading conditions is compared with the stress intensity factor K_I given by

$$K_I = C\sigma\sqrt{a} \tag{5.2}$$

where C = constant, depending on specimen and crack geometry
σ = applied nominal stress
a = flaw size as a critical dimension for a given crack geometry

Some stress intensity factors for common cases are given in Fig. 5.10. A number of additional solutions are given, for example, in Tada et al. (1973) and Carlsson (1976).

The critical crack size for a given stress level is in the static case given by the equation $K_{IC} = K_I$. For a thorough discussion of this method, the reader is referred to Rolfe and Barsom (1977). However, in civil engineering design with low-strength steel grades, the applicability of the method is relatively limited even if some plastification at the tips of the cracks is permitted in an approximate analysis. The limiting condition for the plane-strain state, that is, this method, is given approximately by

$$a, B \geqslant 2.5 \left(\frac{K_{IC}}{\sigma_y}\right)^2 \tag{5.3}$$

where B = thickness of specimen
a = crack length
σ_y = uniaxial yield stress

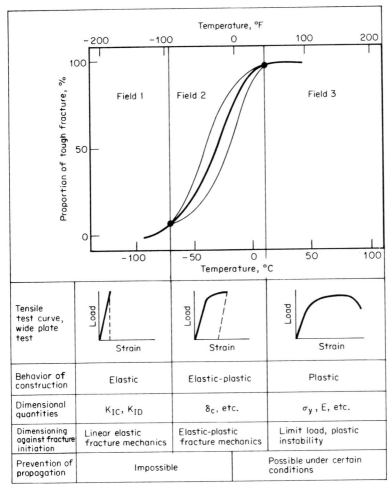

Figure 5.9 Material behavior and fracture mechanical design principles for ferrite-pearlite steel. *(Räsänen, 1982.)*

Elastic-plastic fracture design Elastic-plastic fracture mechanics provides the soundest theoretical basis for the design of steel structures in demanding cold temperature applications. One design approach reviewed at its current state in Harrison et al. (1979) is described below. In the approach the pseudoelastic stress σ_1 and the critical crack opening value δ_c from simple COD tests are used in order to find the maximum design flaw size.

The pseudoelastic stress attempts to estimate the nominal stress or strain state that is relevant for the elastic-plastic fracture. It is given by

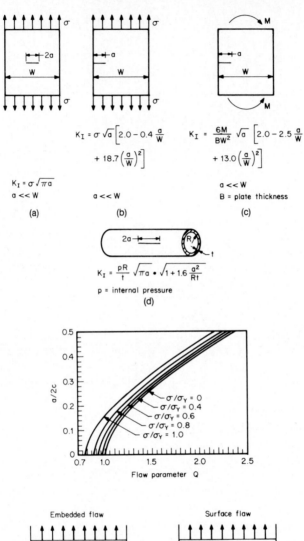

$$K_I = \sigma \sqrt{a} \left[2.0 - 0.4 \frac{a}{W} + 18.7 \left(\frac{a}{W} \right)^2 \right]$$

$$K_I = \frac{6M}{BW^2} \sqrt{a} \left[2.0 - 2.5 \frac{a}{W} + 13.0 \left(\frac{a}{W} \right)^2 \right]$$

$K_I = \sigma \sqrt{\pi a}$
$a \ll W$

(a)

$a \ll W$

(b)

$a \ll W$
$B = $ plate thickness

(c)

$$K_I = \frac{pR}{t} \sqrt{\pi a} \cdot \sqrt{1 + 1.6 \frac{a^2}{Rt}}$$

$p = $ internal pressure

(d)

Embedded flaw

Surface flaw

$K_I = \sigma \sqrt{\pi a / Q}$
$a, c \ll W, B$

(e)

$K_I = 1.1 \sigma \sqrt{\pi a / Q}$
$a, c \ll W, B$

(f)

Figure 5.10 Estimation of the stress intensity factor K_I for several common cases.

$$\sigma_1 = \mathrm{SCF} \cdot \sigma + \sigma_r \qquad (5.4)$$

where SCF = elastic stress concentration or, where localized uncontained yielding occurs, strain concentration factor
σ = computed nominal design stress
σ_r = residual stress due to fabrication process

The stress or strain concentration factor takes into account the effect of the actual local plastic strain that has occurred during the manufacturing process as well as the possible effects of the structural shape and other factors. These effects are known to concentrate on the locations of structural discontinuity, such as the welded connections of structural members and the end, change, or crossing zones of welds. Values from 1 to 8 have been used for SCF. The lowest values correspond to structures that do not have significant stress concentrations, and the high values are for very complex or massive welded intersections, for example, in oil production platforms. The residual stresses due to welding are often equal to the yield strength of steel. Assuming that the allowable stress is 60% of the nominal yield strength of steel and that the yield strength is 30% higher than the nominal values, the pseudoelastic stress may reach $\sigma_1 = 8 \times 0.6\sigma_{yn} + 1.3\sigma_{yn} = 6.1\sigma_{yn}$ at most critical welded intersections.

The design curve is given in Fig. 5.11. The maximum nominal flaw size $\bar{a}_{max}$ (which is equal to the maximum half-length of a through-thickness rectilinear crack) can be directly solved when the pseudoelastic stress and the COD value of, for example, a welded area are known. On the other hand, different flaw types can be converted into nominal flaws, as shown in Fig. 5.12. The design curve has an average built-in safety factor of approximately 2.5, but the test results show considerable scatter. However, tests have revealed about a 95% probability of the predicted maximum crack size being smaller than the critical crack size (Harrison et al., 1979).

Fatigue Fatigue is an important consideration in the design of structures subjected to load fluctuations. In principle, the fatigue process before fracture is composed of two stages, fatigue crack initiation and fatigue crack propagation. The crack initiation has an important role in the fatigue life of structural elements that are free of stress concentrations and flaws and are subjected to relatively low-level stress fluctuations. However, surface irregularities and stress concentrations may reduce the crack initiation period significantly. Welded structures always contain flaws, which may be considered as crack initiation points. There-

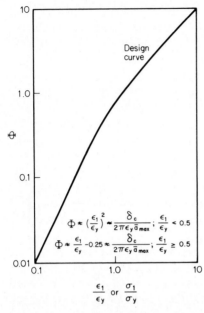

Figure 5.11 Design curve given the maximum nominal flaw size $\bar{a}_{max}$ from the COD test value δ_c and the pseudoelastic stress σ_1. *(Harrison et al., 1979; reprinted with permission of American Society for Testing and Materials.)*

fore crack propagation generally governs the fatigue design of civil engineering structures.

The crack propagation behavior of metals is represented schematically in Fig. 5.13. In region 1 the cracks do not propagate until a critical stress intensity factor fluctuation ΔK_I is reached. In region 2 the crack propagation is governed by the formula

$$\frac{da}{dN} = A(\Delta K_I)^n \tag{5.5}$$

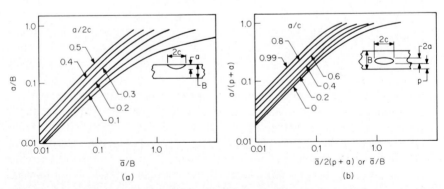

Figure 5.12 Relationship between (a) surface and (b) buried crack dimensions and equivalent through-thickness crack dimension $\bar{a}$. *(Harrison et al., 1979; reprinted with permission of American Society for Testing and Materials.)*

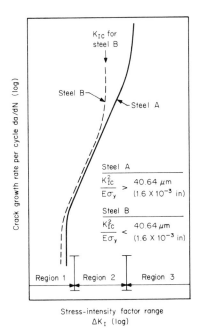

Figure 5.13 Schematic representation of fatigue crack growth in steel. *(Adapted from Rolfe and Barsom, 1977; reprinted with permission of Prentice Hall, Inc.)*

where a = crack length
$\qquad N$ = number of cycles
$\qquad A, n$ = constants

The constants A and n depend very much on the material and on the general character of loading and the environment. At high stress fluctuation levels, Eq. (5.5) underestimates the crack growth, and the fracture toughness of the material is approached.

The propagation of the crack causes deformations and compressive stress at the tips of a crack. That is why the opening of the crack occurs only after a certain crack opening stress σ_{c1} has been exceeded. This is illustrated in Fig. 5.14. The important parameter in the stress intensity factor fluctuation ΔK_I is the effective stress fluctuation $\Delta \sigma_{\text{eff}}$. Examples of effective stress fluctuations for initially stress-free and for welded high- or low-strength steel are shown in Fig. 5.15. It is noted that in welded structures the crack is always open and can propagate more effectively than in a material that is free from initial stresses. Furthermore, high-strength steel may be more resistant to fatigue if the material is free from initial stresses, but in welded structures the effect of yield strength disappears.

The effect of temperature on fatigue near the transition point has not been studied extensively. However, the material behavior becomes

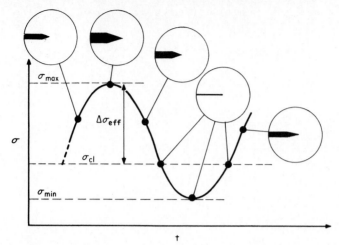

Figure 5.14 Opening of fatigue crack during load fluctuations. *(Alasaarela, 1979.)*

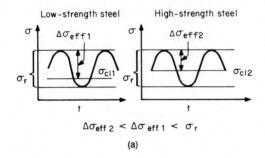

$$\Delta\sigma_{eff\,2} < \Delta\sigma_{eff\,1} < \sigma_r$$

(a)

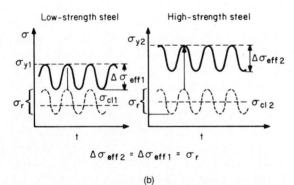

$$\Delta\sigma_{eff\,2} = \Delta\sigma_{eff\,1} = \sigma_r$$

(b)

Figure 5.15 Effective stress fluctuations. *(a)* Material free from initial stresses. *(b)* Welded structure. *(Alasaarela, 1979.)*

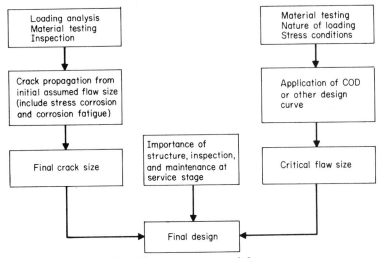

Figure 5.16   Principle of fracture mechanical design.

more brittle, and crack propagation increases at a given stress fluctuation level. Furthermore, the effects of corrosion and stress corrosion (high-strength steel grades) on crack propagation in arctic environments needs to be studied. For some general discussion on these factors, the reader is referred to Rolfe and Barsom (1977).

The design procedure   The principles of a fracture mechanical design procedure are summarized in Fig. 5.16. The allowable initial crack size a_0 by inspection (as least equal to the smallest detectable size) is determined by an iterative process where environmental analysis, stress analysis, COD design curve, and crack propagation estimates are used as basic elements. Because there is considerable scatter in these elements, reliability analysis for the structure could be pursued theoretically by using a statistical approach (Urabe and Yoshitake, 1981). However, the search for reliable statistical elements for the entire analysis would probably be too extensive a task.

5.1.5   General Design Criteria

The basic idea in steel design for cold temperature applications is to define the steel grade and the manufacturing process so that brittle fracture can be avoided and conventional design methods can be applied. Although certain basic elements of fracture mechanics are available, a pure fracture mechanical design approach is still not very widely used in civil engineering practice. This is because not all the necessary design

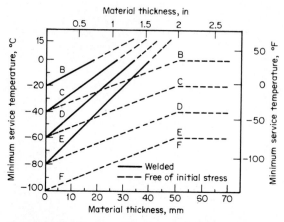

Figure 5.17 Design chart for steel-grade selection. Grades B, C, D, E, and F are for different weldability and fracture toughness criteria. (*Adapted from Länsiluoto, 1977.*)

coefficients are available for welds or even for the steel grade itself. To obtain the required data from a testing laboratory is expensive and time-consuming. At present, most engineering practice still relies on semiempirical design methods and recommendations in an attempt to satisfy the following four design conditions:

1. Prevention of fracture initiation in static loading conditions

2. Prevention of fracture initiation in impact loading conditions

3. Prevention of fatigue fracture initiation in fluctuating loading conditions

4. Prevention of the propagation of an initiated fracture

When the NDT temperature is known, rough estimates for the lowest allowable service temperature can be obtained for conditions 1, 2, and 4 by using empirical formulas giving the relationships between NDT and the important points in K_{IC}, K_{ID}, and CAT curves (Fig. 5.4). Design criteria can also be based empirically on the Charpy V-notch toughness values. A typical design chart is given in Fig. 5.17. The effects of welding, plate thickness, and steel grade are clearly visible.

Because the Charpy V-notch toughness test does not distinguish between the energy required for initiation of the crack and for crack growth, the Charpy V-notch requirement generally increases with increasing steel strength. The requirements of the American Association of State Highway and Transportation Officials (AASHTO) for bridge steels are shown in Table 5.1. The required Charpy V-notch toughness values

TABLE 5.1 Fracture-Toughness Specifications for Bridge Steels

AASHTO designation	ASTM designation	Thickness, cm (in)	Energy absorbed, J at °C (ft·lb at °F)		
			Zone 1 °	Zone 2 †	Zone 3 ‡
M161	A242	≤10 (4)	20 at 21 (15 at 70)	20 at 4 (15 at 40)	20 at −12 (15 at 10)
M183	A36	≤10 (4)	20 at 21 (15 at 70)	20 at 4 (15 at 40)	20 at −21 (15 at 10)
M188	A441	≤10 (4)	20 at 21 (15 at 70)	20 at 4 (15 at 40)	20 at −12 (15 at 10)
M222‖	A588‖	≤10 (4), mechanically fastened	20 at 21 (15 at 70)	20 at 4 (15 at 40)	20 at −12 (15 at 10)
		≤5 (2), welded	20 at 21 (15 at 70)	20 at 4 (15 at 40)	20 at −12 (15 at 10)
		>5 to 10 (2 to 4 in)	27 at 21 (20 at 70)	27 at 4 (20 at 40)	27 at −12 (20 at 10)
M223‖	A572‖	≤10 (4), mechanically fastened	20 at 21 (15 at 70)	20 at 4 (15 at 40)	20 at −12 (15 at 10)
		≤5 (2 in), welded	20 at 21 (15 at 70)	20 at 4 (15 at 40)	20 at −12 (15 at 10)
M244	A514	≤10 (4), mechanically fastened	34 at −1 (25 at 30)	34 at −18 (25 at 0)	34 at −34 (25 at −30)
		≤6 (2.5), welded	34 at −1 (25 at 30)	34 at −18 (25 at 0)	34 at −34 (25 at −30)
		>6 to 10 (2.5 to 4 in), welded	47 at −1 (35 at 30)	47 at −18 (35 at 0)	47 at −34 (35 at −30)

° Zone 1 — minimum service temperature −18°C (0°F) and above.
† Zone 2 — minimum service temperature −18 to −34°C (−1 to −30°F).
‡ Zone 3 — minimum service temperature −35 to −51°C (−31 to −60°F).
‖ If the yield point exceeds 450 MPa (65 kips/in²), the temperature for the Charpy V-notch toughness value for acceptability shall be reduced 8°C (15°F) for each increment of 70 MPa (10 kips/in²) above 450 MPa (65 kips/in²).
SOURCE: American Association of State Highway and Transportation Officials (1982).

359

TABLE 5.2 Example Recommendation for Steel Selection in Ordinary Arctic Construction

Foundations	Welded pipe piles to ASTM A252 grade 3, modified for impacts.
Main structural members	1. Welded plate sections to ASTM A573, A588, and A633 or CSA G.40.21, 44T and 50T. 2. Hollow structural sections to ASTM A500, modified for impacts, or CSA G40.21, 50T and 50A.
Secondary structural members	Cold-formed sections to ASTM A607 or A570.
Fasteners	1. Structural bolts to ASTM A193 or A320, grades B-7 or L-7. 2. Anchor bolts of special-quality deformed bar, modified for impacts.
Cladding	Galvanized sheet to ASTM A446 with G-90 coating. Weathering steel sheet to ASTM A606. Prepainted steel or vinyl-coated steel to manufacturer's specification.

SOURCE: Azmi (1978).

are 20 to 47 J (15 to 35 ft·lb) at temperatures of 17 to 39°C (30 to 70°F) above the lowest service temperature. Loading rates for bridge beams are typically relatively low. One recommendation of steel grades for arctic conditions is given in Table 5.2.

A well-stated selection of the steel grade is an important step in cold temperature construction applications, but it can be nullified if mistakes are made in the general design or manufacturing process. In design, simple smooth stress fields should be favored over complex three-dimensional ones with high stress concentration factors. Heat treatment or preloading, if undertaken, should be pursued according to carefully controlled plans in order to avoid undesirable effects such as plastic deformations, metallurgical changes, or crack growth. Executing critical welded connections in the field should be avoided, because these are difficult to produce, finish, and inspect according to the standards. Bolted connections are more desirable because they are easier to produce and less sensitive to cold temperature effects.

5.2 Concrete

The serviceability of concrete structures for cold regions may be endangered during the construction stage as well as during the service stage. Freezing of fresh concrete has a severe effect on its strength development and final texture in winter construction. On the other hand, when a massive completed structure is exposed to the cold environment, significant thermal cracking may occur as a result of the thermoshock. In the service stage concrete is required to survive the low temperatures and consecutive freezing and thawing cycles, sometimes in a very moist and aggressive environment.

5.2.1 Freezing of Fresh Concrete

One of the main principles of concrete pouring in winter conditions is to prevent the freezing of fresh concrete. When fresh concrete freezes, the volume of the water in the mixture increases by about 9%, which means about a 2% increase in the total volume of concrete. During the freezing process water tends to travel toward the freezing surface, forming ice lenses in a similar fashion as in freezing ground (Fig. 5.18). This has severe effects on the texture of concrete. The frozen concrete is strong and hard, but because the strength is due to the "ice-cementing," it vanishes as the concrete thaws (Fig. 5.19).

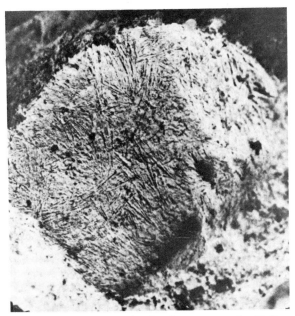

Figure 5.18 Concrete that has frozen too early. Notice needlelike ice-lense formations. *(Courtesy of Technical Research Centre of Finland.)*

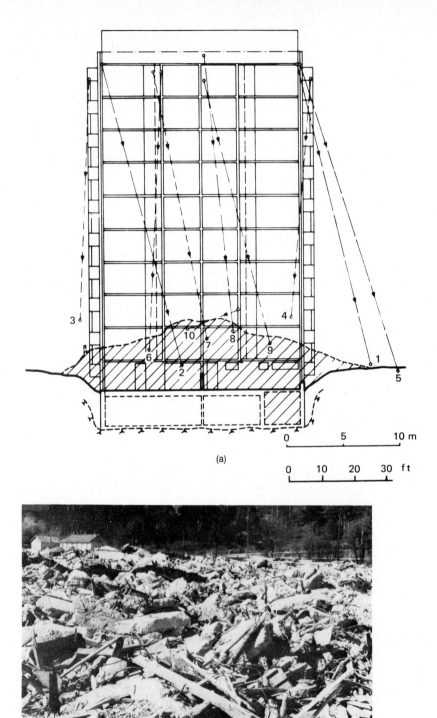

(a)

0   5   10 m

0   10   20   30 ft

(b)

Figure 5.19 Collapse of multistory building frame in the spring after the early frozen concrete had thawed and lost its strength. (*a*) How it happened. (*b*) After collapse. (*Courtesy of Technical Research Centre of Finland.*)

The hydration process, inactive when concrete is deeply frozen, continues after thawing, but because of increased porosity and changes in the texture, the final strength of concrete may not reach even half the design strength (Fig. 5.20). If the water-cement ratio of the concrete is reduced, the situation can be slightly improved. However, the best protective measure is to secure the continuation of the hydration process until the free water content of the concrete is significantly reduced, pores have formed to level the freezing pressures, and the concrete has gained enough strength to resist the pressures. After the concrete has gained the so called freezing strength, generally about 5 MPa (700 lb/in²), it may freeze and still gain the original design strength after the hydration process has recontinued under desirable curing conditions for an adequate period. However, some concrete properties such as the water tightness or frost susceptibility may be damaged because of early freezing.

Estimates of the strength development of concrete are needed to establish the time required to develop the freezing strength of concrete, and the time after which the forms can be removed. (Generally the concrete should have gained 60% or more of its design strength before removal of the supporting forms in order to ensure structural safety, to avoid excessive deformation due to its own weight and working loads, and to provide adequate curing under favorable conditions.) Approximate design charts and tables that give the strength development in concrete as a function of temperature are available for most common

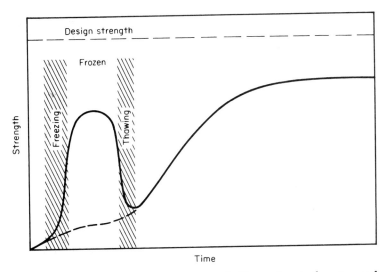

Figure 5.20 Schematic representation of effects of early freezing and thawing on strength development of concrete.

concrete mixtures. They can also be developed based on laboratory experiments (American Concrete Institute, 1978). For example, in Finland the strength development of concrete using ordinary portland cement is estimated according to Fig. 5.21, where the maturity factor N is calculated by

$$N = \sum_{0}^{t} K(T + 10°C)\Delta t \qquad (5.6)$$

where T = temperature of concrete, °C
$\quad\quad t$ = hardening time of concrete at temperature T, days
$\quad\quad K$ = coefficient, depending on the temperature of concrete,

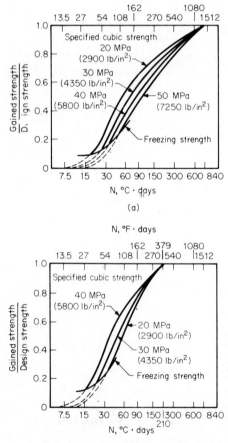

Figure 5.21 Typical design relationships between maturity factor N and relative compressive strength for different specified concrete cubic strengths when using (*a*) ordinary portland cement and (*b*) rapid portland cement. (*Kilpi and Sarja, 1981.*)

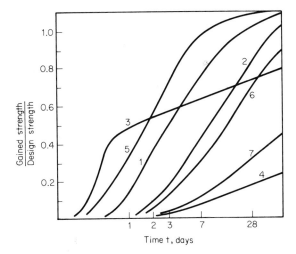

Figure 5.22 Typical strength development curves for concrete mixtures at different curing temperatures. Ordinary portland cement: 1—at $+20°C$ (68°F), 2—at $+5°C$ (41°F), 3—at $+60°C$ (140°F), 4—at $-8°C$ (18°F); 5—rapid portland cement at $+20°C$ (68°F); 6—normal frost-hardening concrete or mortar at $-10°C$ (14° F); 7—special frost-hardening concrete or mortar at $-25°C$ $(-13°F)$.

$$K = \begin{cases} 1.0 & T > 0°C \ (32°F) \\ 0.4 & 0°C \geqslant T \geqslant -10°C \ (32°F \geqslant T \geqslant 14°F) \\ 0 & T < -10°C \ (14°F) \end{cases}$$

According to this estimate the strength development of concrete continues also at temperatures below $0°C$ (32°F). This is due to the fact that not all water in concrete freezes at $0°C$ (32°F) and thus a partial hydration process may continue. The American Concrete Institute (1978) also calculates the maturity factor according to Eq. (5.6), but the value of constant K is somewhat nonconservatively set equal to 1.0 at all temperatures above $-10°C$ (14°F).

Heating, thermal protection, and special frost-resistant concrete mixtures or mortars are used to provide desirable results in concrete construction during periods of freezing temperatures. Some examples for strength development curves of concrete mixtures at different temperatures are shown in Fig. 5.22. If it is necessary to achieve only the freezing strength before the concrete begins to freeze, as in the case of pouring concrete on the ground, hydration heat combined with thermal protection generally provides adequate results. In most cases, however, it is necessary to provide fast strength gain up to the strength where supporting forms can be removed. Heating, combined with thermal protection or special frost-resistant concrete mixtures or mortars, can be used in this case. These concerns are discussed in more detail in Chap. 7.

5.2.2 Admixtures

Admixtures are commonly used in cold region concrete construction to improve the workability, curing properties, or freeze-thaw resistance. Four types of admixtures are especially beneficial in cold region applications:

1. Air-entraining agents
2. Accelerators
3. Plasticizers
4. Antifreezers

The main purpose of the use of air-entraining admixtures is to improve the frost resistance of hardened concrete. They also provide some protection against the negative effects of early freezing. Air-entrained concrete should be used in winter construction even if the structures will not be exposed to freezing conditions in the service stage. Structures may become saturated during the construction stage due to melting snow, and frost damage may thus be experienced.

Frost-resistant concrete should have a low water-cement ratio. Traditional air-entraining admixtures improve the workability of concrete, and the use of plasticizers may thus be avoided. Air bubbles form in the concrete as admixtures decrease the surface tension of water and cause foaming. However, the concrete will experience some loss of strength, the use of other admixtures may be problematic, and the control of air content and air void distribution is difficult when traditional air-entraining admixtures are used. Recently a new method has been developed to introduce protective air pores into concrete. Micropores, polymer bubbles with diameters of about 2 to 5×10^{-5} m (1 to 2×10^{-3} in), are mixed in concrete. The disadvantages of this method are minor compared to those of traditional air-entraining admixtures, but the workability of concrete is not improved either.

Accelerators are used in winter concreting to obtain rapid strength development of the concrete and thus to reduce the time needed for protection, heating, and supporting. Accelerators tend to cause increased shrinkage of concrete. Calcium chloride is the most commonly used accelerator, but it is not an antifreezer because permissible amounts will not decrease the freezing point more than a few degrees.

The U.S.S.R. has gathered a considerable amount of experience in using admixtures in winter concreting. A list of these admixtures (U.S.S.R. design code SNiP III-15-76, 1977) is given below, and the range of their applications is listed in Table 5.3.

1. Plasticizers (SDB, SSB)
2. Plasticizing air-entraining agents (M_1, VLHK, GKZ-10, GKZ-11, TSK, KTSNR, PAS-1)
3. Air-entraining agents (SNV, SDP, TSNIPS-1)
4. Gas-forming admixtures (GKZ-94, PGEN)
5. Accelerators:
   a. Sodium sulfate (SN)
   b. Sodium nitrate (NN_1)
   c. Calcium chloride (HK)
   d. Calcium nitrate (NK)
   e. Calcium nitrite-nitrate (NNK)
   f. Calcium nitrite-nitrate-chloride (NNHK)
6. Antifreezers:
   a. HK with sodium chloride (HN)
   b. Sodium nitrite (NN)
   c. Potash (P)
   d. Calcium nitrate-carbamide (NKM)
   e. HK + NN, NNHK, KH + M, NNK + M, NNHK + M
7. Steel corrosion inhibiting admixtures (NN, NNK)

Reinforced concrete should not contain more than the following percentages of the weight of cement:

1. 2% sodium sulfate (SN)

2. 4% sodium nitrate (NN_1), calcium nitrate (NK), calcium nitrite-nitrate (NNK), and calcium nitrite-nitrate-chloride (NNHK)

3. 3% calcium chloride (HK)

The Soviet experience on the use of antifreezers is of special interest. The amounts of antifreezers for different curing temperatures are listed in Table 5.4, and the strength gain at different temperatures is given in Table 5.5. The effect of antifreezers is based on their capacity to keep at least part of the water in the liquid state and available for hydration at temperatures significantly below freezing.

It should be clear that the side effects of admixtures on concrete behavior and its final properties must be checked. For example, calcium chloride used as an accelerator decreases the frost resistance of concrete and increases shrinkage and risk of corrosion. It is not recommended for rigid massive structures with large spacings between expansion joints, pre-

TABLE 5.3 Range of Uses of Concrete Admixtures According to U.S.S.R. Design Code SNiP III-15-76

Types of structures and their service conditions	HK, HK + HN	SN	NK, NNK, HKM / NK + M, NNK + M	HK + NN	NNHK, HK + NNK / NNHK + M	NN, NN$_1$	P	SDP, SSB, PASTS-1 M$_1$, VLHK, GKZ, NTSK, KTSNR, SNV, SDP, TSNIPS-1, PGEN
1. Reinforced concrete structures with nonstressed reinforcement of:								
More than 5 mm (0.2 in) in diameter	(+)°	+	+	+	+	+	+	+
5 mm (0.2 in) and less in diameter	−	+	+	(+)	(+)	+	+	+
2. Structures and joints without prestressed reinforcement of precast and cast in situ and precast structures having protrudings or inserts:								
Without special protection of steel	−	+	+	−	−	+	+	+
With zinc coating throughout steel	−	−	−	−	−	+	−	+
With aluminum coating throughout steel	−	−	(+)	−	(+)	−	−	+
With combined coating (alkali-resistant varnish and paint or other alkali-resistant protective layers throughout metal sublayer) and joints without inserts and design reinforcement	(+)	+	+	(+)	(+)	+	+	+
3. Prestressed structures and joints (ducts) of precast and cast in situ and precast structures	−	+	(+)	−	−	+	−	+
4. Prestressed structures reinforced by steel of At-IV, At-V, At-VI, A-IV, and A-V classes	−	+	−	−	−	−	−	+
5. Structures from concrete on aluminous cement	−	−	−	−	−	−	−	+

° −: admixture not permitted; +: use of admixture is purposeful; (+): use of admixture is purposeful only as an accelerator for hardening of concrete.
SOURCE: Adapted from Miettinen et al. (1981).

TABLE 5.4 Amounts of Antifreezer (Dry Weight) as a Percentage of the Weight of Cement for Different Curing Temperatures

Curing temperature, °C (°F)	Admixture					
	NN	HK + HN	NK + M, NNK + M	NNHK, HK + NN°	NNHK + M	P
0 to −5 (32 to 23)	4 to 6	0 + 3 to 2 + 3	3 + 1 to 4 + 1.5	3 to 5	2 + 1 to 4 + 1	4 to 6
−6 to −10 (22 to 14)	6 to 8	3.5 + 3.5 to 2.5 + 4	5 + 1.5 to 7 + 2.5	6 to 9	4 + 1 to 7 + 3	6 to 8
−11 to −15 (12 to 5)	8 to 10	4.5 + 3 to 5 + 3.5	6 + 2 to 8 + 3	7 to 10	6 + 2 to 8 + 3	8 to 10
−16 to −20 (3 to −4)		6 + 2.5 to 7 + 3	7 + 3 to 9 + 4	8 to 12	7 + 2 to 9 + 4	10 to 12
−21 to −25 (−6 to −13)				10 to 14	8 + 3 to 10 + 4	12 to 14

° With ratio of components 1 : 1 by dry weight.
SOURCE: Réunion Internationale des Laboratoires d'Essais et de Recherches sur les Matériaux et les Constructions (RILEM) (1980).

TABLE 5.5 Strength Development of Concrete at Different Curing Temperatures with Portland Cement and Various Admixtures

Admixture	Curing temperature, °C (°F)	Concrete strength as a percentage of design strength			
		7 days	14 days	28 days	90 days
NN	−5 (23)	30	50	70	90
	−10 (14)	20	35	55	70
	−15 (5)	10	25	35	50
HK + HN	−5 (23)	35	65	80	100
	−10 (14)	25	35	45	70
	−15 (5)	15	25	35	50
	−20 (−4)	10	15	20	40
NK + M, NNK + M	−5 (23)	30	50	70	90
	−10 (14)	20	35	50	70
	−15 (5)	15	25	35	60
	−20 (−4)	10	20	30	50
NNHK, NNHK + M,	−5 (23)	40	60	80	100
HK + NN	−10 (14)	25	40	50	80
	−15 (5)	20	35	45	70
	−20 (−4)	15	30	40	60
	−20 (−4)	10	15	25	40
P	−5 (23)	50	65	75	100
	−10 (14)	30	50	70	90
	−15 (5)	25	40	65	80
	−20 (−4)	25	40	55	70
	−25 (−13)	20	30	50	60

SOURCE: Réunion Internationale des Laboratoires d'Essais et de Recherches sur les Matériaux et les Constructions (RILEM) (1980).

stressed structures, or structures exposed to rain or aggressive soil conditions. The behavior of traditional air-entraining admixtures depends on the composition of concrete, the manufacturing process, and the transportation conditions. Extensive testing is necessary if relevant previous experience of similar concrete composition and conditions is not available.

5.2.3 Cracking and Drying

The risk of cracking exists at different stages of concrete construction at low temperatures. Excessive drying may occur at the surface of fresh concrete in heated enclosures with typically low relative humidity. Unprotected concrete may also dry due to wind effects, temperature differentials, and sublimation. Both cracking and drying may have a negative effect on the strength, serviceability, or durability of concrete structures.

Concrete must be protected from heat losses and evaporation as soon as possible after it has been poured in cold weather conditions. Otherwise hairline cracks tend to form at its surface due to drying, shrinkage, and temperature differences. Concrete may also experience strength losses due to frost effects and lack of moisture in the hydration process.

When heating, such as wire heating, is used to secure the rapid strength development of concrete, the risk of thermal cracking arises, especially at the beginning of the heating or cooling phases. That is why moderate rates of concrete temperature changes, generally 5 to 10°C/h (10 to 20°F/h), are recommended. The lower values should be used especially in case of massive structures at a relatively early stage of strength development.

Finally, structures are subjected to thermal shock when stripped and exposed to cold air. The U.S.S.R. design code SNiP III-15-76 recommends the use of temporary covers for already stripped structures if the temperature difference between concrete and ambient air is greater than 20°C (36°F) for structures with surface moduli from 2 to 5 or 30°C (54°F) for structures with surface moduli over 5. (The surface modulus is the ratio between the surface area and the volume of the structure.) The U.S. practice (American Concrete Institute, 1978) specifies the maximum allowable gradual temperature drop of concrete within 24 h after the end of protection. The recommended values are 28°C (50°F) for sections less than 0.3 m (1 ft) thick, 22°C (40°F) for sections 0.3 to 0.9 m (1 to 3 ft) thick, 17°C (30°F) for sections 0.9 to 1.8 m (3 to 6 ft) thick, and 11°C (20°F) for sections over 1.8 m (6 ft) thick. Naturally, all of these recommendations should be adjusted to the specific cases and should not necessarily be considered adequate requirements, for example, in the case of watertight structures.

5.2.4 Effects of Low Temperatures on the Properties of Hardened Concrete

The effect of the temperature decrease is not necessarily harmful for concrete if it is not in a fully saturated state. As a matter of fact, a significant increase in compressive strength and tensile strength is experienced as the temperature decreases, as shown in Figs. 5.23 and 5.24. Similar relationships can also be found for the elastic modulus, bending strength, and bond strength of reinforcement (Browne and Bamforth, 1981). The coefficient of thermal expansion decreases slightly and the thermal conductivity increases when the temperature is lowered.

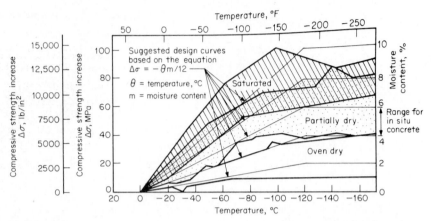

Figure 5.23 Effect of low temperatures on compressive strength of concrete. *(Browne and Bamforth, 1981.)*

The reinforcement also usually behaves in a favorable manner, even in arctic service temperatures, although the notch toughness requirements should be specified following the guidelines given in Sec. 5.1, especially in case of dynamic loading and welded connections. The significant changes in the strength properties of concrete should be considered in the design calculations. However, very little information is available on the long-term strength and creep properties of concrete at low temperatures.

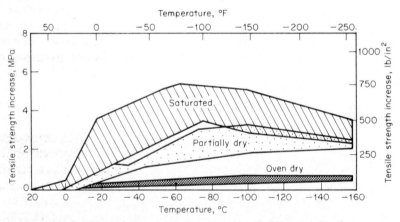

Figure 5.24 Effect of low temperatures on tensile strength of concrete. *(Browne and Bamforth, 1981.)*

5.2.5   Effects of Freeze-Thaw Cycles on the
Properties of Concrete

Gradual deterioration as a result of consecutive freeze-thaw cycles is a potential hazard for concrete structures in cold regions. Concrete always contains some moisture that freezes as the temperature drops below $0°C$ ($32°F$). The freezing depends on the pore structure and moisture content of concrete. It starts from the largest pores. The majority of the free water freezes typically at temperatures above $-10°C$ ($14°F$), but some freezing still occurs in the smallest pores at below $-50°C$ ($-60°F$). The 9% volume expansion of water, its tendency to travel toward the freezing surface, and the osmotic pressures in the presence of salts (concentration differences created by the freezing process tend to level off) cause internal stresses and in unfavorable conditions microcracking during the freezing process.

Numerous factors have an effect on the frost susceptibility of concrete, including air content, characteristics of the pore and void system, degree of saturation, salt content of water, and concrete strength. The concrete paste contains gel pores 1 to 2×10^{-9} m (4 to 8×10^{-8} in) in diameter, capillarity pores with indefinite shapes and sizes of about 10^{-7} to 10^{-5} m (4×10^{-6} to 4×10^{-4} in), and larger voids. The capillarity pores and small air voids are especially important for the frost susceptibility of concrete, the former having a negative effect and the latter a positive effect.

Generally the frost damage occurs in the paste, but in some cases the aggregate may also be vulnerable to frost damage. Vulnerable aggregates typically have large air volumes and small average pore sizes. Standard laboratory tests have been developed to measure the frost resistance of concrete (for example, American Society for Testing and Materials ASTM C666-77 and ASTM C671-77, 1977).

Air-entrained concrete with a low water-cement ratio is generally used for structures subjected to repeated freeze-thaw cycles. The water-cement ratio has an important effect on the general texture of the paste (Fig. 5.25) as well as on the strength and watertightness. At very low water-cement ratios hydration does not occur completely. At high water-cement ratios the volume of capillary water and pores is increased. A continuous capillary pore network may be formed, especially as a result of repeated freeze-thaw cycles in moist conditions. This increases the permeability and frost susceptibility of concrete significantly. The optimum water-cement ratio for frost-resistant concrete is about 0.4 (Mironov, 1956).

Although it is necessary to minimize the amount of capillary pores in the concrete paste, the other kind of pore, isolated air voids, is required

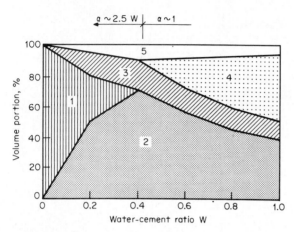

Figure 5.25 Volume portions of components in hardened and totally hydrated cement paste for different water-cement ratios. 1—unhydrated cement; 2—cement gel; 3—gel water; 4—capillary water; 5—air voids.

in frost-resistant concrete (Fig. 5.26). These voids act as pressure-relieving reservoirs in the concrete during freezing and have typically a radius r of between 10^{-5} and 2×10^{-3} m (4×10^{-4} and 8×10^{-2} in). Generally the air content requirement for exposed structures in cold regions is 3 to 8%. (The lower values correspond to applications with coarser aggregates; see Portland Cement Association, 1979.) Because the volume of suitable air voids created during manufacturing and curing stages (shrinkage voids) does not reach the required level, air-entraining admixtures are needed.

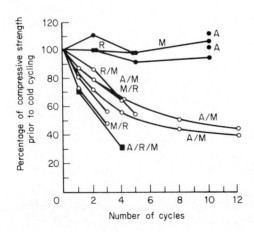

Figure 5.26 Effect of freeze-thaw cycles on compressive strength of concrete. *A*—low air content; *R*—rapid cool down; *M*—high moisture content. *(Browne and Bamforth, 1981.)*

The characteristics of the void system are also important for the frost resistance. This is why the critical spacing factor (average maximum distance from any point in the cement paste to the edge of the nearest air void) is often specified. A typical requirement for freshwater exposure is 0.25 mm (1.0×10^{-2} in) and for saltwater exposure 0.2 mm (8×10^{-3} in). Other factors that are used to describe the character of the air void system are specific surface of air void volume (surface area of air voids per volume unit) and number of voids in a linear unit.

The fictitious spacing factor for a completely dry void system does not describe very well the frost resistance of concrete in the most severe conditions because a considerable part of the voids may be full of water or ice and thus ineffective in relieving freezing pressures. Water absorption of the void system varies drastically in different types of concrete and environments. The steepness and rapidity of the temperature drops also have some effect on the frost susceptibility. Fagerlund (1977, 1981) has presented a thorough discussion of the frost resistance of concretes and described a method to evaluate the frost resistance based on air void distribution, degree of water saturation, and true spacing factor. The time required to reach the critical degree of water saturation is estimated based on freeze-thaw tests.

5.2.6 Special Applications

Concrete pavements, unprotected parking and bridge decks, and hydraulic structures are among the structures exposed to the most severe environmental conditions as far as the frost resistance of concrete is concerned (Fig. 5.27). Concrete is often subjected to abrasions and consequent freezing and thawing in very moist and aggressive environments. Deterioration is a combined result of increased moisture absorption, frost action, mechanical effects, and corrosion of the reinforcement.

Chlorides provide favorable conditions for corrosion of the reinforcement in concrete at levels exceeding about 500 parts per million (ppm). Because the volume of rust is much greater than that of iron, expansive tensile forces will cause eventual cracking and spalling, which greatly increases the rate of deterioration of the structure. The absorption capacity of concrete also increases significantly when the salt content of water increases. Subsequent saturation and drying cycles tend to disturb the original microstructure of the concrete and thus also increase its absorption capability. Finally, attention should be given to certain chemicals that tend to react with concrete, such as sulfates in seawater and in certain soil waters.

There are several ways to protect concrete from abrasion and saturation, including epoxy coatings and steel covering. Special low-permea-

(a)

(b)

Figure 5.27 Examples of deterioration. (*a*) Bridge damaged by combined effects of frost and salt. (*Courtesy of State Research Centre of
Finland.*) (*b*) Splitting of steel pile filled with poor concrete due to frost
action. (*Courtesy of TAMS Engineers.*) (*c*) Poor-quality concrete in
lighthouse being rapidly worn out by combined effects of frost action
and ice abrasion. (*Courtesy of State Research Centre of Finland.*)

(c)

bility and very strong concrete overlays are also often used to protect the ordinary concrete. However, adequate resistance against severe environmental conditions can often be provided also by means of concrete technology.

Strong, well-compacted, and properly cured concretes with a controlled air content and void size distribution and low water-cement ratio (< 0.45) provide good resistance to water penetration and abrasion (ice, traffic, and so on). The reinforcement bars can be protected from corrosion by careful crack control and adequate concrete cover, usually on the order of 30 mm (1.2 in) or more, depending on the bar diameter. Cements with low tricalcium aluminate content (about 5 to 7%) provide protection against sulfates, and finally pozzolana can be used as admixture to improve the sulfate resistance, decrease the permeability, and reduce the problems connected with hydration heat generation in massive concrete structures.

5.3 Wood

Wood is a traditional construction material in the cold environments. It is light and has good workability and insulation properties. The strength properties of wood improve slightly when the temperature is decreased, as shown in Figs. 5.28 and 5.29. Even repeated freezing and thawing

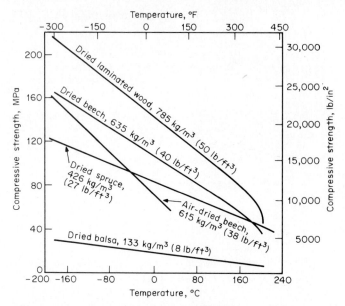

Figure 5.28 Compressive strengths of beech, spruce, balsa, and laminated wood as a function of temperature. *(Kollmann, 1951.)*

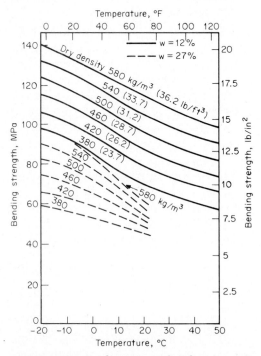

Figure 5.29 Bending strength of pine with moisture contents $w = 12$ and 27% as a function of temperature. *(Thunell, 1942.)*

TABLE 5.6   Compressive Strength (percent) of Test Specimens Frozen 10 Times to −20°C (−4°F) Compared with Specimen kept at Room Temperature

Wood type	Moisture content of wood				
	0%	7–8%	10–11%	19–20%	100–200%
Pine	90	99	95	98	98
Beech	95	98	101	98	99
Oak	95	97	97	97	97

SOURCE: Save (1975).

does not have a very significant effect on the mechanical properties of wood (Table 5.6).

Weather-resistant adhesives that maintain their shear strength through freezing and thawing should be used in cold region construction. Thus the behavior of laminated wood products such as plywood or laminated wood beams should not be affected by the cold environment. However, the problems of rotting of wood due to excess moisture, shrinkage and cracking due to excessive drying, and warping due to

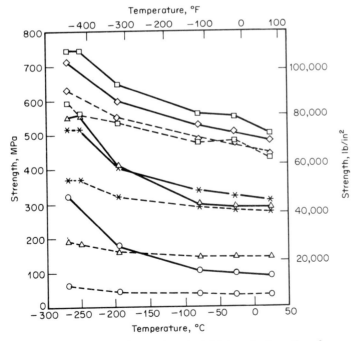

Figure 5.30 Tensile (———) and yield (–––) strengths of some aluminum alloys. ○—1100-0; ◇—2024-T8; △—5083-0; *— 6061-T6; □—7075-T73. (*Kaufman et al., 1978.*)

excessive moisture changes should be considered and controlled in the design.

5.4 Aluminum

The properties of aluminum alloys are generally very favorable for cold weather applications. Aluminum has no ductile-to-brittle transition. As a matter of fact, its mechanical properties, including elastic modulus, static strength, fatigue strength, and ductility, seem to improve slightly with decreasing temperature. Other favorable properties of aluminum include good corrosion resistance, generally good weldability, and low weight-strength ratio. The effects of temperature on the tensile strength of some aluminum alloys can be seen in Fig. 5.30 and on the strength of a bolted connection in Fig. 5.31.

Although aluminum alloys provide good reliability in a wide variety of strength properties, steel has maintained its leading role in arctic metal

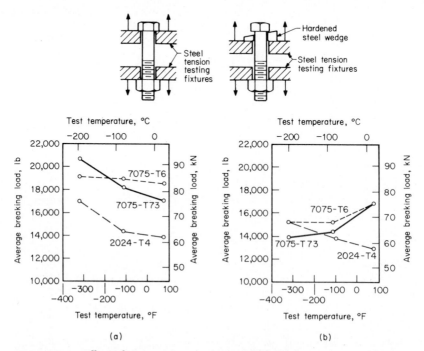

Figure 5.31 Effect of temperature on strength of bolted connections. (*a*) Axial tension test of bolt or machine screw. (*b*) Tension test of bolt with 10° wedge under bolt head (ASTM A370-61T). Aluminum bolts, ⅝-11 × 3-¼ in economy. Each point is the average of at least two tests. (*Kaufman et al., 1978.*)

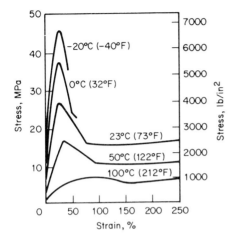

Figure 5.32 Example of stress-strain curves for polyethylene at different temperatures. *(Reike and Clock, 1966.)*

construction mainly because of better economy. It also has a higher modulus of elasticity and a lower coefficient of thermal expansion. Aluminum has been used for window frames, bolts, and corrugated sheets. Special applications include, for example, liquid gas tanks, space trusses, and electric transmission towers. For further discussion on aluminum alloys for cold temperature applications the reader is referred to Kaufman et al. (1978) and on general aluminum design to the Aluminum Association's aluminum standards and data (1976).

5.5 Plastics

Plastics have numerous applications in construction. However, because of the large number of different types of plastics and the significant effects of alloying elements, it is difficult to describe their characteristics in cold region applications in a general fashion. Only some general features are reviewed in this context. More information is available, for example, in Titus (1967).

The mechanical properties of plastics depend very strongly on time and temperature. The viscoelastic behavior of plastics reflects the time dependence. The dependence on temperature is especially profound in the vicinity of transition temperatures. Plastics may have many transition temperatures. Glass transition temperature is the average softening temperature as the plastic is melted. Some plastics also have a transition temperature from ductile to brittle fracture. Notch toughness tests similar to those used for metals can also be used to study the fracture behavior of plastics. Some examples of the mechanical properties of plastics are

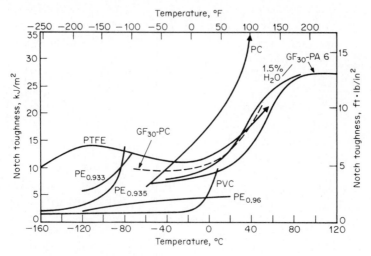

Figure 5.33 Notch toughness of some polymers as a function of temperature. *(Oberbach, 1975.)*

given in Figs. 5.32 and 5.33. Measured values for glass transition temperatures and ductile-to-brittle transition temperatures are given in Table 5.7.

5.6 Other Materials

Various aspects are related also to the use of other construction materials in cold environments. For example, the frost resistance of different kinds of brick and lightweight concrete products depends among other things

TABLE 5.7 Glass Transition and Brittle-to-Ductile Transition Temperatures for Some Polymers

Polymer	Glass transition temperature T_g, °C (°F)	Brittle-to-ductile transition temperature T_b, °C (°F)
PMMA	105 (221)	45 (113)
Polycarbonate	150 (302)	−200 (−328)
Rigid polyvinyl chloride	74 (165)	−20 (−4)
Natural rubber	−70 (−94)	−65 (−85)
Polystyrene	100 (212)	90 (194)
Polyisobutylene	−70 (−94)	−60 (−76)
Polyethylene	−118 (−180)	
Polypropylene	−18 (0)	

SOURCE: Arridge (1975).

on their strength, air content, void distribution, and moisture content. Frost-resistant products should be used and significant moisture buildup should be prevented in order to avoid deterioration in cold environments. Rapid material development has occurred in this respect.

It is possible to carry out lightweight concrete or masonry construction also during the winter without extensive protection and warming measures. The general guidelines given for winter concreting apply also for casting of lightweight concrete. New admixtures, partial hydrophone polymer microparticles, have been developed to provide lightweight concrete with good water tightness and frost resistance properties (Hedberg et al., 1979).

In the case of masonry, the bricks absorb water, and this increases the frost resistance of fresh mortar. The bricks that have a high moisture content due to water absorption from the mortar have to sustain freezing for favorable results in winter masonry. The mortar should have gained sufficient strength and lost enough moisture so that it also can sustain freezing (moisture content $< 6\%$) and carry the loads as thawing occurs without excessive settlements and cracking. Heating of bricks and mortar, using bricks that absorb water rapidly, using mortars with low freezing points, and providing protection are among the methods used to achieve desirable results in the winter.

The major difficulty with asphalt-concrete pavements in cold regions is their susceptibility to cracking due to increased stiffness at low temperatures and the effects of thermal shrinkage. Transverse thermal cracks may develop only on the surface or penetrate through the whole pavement to the subgrade. Numerous factors have an effect on the pavement's susceptibility to thermal cracking, but generally the phenomenon can be reduced by using softer asphalts. However, this may cause rutting in the warm summer period. The design of asphalt-concrete mixtures for cold regions has been discussed in detail in Dempsey et al. (1980) and Johnson et al. (1975). The frost susceptibility of asphalt-concrete mixtures can be studied in laboratory freeze-thaw tests (for example, Lottman, 1978).

The construction of asphalt-concrete pavements becomes increasingly difficult as the air temperature approaches the freezing point. The mixture cools down rapidly and sufficient compaction is hard to achieve with thin lifts. Thick lift construction should be used. One example of the low-temperature effects on the temperature development and compacting of an asphalt-concrete overlay is described in Eaton and Berg (1978).

Insulation materials, such as foams, fibers, and others, are essential parts of cold region construction. Two important cold region material properties, in addition to their cost, building physical function, and construction technical applicability, are their moisture adsorption proper-

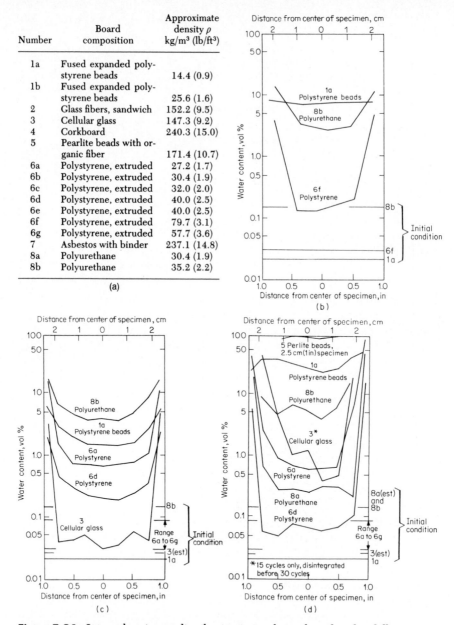

Number	Board composition	Approximate density ρ kg/m³ (lb/ft³)
1a	Fused expanded poly-styrene beads	14.4 (0.9)
1b	Fused expanded poly-styrene beads	25.6 (1.6)
2	Glass fibers, sandwich	152.2 (9.5)
3	Cellular glass	147.3 (9.2)
4	Corkboard	240.3 (15.0)
5	Pearlite beads with or-ganic fiber	171.4 (10.7)
6a	Polystyrene, extruded	27.2 (1.7)
6b	Polystyrene, extruded	30.4 (1.9)
6c	Polystyrene, extruded	32.0 (2.0)
6d	Polystyrene, extruded	40.0 (2.5)
6e	Polystyrene, extruded	40.0 (2.5)
6f	Polystyrene, extruded	79.7 (3.1)
6g	Polystyrene, extruded	57.7 (3.6)
7	Asbestos with binder	237.1 (14.8)
8a	Polyurethane	30.4 (1.9)
8b	Polyurethane	35.2 (2.2)

(a)

Figure 5.34 Internal moisture distribution in insulation board under different test conditions. (*a*) Index for insulating materials. All insulations were plant manufactured. (*b*) After 18 months in water. (*c*) After 34 months in moist silt. (*d*) After 30 freeze-thaw cycles over approximately a 30-day period, with specimen immersed in water.

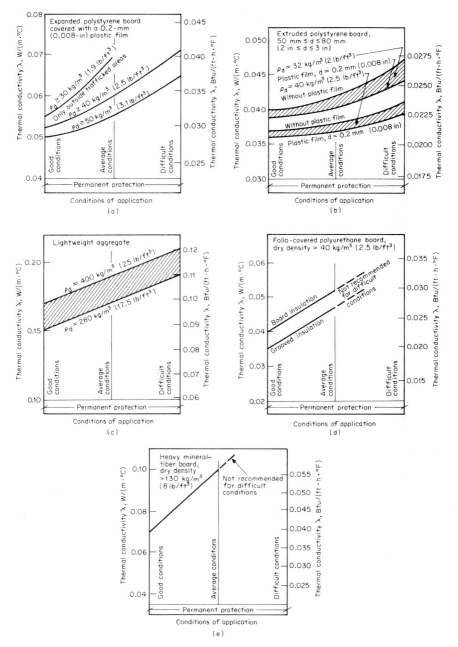

Figure 5.35 Suggested thermal conductivity values for different buried insulation materials from good (insulation remains dry, frost action limited, and no heavy loads on insulation) to difficult (insulation may get wet, ground is frost-susceptible, and loads are heavy) conditions of application. *(Mäkelä, 1982.)*

ties and their resistance to freeze-thaw cycles, especially in moist conditions. Some insulation materials, such as fibrous glass, absorb moisture effectively and must thus be kept dry. Kaplar (1974) has made an extensive investigation on the moisture and freeze-thaw effects on rigid thermal insulations and found that none of the materials was totally resistant to moisture absorption. Subsequent freeze-thaw cycles in the presence of free water generally increased the water absorption (Fig. 5.34) and also caused some deterioration at the saturated parts. Cellular glass, although highly resistant to moisture penetration, experienced surprisingly rapid deterioration. The effect of moisture on the material properties and deterioration should thus be recognized if highly resistant materials such as some extruded polystyrenes are not used. An example of the effect the service conditions have on the thermal conductivities of some insulation materials is given in Fig. 5.35.

Numerous problems related to material applicability for cold regions still remain. Some examples are given below. Ordinary latex paints will lose their emulsion if fresh paint is subjected to freezing, for example during storage or transportation. In roofing, rapid heat losses from hot bitumen make spreading difficult when proper cementing without excess of material has to be guaranteed also. Painting may become ineffective and difficult because a large amount of thinner may be needed already at temperatures above freezing. The curing of many materials such as paints, resins, glues, and coatings does not occur effectively in low temperatures (see Johnson, 1980), and early freezing may prevent proper curing at later stages. The lowest service temperatures have been specified for most roofing membrane systems, because the system or some part of it may begin to show decreased elasticity and increased brittleness at the same time as thermal strains are reaching very high values. A similar phenomenon occurs also with ordinary neoprene bearing plates, and it should be considered in cold region bridge design.

In addition to the above examples there are still numerous cases where cold temperatures have an effect on material applicability. Field experience is the best guide for cold region material selection. If such experience does not exist, extensive laboratory studies and field experiments are necessary in order to establish the necessary design criteria.

CHAPTER
6

OTHER DESIGN CONSIDERATIONS

6.1 Thermal Insulation

In cold regions the main function of thermal insulation is to provide a barrier against heat losses. The modes of heat transfer are conduction, convection, and radiation. Conduction has a dominant role in structural heat losses in a properly designed and constructed building. However, poor design or construction and a lack of understanding the physical functioning of the structure may to a large degree eliminate the investments made on better thermal insulation. The consequences of such mistakes increase in severity as the climate gets colder.

6.1.1 Conduction

The thermal conductance of a structure can be estimated according to

$$C = \frac{1}{R} \tag{6.1}$$

and

$$R = R_o + \frac{d_1}{k_1} + \frac{d_2}{k_2} + \cdots + R_a + \cdots + R_i \tag{6.2}$$

where C = total thermal conductance or transmittance of structure
R = total thermal resistance of structure
R_o = resistance of exterior surface
R_i = resistance of interior surface
R_a = resistance of air spaces in structure
d_i = thickness of layer i
k_i = thermal conductivity of layer i

The temperature T_p in any location of the structure can be calculated at steady-state conditions from

$$T_p = T_i - \frac{R_p}{R}\,(T_i - T_o) \tag{6.3}$$

where T_i = temperature inside structure
T_o = temperature outside structure
R_p = resistance of structure from inside to point in question

The heat flux q in steady-state conditions is obtained simply from

$$q = \frac{T_i - T_o}{R} \tag{6.4}$$

In practice the temperature fluctuates and the heat storage capacity of especially massive structures delays heat flux changes. In spite of these disadvantages, Eq. (6.4) provides an acceptable tool for estimating average conductive heat losses through different parts of the building.

A thorough list of heat transmission coefficients for different materials is contained in the American Society of Heating, Refrigerating, and Air-Conditioning Engineers (ASHRAE) *Handbook of Fundamentals* (1981). Some typical values for thermal conductivities of construction materials in service conditions are given in Table 6.1. However, one should recognize that conductivity values are not constants but depend on numerous factors, including moisture content, temperature, material density, aging effects, and the quality of the installation work of the insulation. The in-place R value does not always live up to the designer's expectations.

The resistance values for surfaces and air spaces depend much on the velocity of the air movements (actually convection). A typical value for outside surface resistance is between 0.02 and 0.05 $(m^2 \cdot {}^\circ C)/W$ [0.1 and 0.3 $(ft^2 \cdot h \cdot {}^\circ F)/Btu$]; for inside surface resistance it is between 0.10 and 0.15 $(m^2 \cdot {}^\circ C)/W$ [0.6 and 0.8 $(ft^2 \cdot h \cdot {}^\circ F)/Btu$]. The resistance of unventilated narrow air spaces of 1 to 10 cm (0.4 to 4 in) is 0.1 to 0.2 $(m^2 \cdot {}^\circ C)/W$ [0.6 to 1.1 $(ft^2 \cdot h \cdot {}^\circ F)/Btu$].

An example of the extreme temperature distributions in a structure is shown in Fig. 6.1. The radiation effects, the heat losses during a clear

TABLE 6.1   Typical Thermal Conductivity Values in Practice

Material	Dry density, kg/m³ (lb/ft³)	Water content, %	Conductivity, W/(m·°C) [Btu/(ft·h·°F)]
Concrete	2300 (144)	2	1.75 (1.0)
Lightweight concrete	500 (31)	5	0.16 (0.09)
	1200 (75)	3	0.5 (0.3)
Brick wall	1500 (94)	1	0.6 (0.35)
Wood, perpendicular to fibers	500 (31)	16	0.14 (0.08)
Fibrous glass board	60 (3.7)	0.5	0.05 (0.03)
Polystyrene, bead board	23 (1.4)	4	0.045 (0.026)
Polystyrene, extruded	40 (2.5)	1	0.035 (0.02)
Polyurethane board	40 (2.5)	2	0.03 (0.017)
Cellular glass	150 (9.4)	—	0.06 (0.035)
Lightweight aggregate	320 (20)	0.5	0.10 (0.06)
Gypsum board	900 (56)	—	0.23 (0.13)
Particle board	600 (37)	10	0.14 (0.08)
Cement mortar	2000 (124)	2	1.2 (0.7)
Lime mortar	1700 (106)	2	0.9 (0.5)
Bitumen	1050 (66)	—	0.18 (0.1)
Sawdust	120 (7.5)	12	0.12 (0.07)
Straw, covered	30 (1.9)	—	0.1 (0.06)
Gravel	1900 (119)	4	1.5 (0.9)
Snow	200 (12)	—	0.2 (0.1)

night with no wind, and the heat absorption during a windless sunny summer day have important temporary effects, especially when the surface is not directly in contact with a high heat capacity substratum such as concrete. These effects should be considered for example in determining the lowest service temperatures of roofing materials or the possible cooling requirements during the summer.

Open insulation joints, door and window frames, and other structural members penetrating through the primary insulation layer form cold bridges that may considerably reduce the total thermal resistance of the structure. One way to estimate the effects of cold bridges or nonuniform layers (Fig. 6.2) is to assume that the different parts do not have an effect on each other (Fig. 6.3). The total thermal conductance C of the structure is given by

$$C = \frac{1}{R} = \frac{A_1/R_1 + A_2/R_2 + \cdots + A_n/R_n}{A_1 + A_2 + \cdots + A_n} \tag{6.5}$$

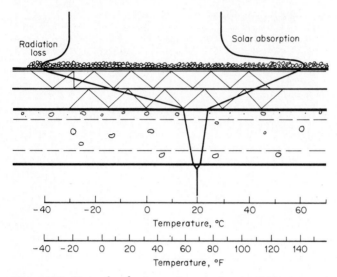

Figure 6.1 Example of extreme temperature variations.

where $A_1, A_2, \ldots, A_n$ = areas of wall parts perpendicular to heat flow
$R_1, R_2, \ldots, R_n$ = total resistances of different wall parts

In another estimate we concentrate on the nonuniform layer. The resistance of this layer j is given by

$$R_j = \frac{d(A_1 + A_2 + \cdots + A_n)}{A_1 k_1 + A_2 k_2 + \cdots + A_n k_n} \qquad (6.6)$$

where d is the thickness of the nonuniform layer. The total resistance R is computed from Eq. (6.2).

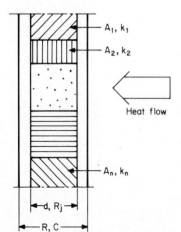

Figure 6.2 Notation for Eqs. (6.5) and (6.6).

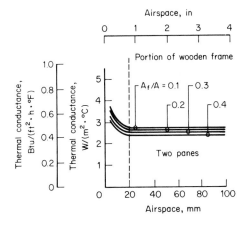

Number of panes	Thermal conductance, $W/(m^2 \cdot °C)$ [$Btu/(ft^2 \cdot h \cdot °F)$]
1	~ 6.0 (~ 1.1)
2	~ 3.0 (~ 0.53)
3	~ 2.0 (~ 0.35)
4	~ 1.5 (~ 0.26)

Figure 6.3 Effect of frame, airspace, and number of panes on thermal conductance of windows.

The actual thermal resistance lies generally between these two estimates, and the average value is usually a reasonable estimate. More accurate results can be obtained using three-dimensional numerical analysis or actual thermal resistance measurements.

6.1.2 Air Leaks

Convective heat transfer may occur through the structure because of the air pressure differences. The pressure differences are created among other things by wind, air-conditioning, and stack effect. The average values are generally small, 5 to 100 Pa (0.1 to 2 lb/ft²), although larger differences can be experienced, especially in highrise buildings. The airflow through a structure q can be computed from

$$q = LA\Delta p$$
$$\frac{1}{L} = \sum_{i=1}^{n} \frac{d_i}{l_i} \tag{6.7}$$

where L = air penetration through structure per area and pressure unit
Δp = pressure difference
l_i = air penetration coefficient of layer i

The theoretical air tightness of structures in cold regions is usually sufficient. However, large amounts of air, moisture, and thermal energy leak through cracks, openings, and unsealed joints. For example, the air penetration coefficient of a well-done brick wall may be about 5×10^{-7} m^2/(s·Pa) [2.5×10^{-4} ft^2/(s·lb/ft^2)]. The leakage through cracks depends on their type and is a nonlinear function of pressure difference and crack width. The leakage through a 1-mm (0.04-in) open crack often exceeds 10 m^3/(h·m) [100 ft^3/(h·ft)]. Rough estimates can be obtained, for example, from the equation (adapted from Baker, 1980)

$$Q = 2.4Ah_W \qquad (6.8)$$

where Q = airflow, m^3/s
$\quad\;\; A$ = area of leakage openings, m^2
$\quad\;\; h_W$ = head, mm of water (1 mm H$_2$O = 9.8 Pa = 0.20 lb/ft^2)

The thermal energy needed to raise the temperature of air from -20 to $+20°$C (-4 to $68°$F) is about 50 kJ/m^3 (1.34 Btu/ft^3). The total air volume change in a poorly constructed house may occur in less than an hour. The heat losses caused by this kind of "uncontrolled ventilation" may be even larger than the actual ventilation losses or conductive losses through the outer surface of a dwelling. Large savings can be achieved with a relatively small effort if the tightening and sealing of cracks and openings is done properly.

6.1.3 Optimum Insulation

Buildings with low surface-area-to-volume ratios and reasonable window areas are advantageous from the energy conservation point of view. The optimum insulation thickness (or, for example, window type or heating system as well) can be found in an analysis where the cost of increasing

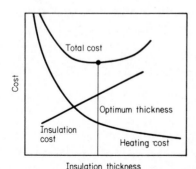

Figure 6.4 Determination of optimum insulation thickness.

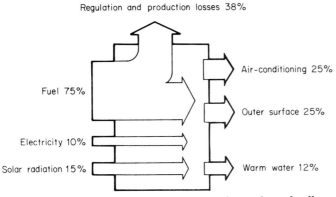

Regulation and production losses 38%

Air-conditioning 25%

Fuel 75%

Outer surface 25%

Electricity 10%

Solar radiation 15%

Warm water 12%

Figure 6.5 Example of the energy budget of a northern dwelling.

the insulation thickness is compared to the cost savings in heating (Fig. 6.4). The interest for the invested capital should be considered in the analysis.

The capital invested in thermal insulation only works effectively if design and construction faults such as large air leaks are eliminated and the thermal function of the building (roof, floor, walls, doors, windows, ventilation, and heating system) is well balanced (Fig. 6.5). Some typical faults are shown in Fig. 6.6. Moisture buildup in structures may also significantly reduce their thermal insulation capacity. This subject is discussed further in the following section.

6.2 Moisture and Condensation

The reason for moisture buildup in structures may be condensation as diffusion or air leakage occurs through a significant thermal gradient or the existence of free water that can penetrate into the structure because of gravity, capillary action, or air pressure. The structure may also contain significant amounts of construction moisture. Moisture and its buildup and transportation may have several negative effects on cold region construction, including deterioration of structural materials and coatings, a decrease in the thermal insulation capacity, and aesthetic and hygienic drawbacks.

6.2.1 Vapor Diffusion and Condensation

Vapor, like air, flows from higher to lower pressure. If no condensation occurs, vapor diffusion through the structure can be estimated from

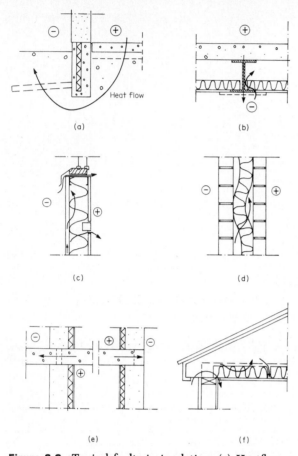

(a)

(b)

(c)

(d)

(e)

(f)

Figure 6.6 Typical faults in insulation. (*a*) Heatflow
should be cut with additional insulation. (*b*) Steel beam
forms cold bridge and air penetrates into uninsulated
space, partly eliminating effect of insulation. (*c*) Cold-
formed section forms cold bridge. Air flow behind corrugated sheet penetrates into soft mineral wool insulation, reducing its effectiveness. Air leaks through joints
and openings mean large thermal losses. A wind protection board should be placed behind corrugated sheet
and a sealed vapor barrier should be placed to warm
side. (*d*) Insulation should be in contact with warm side
because otherwise cold air will enter into space, cooling
the wall. (*e*) Cold bridge formed by concrete slab should
be cut. (*f*) Air leaks and cold bridges should be cut.
Insulation should tightly fill space reserved for it. Additional insulation and wind protection board eliminate
gaps and wind effects.

$$q = k_d \frac{p_i - p_o}{d} = \frac{\Delta p}{r} \tag{6.9}$$

where q = mass of vapor transmitted per unit area
k_d = vapor permeability
p_i = vapor pressure at interior surface
p_o = vapor pressure at exterior surface
d = thickness of structure
r = vapor flow resistance of structure

In a layered structure the vapor pressure can be computed at any point from

$$p_x = p_i - \frac{p_i - p_o}{r} \sum_{i=1}^{x} r_i \tag{6.10}$$

where r is the total vapor flow resistance and r_i is the vapor flow resistance of layer i. Some vapor resistivity values are given in Table 6.2, and the relationship between vapor pressure, air moisture content, and temperature is given in Fig. 6.7.

When the temperatures within the structure are known, the saturation vapor pressure curve can be drawn across the structure using Fig. 6.7.

TABLE 6.2 Typical Vapor Resistivity Values for Construction Materials [Vapor Resistance for Material Thickness of 1 m (3.3 ft)]

Material	Density, kg/m³ (lb/ft³)		Vapor resistivity, (Pa·s·m²)/ng·m [(h·ft²·in Hg)/grain·in]	
Calm air			0.005	(0.007)
Concrete	2300	(142)	0.1–0.3	(0.15–0.44)
Lightweight concrete	500–700	(31–43)	0.02–0.05	(0.03–0.07)
Brick	1200–1800	(74–112)	0.02–0.1	(0.03–0.15)
Wood	500	(31)	0.3–1	(0.44–1.45)
Metals			∞	(∞)
Bitumen	1000	(62)	500–5000	(725–7250)
Polyethylene film	100	(6)	1000–10,000	(1450–14,500)
Lime-cement mortar	1800	(112)	0.05–0.2	(0.07–0.29)
Wood fiberboard	300–1000	(19–62)	0.04–0.2	(0.06–0.29)
Gypsum board	800	(50)	0.06	(0.087)
Fibrous glass board	20–200	(1.2–12)	0.01–0.03	(0.015–0.044)
Polystyrene	15–50	(1–3)	0.1–1.0	(0.15–1.5)
Polyurethane	30–60	(1.9–3.7)	0.4–10	(0.6–15)

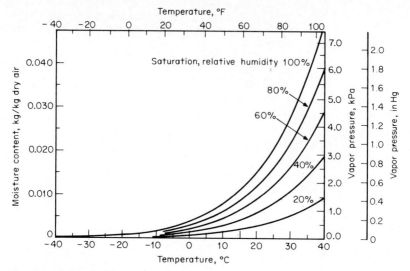

Figure 6.7 Air vapor pressure and moisture content as a function of temperature for different relative humidities.

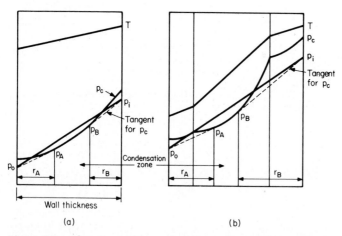

(a)   (b)

Figure 6.8 Glaser's graphic method for evaluating the condensation zone for (a) a uniform and (b) a layered structure. Steps: 1 — Thicknesses of layers are drawn in proportion to vapor resistances. 2 — Temperature curve is drawn for structure assuming that thermal conductivity does not depend on moisture content of material, that is, curve is linear within layers. 3 — Vapor saturation curve is drawn based on temperature curve. 4 — If vapor saturation curve crosses theoretical vapor pressure line between p_i and p_o, condensation occurs at a zone defined by tangents from p_i and p_o to saturation curve.

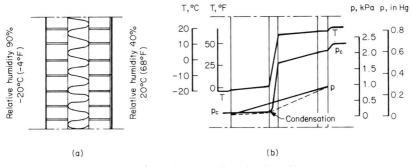

Figure 6.9 Computation of condensation for a brick wall.

On the other hand, the theoretical vapor pressure curve can be found using Eq. (6.10). If these two curves cross, condensation occurs. Quantitative estimates of the condensation can be found by computing the amounts of vapor coming into and going out from a condensation zone. One suggested method to define the condensation zone is given in Fig. 6.8. The rate of moisture condensation q_c becomes

$$q_c = \frac{p_i - p_B}{r_B} - \frac{p_A - p_o}{r_A} \tag{6.11}$$

where p_A, p_B = vapor pressures at points A and B
$\quad\quad r_A$, r_B = vapor flow resistances of corresponding layers

On the other hand, condensation tends to occur at a solid interface in a layered structure.

EXAMPLE 6.1 Estimate the rate of condensation in the cross section shown in Fig. 6.9a. The computations are tabulated on page 398. The thermal profile is computed from Eq. (6.3), and the saturation vapor pressure for different parts of the structure is found from Fig. 6.7. The theoretical vapor pressure distribution is given by Eq. (6.10).
   The theoretical and critical vapor pressure curves can now be drawn following the guidelines given in Fig. 6.8. This is done in Fig. 6.9b. Condensation occurs because the lines cross and the rate of condensation is, according to Eq. (6.11),

$$q_c = \frac{920 - 140}{0.0076} - \frac{140 - 90}{0.005} = 92{,}000 \text{ ng/(m}^2 \cdot \text{s)}$$
$$= 8 \text{ g/(m}^2 \cdot \text{day)} \ [0.0016 \text{ lb/(ft}^2 \cdot \text{day)]}$$

Whether this condensation is dangerous and ventilation or an additional vapor barrier is necessary, depends among other things on the duration of the wetting and drying seasons, the frost susceptibility of brick, and the type of the possible coating material on the outer surface.

Layers	d_i, m (in)	k_i, W/(m·°C) [Btu/(ft·h·°F)]	R_i, (m²·°C)/W [(ft²·h·°F)/Btu]	ΔT, °C (°F)	T, °C (°F)	p_c, kPa (in Hg)	r_i, (Pa·s·m²)/ng [(h·ft²·in Hg)/grain]	Δp, kPa (in Hg)	p, kPa (in Hg)
Air			0.05 (0.28)	0.6 (1.1)	−20 (−4) −19.4 (−2.9)	0.10 (0.030) 0.11 (0.032)			0.09 (0.027)
Brick	0.13 (5.1)	0.6 (0.35)	0.22 (1.25)	2.9 (5.2)	−16.5 (+2.3)	0.14 (0.041)	0.005 (0.29)	0.33 (0.097)	0.42 (0.124)
Mineral wool	0.12 (4.7)	0.05 (0.03)	2.4 (13.6)	31.4 (56.5)	+14.9 (+58.8)	1.7 (0.501)	0.0012 (0.07)	0.08 (0.024)	0.50 (0.148)
Brick	0.13 (5.1)	0.6 (0.35)	0.22 (1.25)	2.9 (5.2)	+17.8 (+64.0)	2.0 (0.590)	0.005 (0.29)	0.33 (0.097)	0.83 (0.245)
Plaster	0.02 (0.8)	1.0 (0.58)	0.02 (0.11) 0.15 (0.85)	0.2 (0.4) 2.0 (3.6)	+18.0 (+64.4)	2.1 (0.620)	0.0014 (0.08)	0.09 (0.027)	0.92 (0.272)
Air			3.06 (17.4)		+20.0 (+68)	2.3 (0.680)	0.0126 (0.73)		

6.2.2   Other Moisture Sources

Condensation may also occur as air and moisture leak through cracks and openings. As a matter of fact, this is usually the dominant mode when good vapor barrier materials are used. Massive ice and moisture buildups of totally different magnitudes than what can be expected by diffusion have been experienced in leaky structures in cold regions.

The moisture content of recently finished structures may be high due to construction moisture. The moisture content may also be increased when the structure is in contact with the ground or during rain due to capillary action. The wind pressure may increase water absorption, especially through cracks and openings or when the surface is saturated. Finally, water may penetrate into the structure if joints and other details are not properly designed and constructed.

6.2.3   Practical Moisture Control

The general rule of moisture control in structures is to provide adequate insulation to prevent surface condensation, and to use a continuous vapor and air barrier at the warm side of the insulation so that the structure will breathe outward. The colder the climate, the more important it is to have the physical functioning of the structure under control, because the consequences of failures increase in severity and the moisture buildup potential increases at the same time as the drying potential decreases.

The possibility of condensation at the inner surface of the structure depends on its temperature and the moisture content of the surrounding air. When the moisture content of air is larger than its saturation content at the surface, moisture condensation or frost buildup occurs. The relative humidity of the inside air is normally low in the winter, and large-scale surface condensation is thus not likely to occur. However, local condensation may occur in connection with cold bridges or at the inner surface of poorly designed windows.

It is important that the air and vapor barrier be located as close to the warm surface as possible. Otherwise condensation may occur at the barrier. Any layer that has high resistance against water vapor permeability can serve as the vapor barrier if it does not contain leaks. Special attention should be given to sealing joints and penetrations through the barrier. Double polyethylene films with overlapped seams or, for special applications, taped seams are commonly used as vapor barriers. Brick walls, concrete layers, or lightweight concrete may also serve the purpose. Careful evaluation of the moisture buildup in the winter, its effects on the thermal insulation capacity and the frost susceptibility of materials, and of the drying of the structure during the summer is necessary.

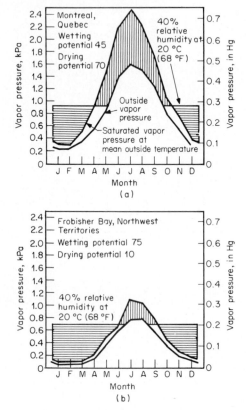

Figure 6.10 Typical wetting and drying potentials of structures in heated buildings in (a) subarctic and (b) arctic environment. (*Baker, 1980.*)

The moisture must have the possibility to dry out. The physical potential for drying decreases and the moisture buildup increases as the climate becomes colder, as shown in Fig. 6.10. Although some drying of the structure may occur in the wrong direction, that is, from cold to warm, proper drying of construction moisture and the use of an effective vapor barrier near the warm surface are essential. Because the vapor barrier is never perfect and construction moisture has to have a way to escape, vapor traps should be avoided. Ventilation of the insulation should be arranged in order to avoid structural damage if the outer surface of the structure is highly resistant to vapor diffusion. Some examples of moisture problems in poorly constructed buildings are illustrated in Figs. 6.11 to 6.13.

6.3 Thermal Stresses

The extreme temperature variations in cold regions are generally large, in some arctic inland areas close to 100°C (180°F). When the radiation effects are taken into account, extreme temperature variations in the outer surfaces of roofs and walls may well exceed 100 °C (180 °F) and the temperature differences between outer and inner surfaces may be close to 70 °C (130 °F). The structures develop large thermal strains at low temperatures, and at the same time the modulus of elasticity may have increased and the ductility of the material may have dropped significantly.

Large nominal thermal stresses are not necessarily very harmful for a structure. Consider the following case of a normal reinforced concrete structure. Analysis of homogeneous cross sections may show considerable thermal stresses induced by extreme temperature gradients or fluctuations. However, when the stiffness of the most stressed parts (the cracking capacity is exceeded) is replaced by that of a cracked cross section in a subsequent iterative step, the magnitude of the thermal stresses will drop to only a fraction of the initial values. With proper placing of expansion joints the thermal stresses and the crack dimensions may be kept under control.

In some cases thermal effects may contribute to the worsening of serviceability, to accelerated deterioration, or even to a loss of structural

Figure 6.11 Ice buildup as evidence of severe air leaks and heat losses. *(Courtesy of W. Tobiasson.)*

Figure 6.12 Massive frost buildup in poorly ventilated attic as a result of air leaks. *(Courtesy of W. Tobiasson.)*

Figure 6.13 Stains on ceiling created by thawing of frozen moisture. *(Courtesy of P. Johnson.)*

stability. Thermal stresses may become significant in the design of steel or aluminum structures. They should be considered also in the design of ordinary or prestressed concrete structures when crack control is important, as for arctic offshore structures (Almazov and Kopaigorodski, 1982).

The thermal effects on simple steel and reinforced concrete structures can be analyzed with reasonable accuracy, but the analysis of many composite systems may be more difficult. Such systems include, for example, sandwich panels, coatings or surface materials and base materials of walls, and roofing membranes, insulations, and supporting structural materials. The coefficients of thermal expansion in composite systems generally differ, and the behavior of most materials is nonlinear depending not only on the stress level but also on time and temperature. The fatigue and aging effects are especially difficult to analyze. Quick estimates of the thermal stresses can be made by using conservative nominal values for elasticity constants in the analysis. For example, Heger (1978) has presented simple formulas to estimate the internal stresses and support reactions of continuous structural sandwich panels subjected to thermal gradients.

Experience, knowledge of the properties of materials at low temperatures, and analysis of thermal stresses are valuable in evaluating the applicability of structural systems for cold environments. However, the general design is of equal importance. A good design often contains proper use of expansion joints and specially designed connections, joints, supports, and other details so that the consequences of thermal movements are under control. It is often practical to have connections or other structural details that have enough flexibility or tolerance to allow the thermal movements to occur freely.

6.4 Fire

There are several reasons why the fire hazard needs special design attention, especially in the arctic environment. The air and materials inside buildings are very dry and the circumstances are favorable to fire ignition and burning. Heavy utilization of the heating system increases the risk of fire, especially if the system is not properly designed and installed. Fire fighting may also be more difficult. Problems may be connected, for example, with transportation and maneuverability, the acquisition of fire fighting water or the operation efficiency in winter conditions.

The consequences of a fire may also be severe, especially in remote arctic circumstances. People have to find shelter, but the situation is difficult if the transportation system or the power delivery system is not functioning. Stringent requirements must thus be set for structural fire safety. A fire alarm system, early extinguishing, escape routes, and readiness and effectiveness of fire fighting also need special consideration. As a matter of fact, fire hazard is one of the main reasons why in remote arctic settlements essentials such as shelter, power, transportation, and communications are often secured with some kind of separate reserve system.

6.5 Some Roof Design Considerations

Roofs have been traditionally vulnerable to damage and poor serviceability in cold regions (see, for example, Tobiasson, 1980; Baker, 1980). General weathering, thermal forces, mechanical damage, consequences of moisture buildup, and poor workmanship of roofing are common reasons for unsatisfactory performance. In some cases ice formations have also contributed to roof damage.

Ice formations may cause splits in roofing membranes that lack high strength and good deformation properties. In the spring meltwater may

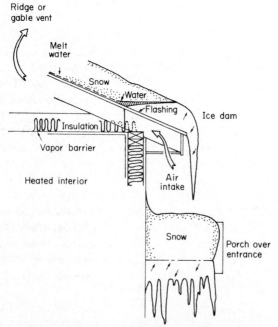

Figure 6.14 Eave icing. *(Tobiasson, 1971.)*

Figure 6.15 Heavy eave icing caused by improper roof insulation. *(Courtesy of J. P. Zarling.)*

stay on depressions of long span roofs or in valleys of roofs with internal drains blocked by ice and slush. In varying weather conditions meltwater freezes, undergoes deformation, and may crack. If the ice has bonded strongly on the roofing membrane and has sufficient thickness, at least 5 to 10 cm (2 to 4 in), splitting may occur. This type of ice formation can be prevented if the roof has adequate slopes to drain depressions or the valley areas. The valley areas can also be heated by using reduced insulation or electric heating cables.

When overlapping water-shedding units such as corrugated sheets, tiles, or composition shingles are used, roofs have generally steep slopes. However, ice dams may form due to uneven thermal distribution on the roof, for example at the eaves (Figs. 6.14 and 6.15) or in the vicinity of warm ventilation ducts. The major problem here, apart from the general deteriorating effect of ice and moisture and the hazard created by falling icicles, is that water backed up behind an ice dam may penetrate into the roof through seams of such a shedding roofing. Adequate insulation and ventilation of roof spaces reduces the ice dam problems. If valleys and internal drains exist on such roofs, proper drainage can only be guaranteed by heating.

In areas of high winds, blowing snow may penetrate into the attic or other roof spaces through ventilation openings. The amount of moisture created by melting snow in such cases tends to be even larger than what frost buildup or water leakages could have created in an unventilated attic. Snow penetration can be prevented by arranging the opening to the attic sufficiently above the eave intake [normally about 20 cm (8 in)], and by using a high ratio of eave inlet to attic opening areas (Fig. 6.14).

6.6 Some General Design Aspects

The cold environment sets many special requirements for structures. The designer and the supervisor have to be very familiar with frost, snow, ice, the behavior of materials, building physics, and related problems because consequences of failures and mistakes in cold regions, especially in the Arctic, are more severe than in more temperate climates. As an example, consider a wall. If the vapor barrier is not complete, heavy moisture buildup, sagging of insulation, and massive ice formation may occur. The thermal insulation capacity is reduced to only a fraction of its intended value, the heating capacity may prove inadequate, and pipes located in the wall may freeze.

Experience of practical problems is also important. Entrances, for example, should be provided with storm locks or hallway shelters with two doors. This is not only because of energy conservation efforts but also to avoid cold wind blowing directly into the house. An entrance with a canopy provides protection against falling snow and icicles. Doors should not have low thresholds because hard packed snow may prevent proper opening and closing. Sometimes doors are designed to open inward to prevent them from being blocked by snowdrifts or broken by high arctic winds. Cold floors are a common nuisance in elevated buildings because the temperatures tend to stratify within the room. This can be avoided with a double floor with insulation in the lower floor and some sort of heating also between the floors.

Economy and serviceability are key elements in cold region construction. The design must recognize the requirements of construction techniques and the possible repair and maintenance needs. By using simple solutions and local experience the risk of mistakes and failures can be minimized. But this is not enough in remote settlements, where social aspects and the well-being of people have proved to be especially important considerations.

Although in general appearance, cold region communities and buildings do not necessarily differ from those in the temperate zone, certain planning aspects connected to microclimate, light, and coziness require special attention. Southerly slopes are often favored in siting communities and buildings. The slopes get maximum amounts of sunlight and solar radiation. They are less windy than hilltops and warmer than valleys and depressions, which gather cold air in very cold weather conditions. (On a calm clear night the air nearest the ground gets colder than the average temperature due to radiation; being denser, it flows downward.)

Grouping of buildings can also be used to create a milder microclimate. Heating energy savings of up to 50% can be achieved by proper siting and grouping of buildings. The psychological effects of long winters can

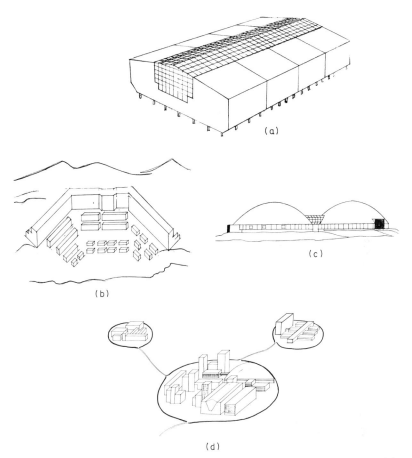

Figure 6.16 Designs that recognize the specific features of cold regions. (*a*) Well-lit recreational space created between two wings of a building. (*b*) Shelter building used to protect a community from high winds and to form a solar trap. (*c*) Energy-efficient arctic activity center. (*d*) Enclosed community.

be counteracted by warm colors and textures in the interior design, by fenestration to maximize sunlight, and by paying special attention to recreational facilities.

Some interesting cold region designs are shown in Fig. 6.16. The first design illustrates how a very stimulating and quite energy-efficient recreational space can be created between two building wings. Note how the corners have been rounded to improve snow control. Such a design has been used in the BP/Arco operations center at Prudhoe Bay. In the second design a long building is used to shield the community from high winds and to form a solar trap. Fermont, Quebec, provides one example of such a community plan. The third design represents energy-efficient planning that fits well into the arctic landscape. In this U.S. Army Corps of Engineers idea, working and living accommodations are separated by a tunnel to make people feel more comfortable and to improve fire safety. Economic erection of such structures in the Arctic may, however, be problematic. Ideas of completely enclosed communities, such as the fourth design, have also been presented from time to time. Although these ideas have not yet been carried out on a large scale, many campus complexes, metropolitan shopping areas, and arctic operation centers where different buildings are connected with tunnels or enclosed walkways represent this ideology. The reader is referred to Van Ginkel Associates Ltd. (1976) and to ASCE specialty conference proceedings (1981) for further discussions on the architectural planning of cold region communities and buildings.

7

COLD WEATHER CONSTRUCTION: TECHNIQUES AND RESTRICTIONS

7.1 Feasibility Considerations

Cold weather construction, although generally more expensive and less efficient than construction in more temperate conditions, is common in subarctic and arctic areas. The construction schedule is very often tight because of high capital costs. In Scandinavia, cold weather construction is encouraged further by society in order to control seasonal unemployment. In the high Arctic, it is simply unrealistic to attempt to execute all construction stages of a major project during the short summer periods.

Low temperature, wind, snow, lack of light, and other environmental conditions in cold weather influence the construction efficiency in several ways. First, the combination of low temperature and wind often makes working conditions unpleasant or even dangerous (Fig. 1.7). The use of special work clothing and wind shelters can provide some possibilities to extend the working period. The temperature of $-40\,°C\,(-40\,°F)$ is generally considered the absolute limit for any type of outdoor construction activity.

Low temperature may also limit the use of machinery and other equipment. Freezing and starting problems and adfreeze effects are often experienced. Hydraulics is commonly a weak point for ordinary machinery at very low service temperatures. The serviceability of machinery may also be endangered because of failures due to brittle fractures. The use of cranes is especially hazardous if sufficient ductility of the crane structure is not guaranteed at the service temperature.

Frost is another phenomenon that must be dealt with in cold weather construction. Conventional excavation methods are usually not very effective on massively frozen ground and compaction of frozen fill seldom produces acceptable results. On the other hand, frozen ground provides good support for transportation and heavy machinery (Table 7.1). That is why it may even be preferable to execute certain construction stages during winter if during other seasons surface soil conditions are wet and soft. Difficult excavation and construction stages below the groundwater table have also been pursued successfully by taking advantage of frost penetration.

Lack of daylight has a considerable effect on the efficiency of construction operations in the Arctic in midwinter. The harmful effects of snow in the form of precipitation or snow transport and accumulation may also be considerable. Generally, snow has to be removed first before construction activities such as placement of soil, pouring of concrete, or roofing can be continued.

Cold temperature affects the properties and behavior of many construction materials in a manner that limits their applicability at certain construction stages. Some materials show brittleness that makes their handling and especially deforming difficult. The rapid cooling of hot bitumen limits its use in roofing as well as in paving at low temperature. Freezing or drying of fresh unprotected concrete may have a serious impact on its mechanical properties. Painting or coating is generally not possible at low temperatures or on moist or icy surfaces. These features and others require the use of different kinds of protective skirtings and heating methods to achieve desirable results in cold weather construction. The protective measures range from removable weather protection structures and wind shelters to enclosures of the entire construction site (Fig. 7.1).

7.2 Earthworks and Foundation Construction

The special problems of cold region earthworks and foundation construction are mostly connected to frost and the freezing and thawing of soil. Because earthworks and foundation construction cannot always be scheduled during the most desirable season, machines and methods must meet strict requirements. Frozen ground is difficult to excavate and compact. On the other hand, the strength and imperviousness of frozen ground can be an advantage when the construction is in soft and wet areas or below the groundwater table.

TABLE 7.1 Short-Term Bearing Capacity of Frozen Marshy Ground According to Military Practice. A Factor of Safety on the Order of 2 Is Recommended

Vehicle type	Total mass of vehicle, tonnes (short tons)	Frost thickness of marshy ground — Grass vegetation, cm (in)	Frost thickness of marshy ground — Turf vegetation, cm (in)	Distance of vehicles, m (ft)	Remarks
Tracked vehicle	4 (4.4)	12 (4.7)	15 (5.9)	10 (33)	When a single tracked vehicle moves on marsh linearly in first gear, the given values can be reduced by 20%. When the temperature is above $+5°C$ (41°F), the given values must be increased by 2–3 cm (0.8–1.2 in).
	6 (6.6)	14 (5.5)	17 (6.7)	15 (49)	
	10 (11.0)	18 (7.1)	21 (8.3)	25 (82)	
	20 (22.0)	20 (7.9)	24 (9.4)	25 (82)	
	25 (27.6)	23 (9.1)	27 (10.6)	30 (98)	
	30 (33.1)	26 (10.2)	30 (11.8)	35 (115)	
	40 (44.1)	32 (12.6)	36 (14.2)	40 (131)	
	50 (55.1)	40 (15.7)	45 (17.7)	45 (148)	
Wheeled vehicle	2 (2.2)	10 (3.9)	12 (4.7)	15 (49)	
	3.5 (3.9)	13 (5.1)	16 (6.3)	18 (59)	
	6 (6.6)	15 (5.9)	18 (7.1)	20 (66)	
	8 (8.8)	17 (6.7)	20 (7.9)	22 (72)	
	10 (11.0)	18 (7.1)	21 (8.3)	25 (82)	
	15 (16.5)	25 (9.8)	29 (11.4)	30 (98)	

(a)

(b)

Figure 7.1 Various protective measures in winter construction. (*a*) Removable weather shelters. (*b*) Home building within an enclosure. (*c*) Air-supported structures used for foundation construction. (*d*) Construction of water-treatment facility under cover. (*Courtesy of A. Ahlström Ltd.*)

(c)

(d)

In seasonal frost areas, construction problems are encountered during and right after the season of freezing temperatures. Penetrating through the thin, seasonally frozen layer and protecting the exposed ground from frost action are principal considerations. Meltwater and the softening of ground may create additional problems at the construction site in the spring. In permafrost areas one works with massively frozen ground which is difficult to excavate or to thaw effectively. Drainage and softening of the active layer are important construction problems during the summer. The fact that permafrost is sensitive to thermal disturbances is also one of the major problems.

7.2.1 Winter Excavation Operations in Seasonal Frost Areas

One of the principal indicators of the excavation properties of soil is its shear strength. The shear strength of frozen soil is generally much higher than that of thawed soil (Fig. 7.2), and therefore its relative excavation resistance is also higher. Only soils with very low moisture content can be excavated in the frozen state by conventional methods.

The depth of a frozen layer also has a great influence on winter excavation. The progress of frost and thaw depends on numerous factors, including the moisture content and the thermal properties of the soil, the insulation provided by snow and vegetation, and the radiation effects. Rough estimates can be made based on Fig. 7.3. Winter excavation is usually based on breaking or thawing the frozen layer. If frost penetration has been slowed down by insulating the ground prior to freezing, operations can be eased significantly.

With mechanical excavating methods frozen ground is broken into pieces manageable for the loading equipment. Normal 20-ton digging machines can break frost layers up to a thickness of 0.5 m (20 in). Hammering with the machine bucket, possibly provided with a frost tooth, although effective, may strain the hydraulic system of some machines. Digging machines may be equipped with a vibrating tooth that breaks the frost by wedge effect. Manual compressed-air jackhammers are effective only in small-scale works.

Heavy-duty bulldozers equipped with single or multitooth hydraulic rippers have been used effectively in large-scale operations (Moore and

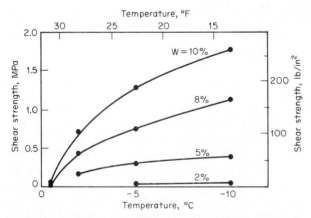

Figure 7.2 Shear strength of silty sandy moraine as a function of temperature and moisture content. (*Adapted from Heiner, 1972.*)

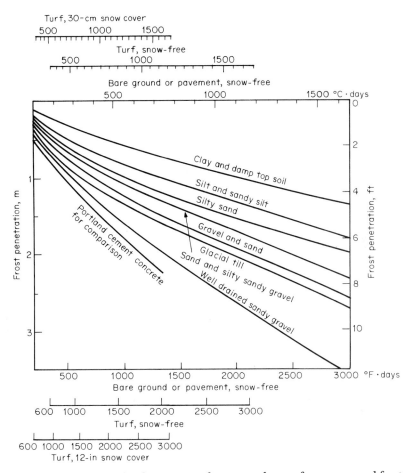

Figure 7.3 Relationship between air freezing index, surface cover, and frost penetration into homogeneous soils. *(Sanger, 1963.)*

Sayles, 1980). The rippability of frozen ground can be estimated based on its seismic velocity. Saturated and dense gravel has proven most difficult to rip at low temperatures. Still another mechanical method is to drop heavy pieces of concrete or rock by a crane or digging machine to break the frozen layer. Figure 7.4 gives some examples of the use of mechanical methods in breaking seasonally frozen ground. The use of explosives is an alternative approach. It is commonly used on thick, strong frost when possible damages to surrounding structures can be avoided. The consumption of explosives is quite high, typically 0.3 to 1 kg/m^3 (0.02 to 0.06 lb/ft^3). It increases with increasing moisture content in the frost.

(a)

(b)

Figure 7.4 Mechanical methods to break frost. (*a*) Dropping a concrete block. (*b*) Hitting with bucket that has frost-breaking tooth. (*c*) Hitting wedge attachment. (*d*) Heavy-duty ripper. (*Courtesy of R. Virtanen.*)

(c)

(d)

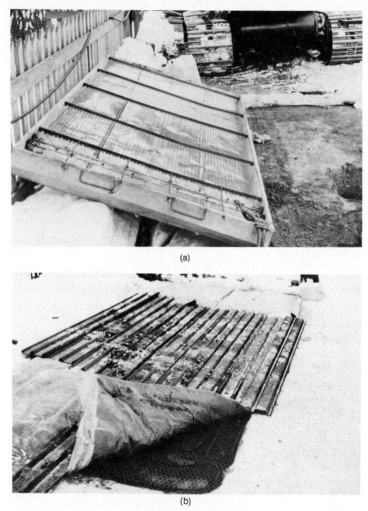

(a)

(b)

Figure 7.5 Methods to thaw frozen ground. (*a*) Thermal radiation. (*b*) Electric heating protected by steel sheets in traffic lane. (*c*) Thawing with steam. (*d*) Direct-fired heater connected to an insulated thawing channel. (*Courtesy of R. Virtanen.*)

(c)

(d)

Thawing of seasonally frozen ground provides an attractive alternative to mechanical methods and explosives, especially in trenching, if the time factor is not critical. It is very practical in areas where buried cables and pipes may be damaged as frozen ground is broken. Thermal energy can be produced on site from electricity, oil, or liquid gas, and it can be transferred to the ground by radiation, conduction, or in the form of heated air, steam, or water. Some examples of prethawing of the frozen ground are given in Fig. 7.5.

The thaw penetration can be estimated roughly from Fig. 7.3 by using the ground surface temperature if it is below 100°C (212°F). In temper-

TABLE 7.2 Insulation Requirements for Frost Protection of Ground and Foundations[a]

Required thermal resistance, $(m^2 \cdot {}^\circ C)/W$ $[(ft^2 \cdot h \cdot {}^\circ F)/Btu]$

Soil type	Allowed depth of frost penetration, m (ft)	Design freezing index, h·°C (°F·days) 10,000 (750)	20,000 (1500)				30,000 (2250)				40,000 (3000)		
		Mean annual temperature °C (°F) +2−+7 (36−45)	+1 (34)	+2 (36)	+3 (37)	+4−+7 (39−45)	+1 (34)	+2 (36)	+3 (37)	+4−+6 (39−43)	+1 (34)	+2 (36)	+3−+4 (37−39)
Clay, silt, w = 20%, γ_d = 1500 kg/m³ (94 lb/ft³)	0.1 (0.3)	1.0 (5.7)	2.5 (14)	2.2 (12)	2.0 (11)	1.8 (10)	3.8 (22)	3.5 (20)	3.3 (19)	3.0 (17)	—	—	5.0 (28)
	0.3 (1.0)	0.6 (3.4)	1.4 (7.9)	1.2 (6.8)	1.1 (6.2)	1.0 (5.7)	2.4 (14)	2.0 (11)	1.7 (9.7)	1.5 (8.5)	3.5 (20)	3.2 (18)	3.0 (17)
	0.5 (1.6)	0.3 (1.7)	0.9 (5.1)	0.7 (4.0)	0.6 (3.4)	0.5 (2.8)	1.4 (7.9)	1.2 (6.8)	1.1 (6.2)	1.0 (5.7)	2.0 (11)	1.8 (10)	1.6 (9.1)
Sand, gravel, w = 8%, γ_d = 1700 kg/m³ (106 lb/ft³)	0.1 (0.3)	1.5 (8.5)	3.9 (22)	3.2 (18)	2.8 (16)	2.5 (14)	6.0 (34)	4.8 (27)	4.3 (24)	4.0 (23)	—	—	—
	0.3 (1.0)	1.0 (5.7)	2.8 (16)	2.3 (13)	2.0 (11)	1.8 (10)	4.5 (26)	3.6 (20)	3.0 (17)	2.8 (16)	—	4.8 (27)	4.0 (23)
	0.5 (1.6)	0.6 (3.4)	2.2 (12)	1.6 (9.1)	1.3 (7.4)	1.2 (6.8)	3.3 (19)	2.7 (15)	2.2 (12)	2.0 (11)	4.5 (26)	3.8 (22)	3.0 (17)

[a] Insulation should be installed before the beginning of the freezing season.

SOURCE: Algaard (1976).

atures above 100°C (212°F), ground moisture vaporizes and efficiency is reduced, especially if this vapor is lost from the system. When steam or water is used for thawing, one must keep in mind the possibility that the increased water content of the soil may cause frost action or loss of bearing capacity. Because the energy cost is about half the total cost of prethawing in seasonal frost areas (the rest being labor and capital costs), energy efficiency is an important consideration. Removal of loose snow and ice and effective insulation of the system would improve the efficiency. The thermal capacity created during the thawing process can also be utilized by stopping the heat slightly before the required extent of thawing has been reached. When thick layers of frozen ground have to be prethawed, the thaw progress can be somewhat accelerated by removing the thawed layers and exposing the frozen ground after thaw has progressed some distance, say 0.6 m (2 ft).

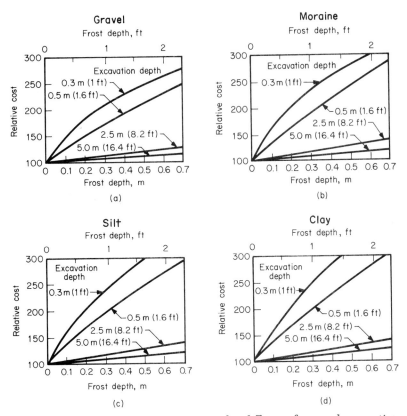

Figure 7.6 Relative cost of excavation for different frost and excavation depths. *(Svenska Byggnadsentreprenörföreningens Produktionsråd, 1963.)*

Insulation provides a convenient way to delay and slow down the penetration of frost and to ease winter excavations (Table 7.2). Fabricated insulations are commonly used; but sawdust, hay, and snow are also used when available. Sometimes all the foundation work including excavation is done inside weather shelters.

Insulation for winter excavation should be installed early in the fall. The thermal reserve of the ground then keeps the ground frost-free long after the freezing season has begun. Snow covering is the most inexpensive approach. If the excavation has not been completed before the thawing season, the insulation should be removed.

Several factors should be considered when selecting the winter excavation method in seasonal frost areas. The type, density, temperature,

TABLE 7.3   Relative Cost Estimates to Break or Thaw a Frost Layer about 1 m (3.3 ft) Thick or to Provide Thermal Insulation°

Method	Relative cost		Remarks
	Mean	Range	
Protection			
Snow	2	0–5	No traffic
Sawdust	4	3–15	No traffic
Fabricated insulation	7	5–10	No traffic
Breaking			
Heavy power shovel	1	0–5	Space
Drop particle	2	1–5	Quaking, flying particles
Frost tooth	>10		Hydraulics
Hitting wedge attachment	5	2.5–10	Noise
Bucket with frost tooth	20		Noise
Heavy ripper	12	3–50	Quaking
Explosives	20	5–50	Quaking, flying particles
Manual jackhammer	100		
Thawing			Reduced efficiency in moist clay and peat
Electric radiator	35	25–50	$<10 \text{ m}^2$ (100 ft²)
Heating cable, 220 V	10	7–15	
Heating cable, 42 V	18	10–30	
Oil-burning direct-fired heater	15	8–50	$>20 \text{ m}^2$ (200 ft²)
Oil-burning indirect-fired heater	20	10–50	$>50 \text{ m}^2$ (500 ft²)
Liquid-gas radiator	15	8–30	
Steam	50		

° Note that the cases are not necessarily comparable.
SOURCE: Virtanen (1978).

and ice content of frozen ground are of importance for the relative effectiveness of different methods. The magnitude and nature of the excavations and the available time and equipment should be considered in the planning. At times the traffic requirements will preclude the possibility of using insulation or prethawing methods. On the other hand, mechanical methods are noisy and may cause damage to pipe and cable installations or even to structures. An example of the relative cost trends of winter excavation for different frost depths and soil types is given in Fig. 7.6. Some examples of relative costs for different methods to break or thaw frozen ground or to prevent it from freezing are listed in Table 7.3.

As was pointed out earlier, although the frozen ground is difficult to excavate, it does provide some advantages that should be considered. Soft and wet soil can carry heavy machinery only when frozen. Walls of a trench stay stable when they are frozen. Dry excavation can be pursued below the groundwater table if the frost layer is not disturbed. On the other hand, if water can leak into the pit, the removal of ice and water may require additional efforts.

7.2.2   Excavation in Permafrost Areas

In permafrost areas the excavation methods are in principle quite similar to those used in seasonal frost areas. They include mechanical methods, drilling and blasting, and prethawing combined with conventional methods. The most extensive excavation works in permafrost areas are often connected with the operation of borrow pits (Fig. 7.7) or mines.

Figure 7.7   Borrow pit at Prudhoe Bay, Alaska.

Figure 7.8 Trenching in permafrost. (*a*) Rocksaw in operation. (*b*) Sections blasted and sections rocksaw cut and blasted. (*Zirjacks and Hwang, 1983.*)

Mechanical methods have been used successfully, especially when excavating relatively warm permafrost. Cross-ripping with heavy rippers has been economical in loosening frozen ground in large areas. Pneumatic machinery and power shovels have also been applied successfully. In the U.S.S.R. a pile driver with a tubular leader has been used together with a power shovel in trench excavation (Tsytovich, 1975).

Drilling and blasting is an effective method to loosen permafrost in large-scale tunneling or trenching (Fig. 7.8). It can also be used to open pits for footings or for soil sampling. Various types of explosives have been used, but according to Mellor and Snellman (1970) liquid or slurry explosives have certain distinct advantages. Test drillings and blastings are usually carried out to determine the optimum blasting arrangement.

Some experiences with mechanical excavation methods and blasting are listed in Table 7.4.

Thawing of permafrost may sometimes be necessary in order to eliminate the possible thaw settlements or viscoelastic permafrost deformations in advance, but its principal usage is to ease the excavation. Velli et al. (1977) and Kudoyarov et al. (1978) have discussed the artificial thawing methods that include hydraulic thawing with hot or cold water, steam thawing, electric heating, and thermochemical thawing. All these methods are quite labor-intensive. Steam and water jets are the most commonly used. Careful control is needed to avoid excessive thawing of permafrost if the methods are applied only locally for the placement of footings or piles.

Solar energy can also be utilized during the summer by first removing the insulative vegetation cover. The exposed ground is then allowed to thaw for a 10- to 15-cm (4- to 6-in) depth before it is scraped away so that the frozen ground is always exposed to thawing. A 7.5-m (25-ft) depth may be excavated by using this method in a 100-day summer operation in areas where the natural active layer is only 0.3 to 0.6 m (1 to 2 ft) (Linell and Lobacz, 1980). This method is widely applied in borrow operations. Flooding large areas has been used in a similar manner to accelerate thawing.

Sufficient knowledge of the subsurface conditions is required in order to evaluate the feasibility of the excavation method. The nature of the work and the schedule are other important factors in selecting the excavation method. In small operations simple compressed-air jackhammers may be acceptable. On the other hand a combination of several methods such as blasting and ripping is often effective.

7.2.3   Earth Handling and Placement

The construction of fills, dams, and embankments in winter conditions has certain restricting features. The earth should be placed on a stable subgrade such as suitable thawed ground, frozen but not frost-susceptible ground, or perennially frozen ground. Compaction of frozen soils to specified densities is usually difficult. Only granular moisture-free frozen soils have been placed and compacted to specified densities successfully in winter conditions. Frozen moisture-bearing soils have also been placed with acceptable results, but recompaction after thaw consolidation is usually required. Finally, special arrangements may be necessary if thawed soil is delivered and placed in winter conditions.

The preparation of borrow pits and stockpiles is important for smooth delivery of suitable material for soil placement. It is generally advantageous to operate borrow pits continuously during the winter to avoid the

TABLE 7.4 Comparison of Methods of Excavating Frozen Ground

Method of disengagement	Details of disengagement mechanism	Material type	Temperature, °C (°F)	Disengagement effectiveness, 10^{-9} m³/J [10^{-6} ft³/(ft·lb)]	$\dfrac{\text{Energy}}{\text{Volume}}$, kPa (lb/in²)	Rate of removal, m³/min (ft³/min)	m/min (ft/min)	Remarks
D-8 bulldozer, no. 8 ripper	Single-tooth ripper digs a furrow 0.225 m² (2.42 ft²) in cross-sectional area.	Frozen clean beach gravel, GW or GP; gravel < 5 cm (2 in), with some sand and clear ice; moisture content slightly above saturation.	−1 (30)	1300 (62.2), short term	790 (114)	13.7 (485)	—	Experimental run: one furrow 40 m (130 ft) long by 0.23 m² cross-sectional area. Effectiveness based on rated power of bulldozer—does not represent cutting power alone. Large cuttings observed, probably accounting for high effectiveness.
D-9 bulldozer, no. 9 ripper		Silty gravel, 2080 kg/m³ (130 lb/ft³); moisture content 6%; 18% silt.	−2 to −1 (28 to 30)	283 (13.6), short term	3450 (500)	—	—	Experimental run: boulders 0.6 m (2 ft) maximum were observed in the ripper cuts. Effectiveness based on rated power of bulldozer.
D-8 bulldozer, two-toothed Easco ripper		Frozen clay.	NR	119 (5.7), day long	8480 (1230)	1.3 (45)	—	Effectiveness based on rated power of bulldozer.

426

Machine	Specifications	Soil						Remarks
Disksaw	Saw diameters varied from 0.8 to 2.5 m (2.6 to 8.2 ft), with peripheral speed up to 15 m/s (49 ft/s).	Very dense clay.	−14 (7)	62 (3.0), highest	—	—	—	Granite chips caused breakage of saw. Stability of large disk saw was low. Cutting with a dull edge has been investigated.
Chain saw	2.5-m (8.2-ft)-deep trench can be made in one pass.	—	—	—	—	—	—	High-wear, expensive materials necessary for parts.
Rotary excavator (FP-7A, FP-2M, FP-4) with special teeth	Trench cut 1.7 m (5.6 ft) deep by 1.2 m (4 ft) wide.	2-m (6.6-ft)-deep frozen layer.			—	0.24 (8.6)	0.12 (0.4) 0.18 (0.6)	60 to 80 m (200 to 260 ft) per working day was excavated. Rotary excavator found to be more efficient than chain saw.
Vibrating hammer attachment to tractor	Weight of falling hammer 1000 kg (2200 lb); vibrating frequency 420 to 725 cycles/min; maximum energy of impact 1500 J (1100 ft·lb); specific impact energy 30 J/cm (55 ft·lb/in) of cutting edge.	Clay soil.	−4 (25)	354 (17.0), average	2760 (400)	0.48 (16.8)	—	Based on rated power of electric motor [22 kW(30 hp)]. Cost of loosening soil with this machine is approximately half that of drilling and blasting.

TABLE 7.4 (*Continued*)

Method of disengagement	Details of disengagement mechanism	Material type	Temperature, °C (°F)	Disengagement effectiveness, 10^{-9} m³/J [10^{-6} ft³/(ft·lb)]	$\dfrac{\text{Energy}}{\text{Volume}}$, kPa (lb/in²)	Rate of removal, m³/min (ft³/min)	Rate of removal, m/min (ft/min)	Remarks
Vibrating hammer on excavator bucket	—	—	—	—	—	1.4 (50.5)	—	
Combination disk cutter and shovel excavator	Narrow slits 60 to 80 cm (2 to 2.6 ft) apart and 0.8 m (2.6 ft) deep are cut with a 2.5-m (8.2-ft)-diameter disk rotating at a circumferential velocity of 3.5 to 20 m/s (11 to 60 ft/s). Frozen ground excavated by a "straight shovel" excavation.	—	—	—	—	—	—	High power consumption for disk cutting process. High cutting speed causes rapid wear of cutting edge.
Wedge-shaped hammer	50-cm (20-in)-wide wedge-shaped hammer splits 30- to 40-cm (12- to 16-in)-wide pieces of frozen soil.	—	—	—	—	0.25 (9.0)	—	Performance data not available.

Cutting of total volume of excavated ground	85-cm (33-in) width of cutter containing 9 cutters 3 cm (1.2 in) wide; cutters staggered so that 3 cutters engaged at one time; followed by 7-cm (2.8-in)-wide cutter wedge, which breaks up large particles; cutting speed 0.6 m/s (2 ft/s).	—	—	52 (2.5) 56 (2.7)	17,200 (2500) 20,700 (3000)	—	5–10 (16–33)	Force required to cut 1-m (3.3-ft)-deep trench, 8.5 cm (3.3 in) wide at an advancing speed of 10 m/h (33 ft/h) is 15 kN (3300 lb).
Rotary excavator	Excavator cuts a trench 1.1 m (3.6 ft) wide by 1.7 m (5.6 ft) deep; cutting speed 1 m/s (3.3 ft/s).	Frost line 50 cm (1.6 ft) deep.	—	—	—	2.5 (89.0)	1.2 (4.0)	Cutting speed of 1 m/s (3.3 ft/s) was too high, resulting in excessive wear of cutting surface.
		Frost line 120 cm (4 ft) deep.				1.7 (60.0)	0.82 (2.7)	
	Excavator cuts trench 1.1 m (3.6 ft) wide by 1.7 m (5.6 ft) deep; cutting speed 0.5 m/s (1.6 ft/s).	Frost line 100 cm (3.3 ft) deep.	—	390 (18.8)	2550 (370)	1.7 (60.0)	0.82 (2.7)	Modified bucket teeth design. Cutter speed reduced to 0.5 m/s (1.6 ft/s).

TABLE 7.4 (*Continued*)

Method of disengagement	Details of disengagement mechanism	Material type	Temperature, °C (°F)	Disengagement effectiveness, 10^{-9} m³/J [10^{-6} ft³/(ft·lb)]	$\dfrac{\text{Energy}}{\text{Volume}}$, kPa (lb/in²)	Rate of removal, m³/min (ft³/min)	Rate of removal, m/min (ft/min)	Remarks
Wedge impact	Wedge width 18 cm (7.1 in).	Frost line 1 to 1.2 m (3.3 to 4 ft) deep.	—	260 (12.5)	4000 (580)	0.21 (7.4)	—	Based on an average value of 100 m³ (3530 ft³) per day of 1.0- to 1.2-m (3.3- to 4.0-ft)-deep frozen ground requiring 10 impacts/min of 80 kJ (59,000 lb·ft) impact energy.
Tractor-pulled ripper	Tractor weight 400 kN (90,000 lb); developed 280-kN (63,000-lb) pull at ripper.	Sandy clay and cobbles; ground frozen 1.5 m (5 ft) deep.	−21 (−6)	83 (4.0), day long	12,000 (1740)	1.1 (40.0)	—	430 m² (4600 ft²) of area excavated. Area cut up into a 1.2-m (4-ft) square grid of kerfs cut by the ripper penetrating to a depth of 1.2 to 1.5 m (4 to 5 ft). Cutting pattern breaks the entire mass which is then removed by normal shovel or bucket excavators.

Trench excavator	Scoops on excavator chain replaced by alternate cutters and chopping wedges; trench cutters consist of blades reinforced with a hard alloy, attached to rotary supports; wedges for chopping clods also attached to rotary supports; cuts trenches 1.8 m (5.9 ft) deep by 0.85 m (2.8 ft) wide.	Ground frozen to a depth of 1.3 m (4.3 ft).	—	52 (2.5)	17,000 (2500)	0.13 (4.6)	0.08 (0.25)	Trench excavation made by cutting two narrow trenches and chopping frozen ground in between. Method appears to be successful.
Explosives	Ammonium nitrate in diesel fuel.	Either frozen sand or gravel.	—	1000 (48.3)	—	—	—	Explosive energy only, does not include shot hole drilling.
Explosives	Experimental craters with single explosive charge.	Frozen till, Ft. Churchill, 2370 kg/m³ (148 lb/ft³).	—	81 (3.9)	—	—	—	—

SOURCE: Foster-Miller Associates, Inc. (1973), from various sources.

thickening of the frozen layer that is to be removed. Good drainage of the borrow area is necessary so that the moisture content of the soil can be kept low.

In permafrost areas borrow material for winter use can be excavated during the summer. After possible draining and screening or crushing, it is stockpiled. The penetration of frost can be minimized by mixing salt to the surface layer or by providing thermal insulation (Tsytovich, 1975). Even if the material is used in the frozen state, it may be necessary to excavate and stockpile during the summer so that the material is properly drained.

The tendency of frozen lumps to freeze onto the equipment and to each other poses some handling problems (Fig. 7.9). Wet thawed soils also tend to freeze onto cold surfaces. Preventing the thawed soil from freezing during handling and placement is an important consideration. Mixing salt with the soil eliminates the freezing and adfreezing problems to some degree.

Placement and compaction of unfrozen snow and ice-free soil in freezing conditions have proved to be generally feasible down to about $-10°C$ ($14°F$), provided that it will not freeze during compaction. The compaction effort, however, increases moderately as the soil temperature decreases.

Placement and compaction should occur rapidly with small lifts in restricted areas and should be interrupted during snowfall. Snow and

Figure 7.9   Earth handling in freezing conditions. *(Courtesy of F. E. Crory.)*

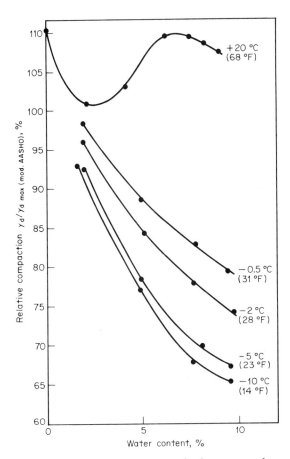

Figure 7.10 Compaction curves for frozen granular material (silty sandy moraine). Vibrating tamper, two layers, 2 min per layer. *(Adapted from Heiner, 1972.)*

frozen soil that have formed during interruptions should be removed before additional fill is placed. Salts such as calcium chloride have been used to decrease the freezing point and to improve the compactibility of soils (Alkire et al., 1975).

The compaction of frozen soil is usually very difficult because the voids are filled with ice that cements the soil into lumps and resists compaction (Fig. 7.10). Excessive settlements and loss of stability may result as the improperly compacted frozen soils thaw. Only low-moisture-content (less than 2 to 5%) granular soils and crushed rock can be compacted effectively to specified densities in their frozen state (Botz and Haas, 1980), as shown in Fig. 7.11. However, acceptable results can often be

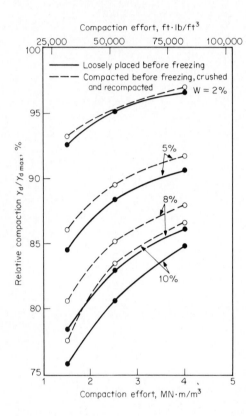

Compaction effort, ft·lb/ft³

Relative compaction $\gamma_d/\gamma_{d\,max}$, %

Compaction effort, MN·m/m³

—— Loosely placed before freezing
---- Compacted before freezing, crushed and recompacted

W = 2%

5%

8%

10%

Figure 7.11 Effect of compaction energy on relative compaction of silty sandy moraine at −5°C (23°F). *(Adapted from Heiner, 1972.)*

obtained when the recompaction and placement of final lifts are pursued after the fill has completely thawed.

Placement and compaction of frozen soils should not be done during a snowfall. Large chunks of frozen soil, snow, and ice should be removed before the material is placed and compacted. Drainage during thawing may be improved with a gravel filter between fill and subgrade.

The construction of embankments, gravel pads, and even dams in permafrost areas has proven to be often quite feasible during the winter. Construction of embankments may occur during the summer by end dumping on the preserved organic mat, but the construction period is short and gravel roads are usually needed to provide access to the borrow pits. During the winter, on the other hand, the tundra is frozen and access to the borrow pits can be rapidly provided by ice and snow roads. However, materials are seldom sufficiently dry and clean for winter placement. Thawing, drying in stockpiles, and screening may be required to obtain the predetermined quality. The right-of-way is carefully cleared without removing the organic mat that even in the compressed state provides thermal protection. Snow is removed and soil layers are dumped and compacted.

If proper granular material is in short supply, thermal insulation can be placed in the fill to cut the penetration of thaw during the summer (Fig. 7.12). The poor-quality subgrade will stay frozen and the thickness of the gravel pad can be cut from a typical value of 2 m (7 ft) to less than 1 m (4 ft). Another method to save good-quality granular material is to use marginal soils in the lower parts of the fill together with filter fabrics that provide sufficient strength for the fill (Tart and Luscher, 1981).

Artificial island construction by end dumping frozen gravel from ice represents a new problem area. Cold gravel may actually gather more ice as it gets in contact with water at the freezing point. On the other hand, the ice in the fill may remain frozen in the cold arctic seawater. As a matter of fact the largest settlements occur in the normally compacted

Figure 7.12 Construction of an insulated gravel pad. *(Courtesy of F. E. Crory.)*

(a)

(b)

(c)

Figure 7.13 Foundation construction in winter conditions. (*a*) Foundation construction can be a mess, (*b*) well organized, or (*c*) done in normal fashion, within a heated enclosure. (*Courtesy of H. Mäkelä.*)

above-water portions of the island during the summer thaw, whereas the partially frozen underwater parts consolidate at a much slower rate (Tart, 1983). The long-term behavior of winter-constructed gravel islands in arctic seas as well as the behavior of winter-constructed gravel islands in more temperate waters still needs further addressing.

7.2.4   Foundation Construction and Seasonal Frost

In seasonal frost areas foundations are not generally made on frozen soil because it cannot be compacted and thaw settlements will take place. Frost also penetrates very quickly into thawed soil, for example, at the bottom of an excavation during the foundation construction stage. Frost protection is thus an important consideration (Fig. 7.13).

Excavation should not be extended to its final level too early. A thawed granular fill placed and compacted at the bottom of the excavation will provide some thermal protection for the subgrade, and thermal insulation materials can be used for additional protection. An alternative is to thaw and compact the frozen soil layer. If cast-in-place foundations are used, the thermal protection and heating of concrete can also keep the soil thawed until the backfill is placed. Ice and water problems may in some cases prove to be quite a nuisance when proper drainage has not been arranged (Fig. 7.14).

Figure 7.14   Ice and water may become quite troublesome for foundation construction in improperly drained sites. *(Courtesy of H. Mäkelä.)*

Figure 7.15 Frozen ground provides good support for heavy piling machinery. *(Courtesy of Finn-Stroi Ltd.)*

It is very important to continue foundation construction in winter conditions without interruption until backfill has been placed. This is the most effective way to control frost problems and to avoid frost damage. A non-frost-susceptible backfill should generally be used to avoid movement of unloaded foundations due to frost action. If the foundations are at shallow depth, additional frost protection may be required during the construction stage to prevent freezing of soil under the foundations (Table 7.2).

The construction of large concrete slabs placed on the ground may be especially laborious in winter conditions. In addition to the problems of pouring concrete, the frozen soil layer has to be replaced with a non-frost-susceptible fill or at least must be thawed and compacted before the granular base material is placed and compacted. After slab construction has been finished one should still make sure that frost does not reach the frost-susceptible subgrade. Many of these slabs, however, can be constructed after the structure has been enclosed and heated.

Piling in winter conditions does not pose any major problems. The frozen ground provides good support for the heavy machinery (Fig. 7.15). Sometimes special arrangements are needed to penetrate through a thick frost layer. The frost heave and adfreezing forces should be considered, especially in case of short temporarily unloaded piles. Several methods are available to eliminate the possibility of uplift, including thermal protection of the ground, use of non-frost-susceptible materials around the structure in the active layer, and isolation of the structures from the uplift with casings or plastic membranes.

7.2.5 Foundation Construction in Permafrost Areas

In permafrost areas construction problems connected with the stability of soil occur mostly during the thawing season. If the foundations are placed on a gravel pad constructed during the summer, the eventual freezing of thawed subgrade may cause excessive heaving. In case the foundations are to be buried in the permafrost, the excavations will experience rapid thawing during the summer, even if a gravel fill is placed quickly at the bottom, precast foundations are used, and the non-frost-susceptible backfill is placed without delay. Water entering the excavation through the active layer may often further complicate the foundation construction. Thawing of ice-rich soils may lead to unacceptable foundation movements (Fig. 7.16).

These problems can be avoided during the winter, although excavation and compaction may require additional effort. If cast-in-place concrete is used for massive foundations, the heat of hydration may cause the thawing of permafrost, and consequently thaw settlement and frost heave will occur. Gravel pads and insulating materials of adequate strength can be used for the thermal protection of the subgrade.

The bearing capacity of piles in permafrost is normally based on the adfreeze bond. The installation methods include driving, dry augering

Figure 7.16 Ice-rich permafrost has thawed under a footing, and the subsequent settlement has caused a structural failure.

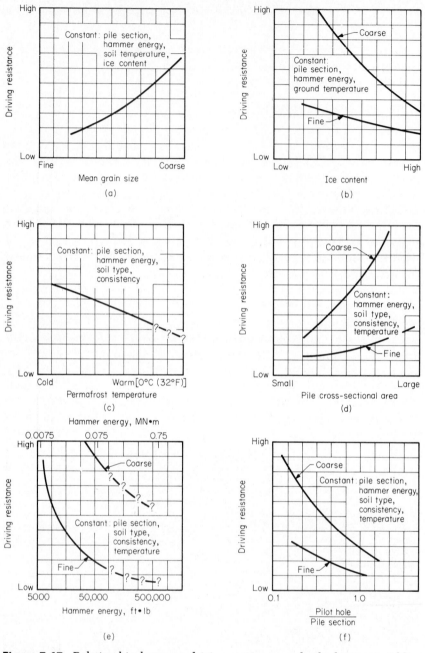

Figure 7.17 Relationship between driving resistance and pile-driving variables. *(Davison et al., 1978.)*

and slurring, drilling, and thawing and driving, as well as combinations of these methods. Selection of the most feasible pile installation method and estimation of the freeze-back time are among the important construction considerations (Crory, 1980).

Open-ended steel pipes and H piles have been driven successfully into warm fine-textured permafrost with conventional or vibratory hammers. Sometimes an undersized pivot hole or a tight hole has been drilled to help guide and ease the penetration. Some factors affecting the driving resistance are illustrated in Fig. 7.17. Because the heat input into the ground is minimal, refreezing occurs rapidly.

Earth augers are commonly used to drill oversized holes for all kinds of piles. The augers designed for frozen soils may penetrate into fine-textured permafrost at rates up to about 0.5 m/min (2 ft/min). However, they are not usually effective in very coarse soils where rotary or churn drills may be required. After the pile has been installed, a backfill slurry is placed and vibrated to the annulus, typically 5 to 10 cm (2 to 4 in), between the pile and the wall of the hole (Fig. 7.18). The gradation and the moisture content of the slurry are usually controlled in order to avoid ice segregation and to achieve good adfreeze strength. Saturated silty sands, sands, and gravelly sands are commonly used. The soil and water are generally mixed in portable concrete mixers and the slurry should have the consistency of a 15-cm (6-in) slump concrete. Piles can also be driven into prefilled holes, but especially in this case wooden and closed-end pipe piles may have a tendency to float (Linell and Lobacz, 1980).

Figure 7.18 Construction of piled foundations on permafrost. *(Courtesy of F. E. Crory.)*

The temperature of the backfill slurry should be only slightly above 0°C (32°F) when placed. Anyway, the freeze back takes time and the temperature of the permafrost is temporarily increased because of the latent heat of fusion of the slurry (Fig. 7.19). This should be considered if the piles are loaded in the early stage. The freeze-back time can be estimated based on Fig. 7.20. During the summer, surface water may enter the holes and further increase the refreezing time. Artificial freezing provides rapid and reliable refreezing, especially in warm permafrost areas or when piles are tightly placed. If thermopiles are not used, this can be accomplished by circulating refrigerant through tubing attached to the pile.

In the U.S.S.R. concrete is commonly used to backfill the annulus between the pile and the hole (Fig. 7.21). As a matter of fact, the augered holes can also be completely filled with fresh concrete. Careful control of the soil temperatures and the strength development of concrete is re-

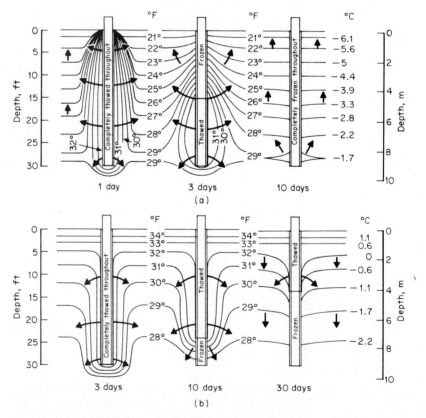

Figure 7.19 Natural freeze back of piles in permafrost during (a) late winter and (b) late summer. *(Crory, 1963.)*

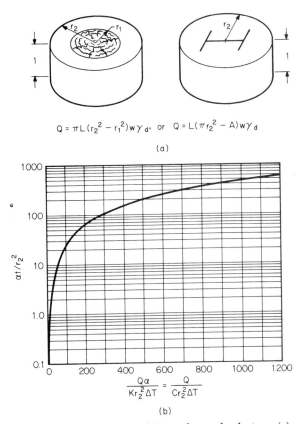

$$Q = \pi L (r_2^2 - r_1^2) w \gamma_d \quad \text{or} \quad Q = L(\pi r_2^2 - A) w \gamma_d$$

(a)

$$\frac{Q\alpha}{K r_2^2 \Delta T} = \frac{Q}{C r_2^2 \Delta T}$$

(b)

Figure 7.20 Estimation of slurry freeze-back time. (a) Latent heat of slurry backfill. (b) General solution for freeze back. Q—latent heat of slurry per pile length; L—latent heat of water; r_1—pile radius; r_2—hole radius; w—water content of dry weight; γ_d—dry density of slurry; A—cross-sectional area of H pile; t—freeze-back time; K—conductivity of permafrost; C—volumetric heat capacity of permafrost; α—diffusivity of permafrost, $= K/C$; ΔT—initial temperature of permafrost before freezing. (*Crory, 1963.*)

quired. Furthermore, concrete piles should be reinforced to resist the possible uplift caused by frost heave in the active layer.

The U.S.S.R. has a history of using different kinds of controlled thawing methods to improve the efficiency of pile driving into permafrost. Recently a method of controlled perimeter thawing and pile driving into thawed ground has gained popularity also in Alaska (Manikian, 1983; Nottingham and Cristopherson, 1983). A 15- to 20-cm (6- to 8-in) pilot hole is drilled into permafrost and filled with hot water. After the soil

(a)

(b)

Figure 7.21 Foundation construction in Norilsk, U.S.S.R. (a) General view of site. (b) Augering machine. (c) Wire machine using mechanical energy and hot water to penetrate into permafrost. (d) Backfilling annulus with cement grout.

(c)

(d)

445

around the hole has thawed, piles can be easily driven by conventional impact or vibratory hammers. The bearing capacity of these "naturally" slurried piles does not appear to differ significantly from that of ordinary slurried piles, but improved piling efficiency and reduced freeze-back times have been reported.

7.3   Concrete Construction in Winter Conditions

Special measures in concrete construction become necessary when the daily mean temperature drops below about $+5\,°C$ ($41\,°F$). That is when the strength development of concrete begins to drop considerably and frost may damage fresh concrete. Concrete construction is feasible at temperatures down to about $-20\,°C$ ($-4\,°F$) with proper protective measures and at much lower temperatures within enclosures, but careful planning is required.

Good statistics of the local climate give a sound basis for preparations. When the construction methods and the concrete strength requirements have been decided, a plan can be made for concrete construction in the winter. Such a plan should cover at least the following:

1. The quality requirements, delivery, protection, heating, and processing of concrete ingredients

2. The impact of air entrainment, strength development requirements, and the extreme concrete temperatures on the mixture design

3. Preparations such as removal of snow and ice from the forms, heating of contact surfaces and frozen ground, or protection of permafrost subgrade

4. Control of concrete temperatures during the pouring process and the necessary protection and heating measures

5. Requirements for curing of concrete, stripping of forms, and removing of construction supports

6. Recording system for the weather conditions and concrete temperatures and control of concrete strength development

The plan should also include considerations for unexpected disturbances, such as extreme weather conditions, power failures, or equipment failures.

7.3.1   Manufacturing

Concrete manufacturing begins with proper storage and handling of the materials (Fig. 7.22). Concrete should be fresh and stored in a slightly heated space so that the relative humidity is low and the risk of moistening can be avoided. Aggregates should be thawed before screening and usually heated before mixing (see, for example, Sadovsky et al., 1978). Hot water, steam, air, and electric resistance heating have been used. Control of the temperature and moisture distribution is important, and this is one reason why open steam jets are troublesome compared to steam confined to a pipe system. Finally, an adequate supply of hot water is necessary.

The temperature of the concrete mixture should generally be increased at low ambient temperatures. As a rule of thumb, the temperature of the mixture should be increased by at least 1° from $+15\,°C\,(59\,°F)$ for each degree of freezing temperature, but not above $+40\,°C\,(104\,°F)$. Governing factors include heat losses during delivery and pouring, massiveness of structures, means of protection and heating, and the desirable strength development.

In moderately cold weather the required mixture temperature can be obtained by heating only the water. When the temperature drops significantly below $0\,°C\,(32\,°F)$, it is necessary to heat also the aggregate or at least the finer part of it. The temperature of the mixture can be calculated simply from

$$T = \frac{0.22(T_s W_s + T_a W_a + T_c W_c) + T_w W_w + T_s W_{ws} + T_w W_{wa}}{0.22(W_s + W_a + W_c) + W_w + W_{wa} + W_{ws}} \quad (7.1)$$

where T = final temperature of concrete mixture
T_c, T_s, T_a, T_w = temperatures of cement, fine aggregate, coarse aggregate, and water
W_c, W_s, W_a = weights of cement, fine aggregate, and coarse aggregate, excluding their moisture contents
W_w, W_{ws}, W_{wa} = weights of water and of free water on fine and coarse aggregates

If the temperature of the aggregate or part of it is below $0\,°C\,(32\,°F)$ and the aggregate contains moisture, it is necessary to take into account the heat required to raise the temperature of ice to $0\,°C\,(32\,°F)$ and above so that the ice is thawed. The specific heat of ice is about half and the heat of fusion 80 times that of water. However, lumps of ice, snow, or aggregate cannot be tolerated because they may survive mixing and weaken the concrete.

If hot water contacts directly with cement, flash setting may occur. That is why it is recommended to mix part of the water with a temperature above $+60$ to $+70\,°C\,(140$ to $160\,°F)$ first with coarse aggregate. Then finer aggregate, the rest of the water, and cement can be added. The concrete mixing time is increased in winter conditions, and careful control of material temperatures, moisture contents of aggregates, and the temperature of the mixture is required.

Figure 7.22 Aggregate proceeding from borrow pit through crushing and screening to batching plant. *(Courtesy of Finn-Stroi Ltd.)*

7.3.2 Delivery and Pouring

The temperature of the concrete mixture should account for the temperature losses that occur during transportation as well as during handling at the site and in place before the fresh concrete can be protected. The temperature drop for a delivery time of 1 h can be estimated according to Peterson (1966),

$$
\begin{array}{lll}
\text{Revolving drum mixers} & T = 0.25(t_r - t_a) \\
\text{Covered dump body} & T = 0.10(t_r - t_a) & (7.2) \\
\text{Open dump body} & T = 0.20(t_r - t_a)
\end{array}
$$

Figure 7.22 *(Continued)*

where T = temperature drop
t_r = required concrete temperature at site
t_a = ambient air temperature

The cooling of concrete during handling can be approximated from Fig. 7.23 when buckets, barrows, and buggies are used for conveying. Finally, it is noted that concrete cools down very quickly in place without protection, as shown in Fig. 7.24.

The temperature distribution in the concrete body is uneven at all stages. Factors such as wind speed, concrete volume, equipment characteristics, and protective measures are of importance. Freezing of any part of concrete should not be allowed. Thermal insulation is a very

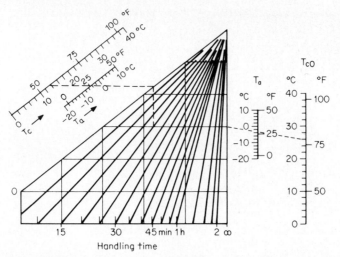

Figure 7.23 Nomogram to estimate the cooling of concrete during site handling. T_{c0} — temperature of concrete at receiving station; T_a — air temperature; T_c — temperature of concrete placed in forms. *(Pihlajavaara and Syrjälä, 1960.)*

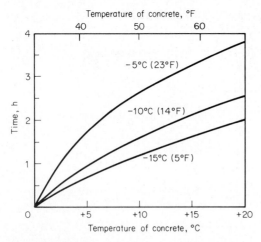

Figure 7.24 Typical time required to freeze upper surface of unprotected concrete slab when initial temperature of concrete is 0 to $+20\,°\text{C}$ (32 to 68 °F) and ambient temperature is -5 to $-15\,°\text{C}$ (23 to 5 °F). *(Nykänen and Ahtola, 1967.)*

(a)

(b)

(c)

Figure 7.25 Stages of winter concreting. (*a*) Snow and ice must be cleared from forms and reinforcement. (*b*) Delivery and pouring. (*c*) Thermal protection of slab. (*Courtesy of E. Kilpi.*)

convenient way to reduce thermal losses. If belt conveyers or pumping is used to transfer concrete, it is common practice to use heated enclosures or insulation along the line. Also concrete tanks and buckets can be insulated to cut the thermal losses during transportation and handling.

The cooling of concrete during long-distance transportation in cold weather can be kept at a low rate, that is, less than 5 to 6°C/h (10°F/h) if proper insulation is used. The stiffening of warm concrete and segregation during long-distance transportation may remain limiting factors. Transportation times up to 2 h can be accepted when mixing trucks are used. Adjustment should also be made for the reduction of the air content of concrete during transportation, typically 0.5 to 1.0% of the volume of concrete per hour. A reduction of air content is also experienced during pumping of concrete.

Preparations for winter pouring begin at the site by ensuring that all surfaces to be in contact with newly placed concrete will not cause freezing or seriously prolong hardening. Heating is applied to cold contact surfaces. Seasonally frozen subgrade is thawed and compacted. In permafrost areas thawed gravel can be placed to provide insulation between the concrete and frozen ground.

All snow and ice should be removed from forms and steel bars, for example with hot air or steam. If the forms and bars are stored under cover so that direct contact with the ground is avoided, many handling and thawing problems can be avoided. The pouring is carried out in a manner similar to that of more temperate climates. Protection from high winds and heavy snowfall may be necessary. Interruptions should be avoided. Quick protection of placed concrete is important (Fig. 7.25).

7.3.3 Protecting, Heating, and Curing

The strength development of newly placed concrete depends considerably on the temperature (Fig. 5.21). Freezing strength of concrete should be achieved to secure the quality. Rapid strength development up to the point when forms can be removed is desirable from the construction point of view. The average concrete temperature T_t at instant t during the hydration process can be computed from

$$T_t = T_c + \frac{Q_c}{c_c \gamma_c} W_c + \sum_0^t \frac{P}{V c_c \gamma_c} \Delta t - \int_0^t \frac{A(T - T_o)}{R V c_c \gamma_c} dt \qquad (7.3)$$

where T_c = initial temperature of concrete
T = temperature of concrete
Q_c = amount of cement in concrete
W_c = heat of hydration occurring before time t
c_c = specific heat of concrete, ≈ 1.05 kJ/(kg·°C) [0.25 Btu/(lb·°F)]

γ_c = density of concrete
P = net efficiency of heating
V = volume of concrete
A = area of surface subjected to thermal losses
R = thermal resistance
T_o = outside temperature

Manufacturers generally provide information concerning the heat of hydration for different types of cement. A typical practical formula to describe the heat development for ordinary portland cement is

$$W_c = 8.4 + 379e^{-120/(24N)^{0.8}} \quad \text{[kJ/kg of cement]} \quad (7.4)$$

where N is the maturity factor given by Eq. (5.6).

The heat of hydration together with proper insulation may provide adequate strength gain. Typical cases include pouring on the ground under marginal weather conditions, massive structures, and structures with moderate requirements for strength development during the construction stage. Some design charts for insulation requirements in this case have been provided by the American Concrete Institute (1978). Commercial insulating blankets or bat insulations are commonly used for thermal protection. Forms built for repeated use are often provided with insulation. Upper surfaces of slabs may be protected by large steel-framed insulation boxes with plywood against fresh concrete so that a good quality of the concrete surface can be maintained.

In most cases a rapid circulation of forms or even adequate strength gain cannot be guaranteed in winter pouring without heating. Several methods are available, including space heaters, embedded electric resistances, electric heating of forms, and radiators. Some examples of protection and heating arrangements are shown in Fig. 7.26.

If high temperatures are used, the concrete should be allowed to set a couple of hours before heating starts in order to prevent strength losses. Heating is generally applied only during the first 1 to 5 days of rapid strength gain.

Enclosures heated with hot air or steam are commonly used for thermal production of concrete slabs and foundations and also in many special applications. Enclosures range from overlapping canvas hoods and wood-framed polyethylene sheetings to self-supporting inflatable shelters and weather halls. Sufficient space should be provided between the concrete and the enclosure to permit free air circulation. Steam heating provides ideal curing conditions for concrete, but the working conditions within the enclosures are naturally poor. On the other hand, air heaters that produce carbon dioxide should be vented to remove the products of combustion to the outside, because carbon dioxide reacts with calcium hydroxide in fresh concrete, causing surface damage. Ther-

(a)

(b)

Figure 7.26 Protection and heating arrangements in winter concretizing. (*a*) Heating foundations with steam. (*Courtesy of Finn-Stroi Ltd.*) (*b*) Heating with electric resistances in Norilsk, U.S.S.R. (*c*) Slipforming within heated enclosure. (*Courtesy of Interbetoni Oy.*) (*d*) Application of infrared heating. If tarpaulins were longer, air circulation could be prevented and heating efficiency improved. (*Courtesy of Finn-Stroi Ltd.*)

(c)

(d)

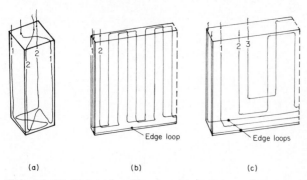

(a)   (b)   (c)

Figure 7.27 Principle of heating (*a*) a column, (*b*) a thin wall, and (*c*) a thick wall with embedded resistances.

mal leaks and losses are quite large when space heating methods are used, and therefore these methods are not generally considered very economical (Bennett, 1976).

Embedded, coiled insulated electric resistances provide a convenient method to heat internally small separate objects such as footings, columns, and beams. Electric heating can also be applied together with other heating methods in areas where heat losses are large. The principle is illustrated in Fig. 7.27. Low-voltage current, typically 9 to 42 V, is passed through the coils, usually plastic-coated steel wires, embedded near the surface of the concrete. The spacing of the coils is on the order of 20 to 30 cm (8 to 12 in) and the power requirement is approximately 40 to 100 W/m [40 to 100 Btu/ft·h)]. Large spacings and powers should be avoided because the heat distribution within the concrete becomes quite uneven (Fig. 7.28).

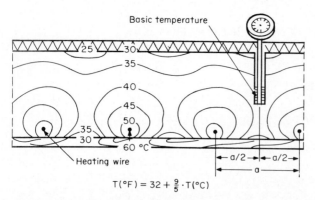

$$T(°F) = 32 + \frac{9}{5} \cdot T(°C)$$

Figure 7.28 Typical heat distribution in slab heated by electric resistances. (*Hyvärinen, 1980.*)

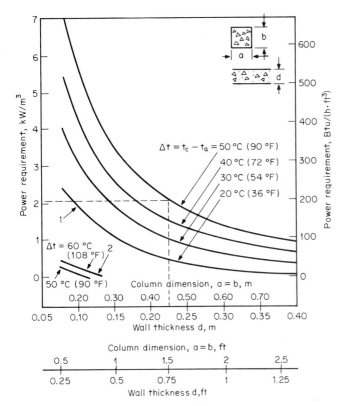

Figure 7.29 Typical design chart for power requirements at early stages of concrete hardening for concrete with about 300 kg/m³ (19 lb/ft³) of ordinary portland cement. 1 — for bare woodform; temperature difference greater than 50°C (90°F) between ambient temperature and basic concrete temperature is not considered feasible; 2 — for insulated form with thermal conductance 0.7 W/(m²·°C) [0.12 Btu/(ft²·h·°F)]. (*Hyvärinen, 1980.*)

The design of the heating requirements can be made according to the principles given by Eq. (7.3). One example of a design chart is given in Fig. 7.29. Concrete temperatures should be monitored during the heating, and at least two circuits should be provided.

Electric resistances can also be used to heat concrete through the surface of insulated forms. The principle is shown in Fig. 7.30. Typical power requirements range up to 300 W/m² [100 Btu/(ft²·h)] for concrete walls heated from both sides and double the amount for concrete slabs, depending on the ambient temperatures, the strength development requirements, and the quality of thermal protection. Additional

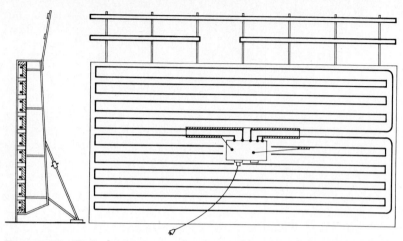

Figure 7.30 Typical arrangement for electrical heating of an insulated form.

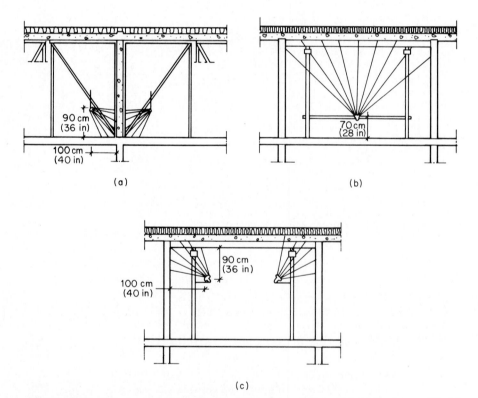

Figure 7.31 Recommended aiming directions for radiators. (*a*) Angle forms. (*b*), (*c*) Forms for slab. (*Penttala and Törmi, 1977.*)

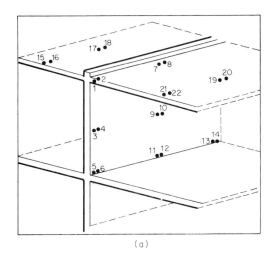

(a)

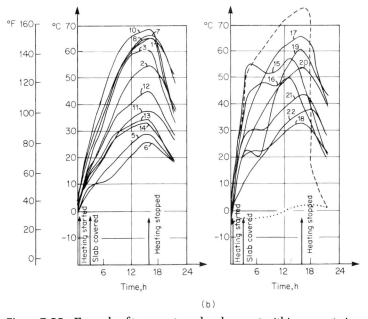

(b)

Figure 7.32 Example of temperature development within concrete in case *a* of Fig. 7.31. Thermoelements are located in slabs and wall. Odd numbers correspond to elements close to form surfaces, even numbers to those in center of structure. — — — temperature in heated space; · · · outside temperature. (*Lehtinen et al., 1979.*)

heating is generally required near the edges of the form. Electric blankets provide another possibility for external heating of concrete.

In infrared heating the heat rays sent by a radiator are absorbed by the object. This heating method is best applicable in heating wide surfaces and offers an alternative to electric heating of forms. The radiators operated with oil, gas, or electricity are generally used with bare steel forms because they have good thermal conductivity and pose no fire hazard, but other types of forms have also been used. Radiation can be applied within enclosures to newly placed concrete through a plastic film that prevents moisture loss.

Some typical examples of infrared heating in residential building construction are shown in Fig. 7.31. The radiators are aimed at the area where concrete is in contact with the cold parts of the structure. Other parts of the form and concrete are heated by conduction and by warming air in the enclosed space. The temperature distribution becomes quite uniform, as shown in Fig. 7.32. In this case, though, overefficient heating may have caused some loss in final concrete strength. A typical power requirement is 1 kW/m^2 [$300 \text{ Btu/(ft}^2 \cdot \text{h})$] of the total surface to be heated. Design charts for infrared heating are provided in Lehtinen et al. (1979).

Although the concrete strength requirement for form removal is typically 60% of the design strength, form circulation periods as short as 1 day have proved to be quite feasible, even in severe winter conditions. This is possible with rapid cement, a concrete mixture that provides a nominal margin of 5 to 10 MPa (700 to 1500 lb/in²) to the design strength requirement, and with effective heating and insulation measures. The effectiveness of winter concreting is in this case quite comparable to that of summer concreting.

7.3.4 Quality Control

Preheating of cold surfaces that will be in contact with concrete should first take place. Newly placed concrete should be protected as quickly as possible in order to prevent it from freezing in the saturated stage.

When the temperature of concrete is much higher than that of the air or when curing of concrete occurs within a heated enclosure with low relative humidity, excessive drying of concrete may occur, especially near the surface. As a result, hydration will not be adequate and strength loss will be experienced. Rapid drying may also cause cracking. A moisture barrier is often necessary. Unformed surfaces such as floors and slabs are especially prone to excessive drying. Thermal insulation may be covered with watertight tarpaulins. Polyethylene sheets may be used within enclosures to prevent moisture from escaping from the concrete. Generally the forms provide adequate curing conditions as long as they are in place. If heating is continued after an early removal of forms, some kind of protective measure may be required, for example, application of a curing compound, an impervious cover, or steam.

The temperatures of newly placed concrete should be monitored closely, especially at the critical locations of the structure and at areas where large heat losses can be expected. Embedded temperature-measuring devices are ideal, but acceptable results for most cases can be obtained by surface thermometers protected by insulation. Temperature records form a basis for estimating the strength development of concrete.

Temperature monitoring also gives a safeguard against overheating, which may be caused by wire heating, radiation heating, or heat of hydration in massive structures. Overheating causes cracking and reduces the final strength of concrete, especially at the early stages of strength development. If high power is required, heating should start gradually or after a short period of setting time. Heating is also often stopped gradually and structures are allowed to cool slowly before being exposed to winter temperatures.

The strength development of concrete varies considerably at the early stages of the hardening process, especially before the time when half of the design strength has been reached. Test specimens cured at the same conditions as the actual structure are useful in verifying the strength development, but this is not always easy to arrange in a reliable manner. Safety supporting is often desirable if forms and shoring are removed from structural members while the latter are carrying significant portions of the design load.

7.3.5 Joining Precast Concrete Elements

Element construction in winter conditions has proved to be quite feasible (Fig. 7.33), but it has its own problems. The joints of precast concrete elements are usually filled with cement mortar. The hardened mortar will take the compressive and the shear loads of the joint at the service stage. Rapid freezing of the mortar in joints is the main problem in concrete element construction in winter conditions. For example, at $-15°C (5°F)$ the mortar begins to freeze within 15 min in a typical joint, and within half an hour the joint is totally frozen. Thermal protection and heat of hydration are of no significant help. The strength development of the mortar is usually secured by heating.

Two methods of heating are commonly used, electric wire heating and space heating. Some examples of the application of electric wire heating are shown in Fig. 7.34. If the heating wires are embedded in the elements, heating can begin before the placing of mortar. If the wires are placed in the joint, the elements have to be cleaned, for example with

Figure 7.33 Element construction in winter conditions. *(Courtesy of Finn-Stroi Ltd.)*

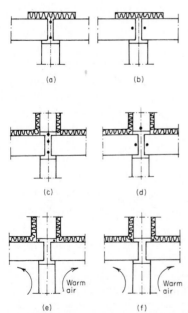

Figure 7.34 Examples of heating the joints of precast concrete elements.

steam. The filling of the joint should be done carefully with fluid mortar so that the wires will not move and no air pockets will be left. Otherwise the heating wires may burn. The heating should begin right after the fill has been placed, and the power requirement is typically 200 to 300 W/m [200 to 300 Btu/(ft·h)] of the joint. The heating should be checked regularly until the mortar has reached the required strength.

Space heaters can be used if the elements are supported so that an enclosure is created. If adequate protective measures are undertaken, the freezing risk of fresh mortar can be avoided. Space heating is quite an expensive method when only the mortar has to be heated. However, heating is also often needed for drying the structures before finishing phases can be accomplished.

7.3.6 New Developments

The need to reduce the costs of winter pouring of concrete has led to numerous innovations in material technology, the creation of new working methods and adoption of unconventional structural systems. The antifreeze compounds studied and used especially in the U.S.S.R. certainly represent a very interesting development in winter pouring, which is gaining ground also in western Europe. Because hydration may continue even at $-20\,°C\ (-4\,°F)$, heating may be avoided in numerous applications, especially in element construction.

The use of rapid cements and accelerators to speed up hydration has been a common practice for some time. Recently the increased use of hot concrete has meant additional savings in heating costs (Fig. 7.35). The concrete is heated to $+40$ to $60\,°C$ (104 to 140 °F) at the batching plant or at the reception station of the construction site, and the resulting strength loss, typically 10 to 25%, is offset by the increased use of cement. Effective thermal protection measures are naturally important after pouring. The method may be well justified if heating can be avoided or reduced because of the increased early heat generation and strength development. Its main disadvantage is reduced workability of concrete.

Allowing the fresh joint mortar to freeze is an unorthodox new approach in precast concrete element construction. The negative effects of freezing are kept under control through admixtures which, among other things, reduce the water-cement ratio. After an enclosure has been created, heating can take place, starting the hydration process. However, new mortars that include antifreeze compounds may reduce the need for such arrangements. Furthermore, element systems based on friction and bolt connections are not very sensitive to weather conditions. Thus precast space elements offer quite an attractive alternative for winter construction.

(a)

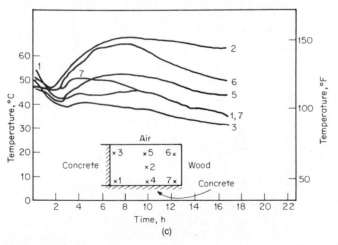

(b)

(c)

Figure 7.35 Hot concreting. Buckets (*a*) and pumping (*b*) are used for concrete delivery. (*c*) Temperature development in wall of a residential building cast in place. Hot concrete was used with insulated forms. Ambient temperature was -3 to $-5°C$ (27 to 23°F), brisk winds. *(Koivupalo, 1981.)*

7.4   Other Considerations

7.4.1   Steel Construction

As long as steel construction is limited to assembling prefabricated components with bolted connections, there are few technical problems. Working conditions and work efficiency will generally have a greater impact on the construction.

In principle it is also possible to produce reliable welded connections at very low temperatures (Fig. 7.36). The risk of hardening increases with decreasing temperatures. Wind may contribute significantly to the cooling rate and hardening of steel. However, from the material point of view, high-quality welds can be produced even in arctic circumstances with proper selection of steel grade, electrodes, preheating practice, and welding process (Cotton and Macaulay, 1976). It is the human factors that have proved to be most critical because the quality standard is more difficult to control. Field welding, therefore, is quite restricted. On the other hand, welding can be carried out inside shelters when the situation calls for it (Fig. 7.37).

7.4.2   Masonry Work

Masonry construction is quite feasible in moderately cold weather without extensive protective measures. Winter work can be based on either one of two principles:

1. The mortar should not freeze until it has gained the "freezing strength," that is, the strength where freezing does not damage the internal structure of the mortar.

Figure 7.36   Pipe welding in Siberia. *(Courtesy of F. E. Crory.)*

2. The mortar should not freeze until the bricks have absorbed sufficient moisture from the mortar so that freezing will not damage it.

The first principle is based on the assumption that the mortar is affected by cold weather in much the same way as concrete. The temperature requirement for masonry materials is at least $+5°C$ ($41°F$) during construction, and masonry should be maintained above freezing temperatures for at least 2 days. The strength development of the mortar can be accelerated by replacing some of the lime with cement or by using calcium chloride as an accelerator. The side effects of admixtures such as corrosion of metals should, however, be recognized.

Winter masonry construction has been carried out with excellent results even when the heating requirements have been neglected after the masonry has been laid up. This is because bricks absorb the moisture from fresh mortar quite rapidly. After the moisture content of the mortar has dropped below 6%, freezing can no longer damage the mortar. When lean mortar and porous brick are used, the time required to reach this value is on the order of 30 min or less. On the other hand, it may take several hours before the moisture content of rich mortar with hard burned bricks has dropped below 6%.

The water absorption properties of different brick-mortar combinations can be tested with simple methods. Test specimens are broken 10, 30, and 60 min after the mortar has been placed. The weights of the mortar samples are then measured in the natural condition and after they have been dried in the oven.

Proper storage for masonry units and mortar materials is important. The materials should be stored on platforms and covered to prevent the

Figure 7.37 Use of weather shelter in welding. *(Courtesy of F. E. Crory.)*

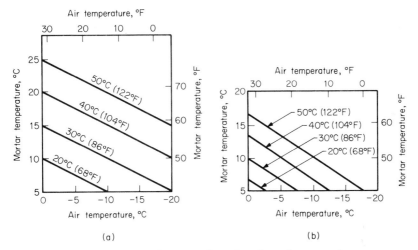

Figure 7.38 Temperature of mortar when placed as a function of air temperature and initial mortar temperature. Handling times (*a*) 30 min and (*b*) 60 min. (*Jonson, 1977; reprinted with permission of Statens råd för byggnadsforskning, Stockholm.*)

absorption of moisture from ground or air. In moderately cold weather it is often necessary to heat only the mortar. With masonry units above freezing and heated mortar at $+20\,^{\circ}\text{C}$ ($68\,^{\circ}\text{F}$), the masonry should remain in the unfrozen state for more than half an hour. However, the temperature drop of mortar during handling should be considered (Fig. 7.38). At lower temperatures the masonry units should also be heated. A warm storage space is obviously ideal, but if the units are stored under a canvas tarp, a hot-air blower will suffice.

It is a common practice to use some kind of windbreak to provide convenient working conditions. The scaffolds can be used as a frame for the windbreak in winter conditions. An enclosure also provides some protection against heat losses, and space heaters can be used (Fig. 7.39). After the work has been finished, the masonry should be protected from rain and snow with tarpaulins.

When the mortar freezes, its hydration slows down and eventually stops. However, the mortar is very strong in the frozen state. As temperatures rise again, hydration recontinues normally, but during thawing the mortar loses the strength provided by the frozen moisture. If the mortar has not gained enough maturity, considerable settlement may occur. This is a very important consideration in cold weather masonry construction. The mortar may not have gained half of its design strength at the moment of thawing if it has been allowed to freeze at a very early stage.

Figure 7.39 Masonry construction within a weather hall. *(Courtesy of Telinevuokraus Kataja.)*

In principle it is possible to let the mortar freeze right after it has been used, even if its moisture content has not dropped sufficiently. Freezing will not damage the fresh mortar if there is no hardening and the internal structure has not had time to form. However, considerable settlement may occur during thawing; the final strength of the mortar will be reduced, and the bond between mortar and brick will be weak. This method is not recommended for load-bearing structures or for structures exposed to severe environmental conditions.

7.4.3 Roofing

Cold weather roofing is generally feasible as long as the windchill effect remains tolerable and the snow does not interfere with the work (Fig. 7.40). Moisture protection, material handling, and winter construction techniques are among the important considerations.

Moisture trapped in the roof system during construction can cause such defects as buckling and blistering of the roofing membranes, reduction in the thermal insulation capability, rotting of wood, and poor adhesion. Proper protection of roofing materials and removal of snow, ice, and frost from the deck are important. All roofing materials should be dry stored, preferably in a heated enclosure.

There are few problems with mechanical roofing methods in winter conditions in addition to moisture control, reduced productivity, and working safety on the slippery roof. Application of bituminous membranes, on the other hand, is more sensitive to weather conditions. Felts become brittle at low temperatures and may be damaged by impacts during handling or careless unrolling. The application of bitumen also becomes difficult because of rapid heat losses during handling, and when bitumen gets into contact with cold surfaces having a large heat capacity, it becomes difficult to spread. Some cementing action may be lost and heavy layers of bitumen may create a danger of slippage. The problems of rapid cooling can be somewhat offset by using higher temperatures of application than normally. Too much heating may, however, damage the bitumen and cause it to burst into flames.

Application of bituminous membranes is quite feasible during calm sunny winter days even below $-20\,°C$ $(-4\,°F)$. The problems can be managed with special materials and techniques. Economy and productivity are important considerations.

7.4.4 Interior Work in Buildings

Heating is typically associated with interior work such as plastering, coating, painting, laminating, and finishing walls and floors. The purpose of heating is to remove moisture from the structures and to create accept-

Figure 7.40 Winter roofing. *(Courtesy of Finn-Stroi Ltd.)*

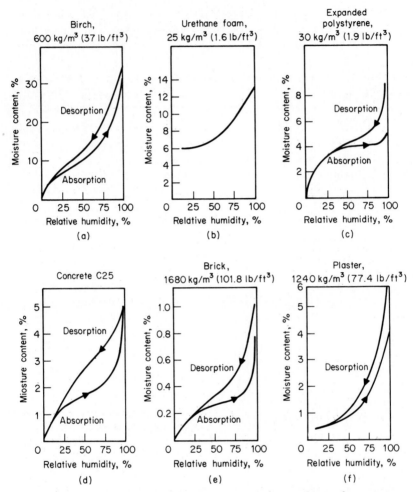

Figure 7.41   Typical examples of balanced moisture conditions for construction materials. *(Ahlgren, 1972.)*

able conditions for the application of finishing materials. When the temperature and the moisture content of structures have come close to those experienced at the service stage, favorable conditions usually exist for interior finishing work.

Construction moisture is generally absorbed water or water used in manufacturing materials such as concrete, mortar, and plaster. Porous materials have a natural moisture content that depends on the surrounding conditions and on the properties of the material in a complex manner. The construction moisture tends to decrease until the natural moisture

TABLE 7.5 Correction Factors Accounting for Different Concrete Properties or Drying Circumstances to be Applied to Drying Time of a Concrete Structure when It Is to Be Coated with an Impermeable Material*

Basic case: Cubic strength of concrete 25 MPa (3625 lb/in²); drying prevented downward; age of concrete 1 month; cured under plastic sheet; slab poured on ground on a plastic sheet. Required time for drying, 60 days.

Property	Correction factor	Remarks
Cubic strength of concrete, MPa (lb/in²)		
15 (2175)	2	Correction factors are not valid
25 (3625)	1	when water curing or rain and
25 (3625), air entrainment	0.5	water damage has increased the
40 (5800)	0.5 – 0.6	moisture content of concrete.
40 (5800), air entrainment	0.3	The air content has been increased from 4 to 8%.
Age		
slab thickness < 150 mm (6 in)	0.7	Age of concrete is 1 week when drying begins
slab thickness > 150 mm (6 in)	1	
Relative humidity		
20 – 50%	1	
60%	1.2	
80%	1.5	
Temperature		
10°C (50°F)	1.3 – 1.4	
20°C (68°F)	1	
30°C (86°F)	0.6 – 0.7	
Slab thickness L		
60 mm (2.4 in)	0.4	The correction factors correspond
80 mm (3.1 in)	0.7	to drying in one direction.
100 mm (3.9 in)	1.0	If drying occurs to both directions,
120 mm (4.7 in)	1.4	L = half the slab thickness.
140 mm (5.5 in)	1.8	
160 mm (6.3 in)	2.3	
200 mm (7.9 in)	3.3	
Material under slab		
50 mm (2 in) of polystyrene	0.9 – 1.0	No plastic sheet between concrete
150 mm (6 in) of lightweight aggregate	0.7 – 0.8	and insulation. Valid at a temperature difference of
50 mm (2 in) of mineral wool	0.6 – 0.7	2°C (3.6°F) between slab and ground.

* The relative humidity of the cooling-concrete interface should not exceed 90%.
SOURCE: Adapted from Nilson (1976).

content has been reached. Some typical curves for natural moisture content are given in Fig. 7.41.

The drying rate of the structure depends among other things on the material, its moisture content, its massiveness, and the drying conditions. The point in time when the structure is sufficiently dry for finishing depends very much on the type of surface material. The requirements are not as strict for painting or plastering as they are for coating with very vapor-tight materials. Concrete structures typically dry very slowly. It may take a couple of months before a concrete slab is ready to be coated with an impervious material, as shown in Table 7.5.

As soon as an enclosure has been created, heating can be started with space heaters or with the building's own heating system. The moisture escaping from the structures is usually carried away with ventilation. In relatively airtight enclosures air dryers can be applied to reduce the heat losses that would be caused by adequate ventilation.

After the moisture content of structures has decreased to acceptable levels (the decision can be made based on experience or measurements), finishing work can begin. Although acceptable results may be obtained even at temperatures below $+5\,°C$ ($41\,°F$), higher temperatures are generally recommended. For example, paint manufacturers often recommend an application temperature of about $+10\,°C$ ($50\,°F$) or more, a relative humidity below 80%, and that the painted surface be not more than $3\,°C$ ($5\,°F$) colder than the air.

Failures in interior works are unfortunately common in winter construction. For example, large changes in the temperature and moisture content of drywall may cause deformations and cracking of the finished surface. Low temperatures or exterior moisture may damage the plaster and cause peeling of painted surfaces. Construction moisture or low application temperatures often also cause the separation of wallpaper, linoleum, parquet, or lamination from its base material. Hence the heating and drying requirements need special attention in the scheduling.

8

CONSTRUCTION PROJECTS IN COLD ENVIRONMENTS

Year-round construction is common in subarctic areas with a well-established infrastructure. It can also be applied in arctic areas, but the nature of activities is different. In developed subarctic areas, year-round construction does not require major arrangements; in the arctic wilderness, however, the environment governs not only the design, but also the logistics and the pace of construction.

8.1 Year-Round Construction in Developed Subarctic Areas

Winter construction means additional direct costs, because the work efficiency decreases. Furthermore, winter construction requires more energy, materials, and equipment.

On the other hand, winter construction means a steady, reliable labor force for the industry. This is a major advantage compared to the social and economical consequences of sharply fluctuating seasonal hiring and firing. In addition, the contractor saves on overhead costs and the project can be completed earlier, which saves capital. The advantages of winter

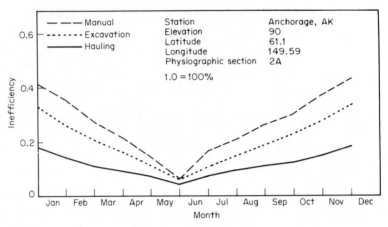

Figure 8.1 Example of subarctic earthwork inefficiency based on temperature, lighting, and precipitation. *(Roberts, 1976.)*

construction are widely considered to outweigh its disadvantages in developed northern areas.

Working conditions have a drastic effect on the construction efficiency. The combined effect of temperature and wind on human activities can be estimated using the windchill factor given in Fig. 1.7. The limit for outdoor activities is around 1500 kcal/(m$^2 \cdot$h). This corresponds to about $-10°$C ($14°$F) and a 20-m/s (45-mi/h) wind velocity or $-40°$C ($-40°$F) and a 1-m/s (2-mi/h) wind velocity. However, work efficiency begins to decline at much smaller windchill factors, and protective measures become necessary.

An example of earthwork inefficiency in subarctic conditions is given in Fig. 8.1 (inefficiency $= 1 -$ efficiency). Full efficiency corresponds to an optimum working temperature of $+10°$C ($50°$F), indirect sunlight, and no precipitation. For example, manual labor efficiency is estimated to drop from 0.73 at $-20°$C ($-4°$F) to between 0.35 and 0.53, depending on the climatic zone, at $-30°$C ($-22°$F), according to a study by Associated General Contractors and House Builders of Sweden (1963). In addition to climatic conditions, local frost conditions have a very significant effect on earthwork efficiency.

Different work phases have practical and absolute cut-off temperatures. The limit may be connected to the working efficiency or to the work phase itself as described in Chap. 7. Activities can be maintained with reasonable efficiency and with extra efforts until the practical limit is reached. Activities can still be continued in special cases with additional arrangements, but the efficiency drops and the costs increase rapidly. The results of a questionnaire, given in Table 8.1, reflect the

TABLE 8.1 Summary of Responses of Alaskan Contractors to the Question: What Is the Lowest Temperature at Which the Following Types of Activities Can Be Performed?

Activity	Absolute low, °C (°F)			Lowest for efficient operation or within enclosure, °C (°F)		
	Number of responses	Range	Average	Number of responses	Range	Average
Machine excavation	15	−57 to −29 (−70 to −20)	−38 (−37)	10	−34 to 0 (−30 to +32)	−18 (−1)
Hand excavation	11	−57 to 0 (−70 to +32)	−30 (−22)	12	−34 to 1 (−30 to +33)	−8 (+18)
Earth moving and grading	11	−43 to −29 (−45 to −20)	−37 (−35)	9	−29 to 0 (−20 to +32)	−11 (+13)
Paving	14	−9 to 7 (+15 to +45)	0 (+32)	4	−4 to +16 (+25 to +60)	+5 (+41)
Concrete formwork	11	−46 to −23 (−50 to −10)	−32 (−25)	7	−23 to +1 (−10 to +33)	−12 (+10)
Concrete placement	6	−46 to −32 (−50 to −25)	−35 (−31)	12	−7 to +4 (+20 to +40)	0 (+32)
Steel erection	10	−46 to −29 (−50 to −20)	−36 (−32)	8	−34 to −1 (−30 to +30)	−18 (−1)
Block and stone masonry	5	−46 to −18 (−50 to 0)	−32 (−25)	12	−18 to +4 (0 to +40)	−1 (+31)
Roofing	7	−34 to −9 (−30 to +15)	−25 (−13)	8	−18 to +16 (0 to +60)	−4 (+24)
Finish carpentry	5	−46 to −23 (−50 to −10)	−33 (−27)	9	−9 to +10 (+15 to +50)	+2 (+36)
Painting	7	−46 to +4 (−50 to +40)	−11 (+13)	10	0 to +16 (+32 to +60)	+6 (+42)
Electrical	14	−57 to −23 (−70 to −10)	−35 (−31)	7	−7 to +4 (+20 to +40)	−2 (+28)
Piping and mechanical	10	−46 to −23 (−50 to −10)	−33 (−27)	5	−7 to +4 (+20 to +40)	−2 (+29)
Pipe welding	9	−46 to −18 (−50 to 0)	−29 (−21)	3	−34 to +1 (−30 to +33)	−13 (+8)
Prebuilt components	13	−46 to −23 (−50 to −10)	−33 (−28)	6	−34 to +1 (−30 to +33)	−13 (+9)
Layout and surveying	17	−46 to −18 (−50 to 0)	−36 (−33)	4	−34 to −7 (−30 to +20)	−17 (+2)
Loading and unloading	14	−51 to −23 (−60 to −10)	−39 (−39)	4	−37 to −7 (−35 to +20)	−18 (−1)
Subsurface exploration	1		−40 (−40)			

SOURCE: Bennett (1976).

attitudes among Alaskan contractors. In the development of Siberia some major operations have continued at temperatures considered inefficient among Alaskan contractors.

Winter construction means additional work, equipment, material, and energy supplies besides the effects of reduced efficiency and temporary interruptions caused by the weather. The additional winter phases include snow and ice works, thermal protection, heating and drying, and the erection of temporary structures. Snow and ice works are in particular connected to earthwork, foundation construction, and frame erection stages. They are also related to site maintenance efforts. Providing thermal protection, heating, drying, and lighting is essential not only for regular work but often also in temporary structures, as described in Chap. 7.

The additional expenses for equipment are due to the increased capacity required in regular work phases and the need for special winter operation equipment. Equipment and machinery break down more frequently in the winter and do not work as efficiently in earth or concrete works. Special equipment needs in winter operation include frost-melting or excavating equipment, concrete-heating equipment, space-heating equipment, steam generators, and plows.

The increase in material expenses during winter operations results from increased losses, especially because of snow, special requirements for material properties, and additional material needs for protective purposes.

Finally, winter operations require energy supply for heating of materials, structures, and buildings, for melting of frost, ice, and snow, and for lighting and machinery. In Finland, where average frost indexes range from 400 to 1600°C·days (750 to 3000°F·days), typical energy consumption in the construction of a multistory residential building is 30 to 130 kWh/m^3 (3000 to 13,000 Btu/ft^3) (Kokki and Mäkelä, 1980). The time required for construction ranges from 9 to 12 months. Construction methods and timing are among the major factors affecting energy consumption, but savings can also be achieved with careful planning and execution of heating and protective measures.

The additional costs of winter construction have been extensively studied by Tynkkynen and Ojala (1979). An example of the effect the starting date has on the additional winter expenses at three Finnish locations is given in Fig. 8.2. A breakdown of the additional expenses is listed in Table 8.2. The construction time of the typical cast-in-place multistory residential building is assumed to be 9.5 months. The location and starting date have a significant effect on the additional winter expenses. Other important variables include construction methods, project size, length of the construction season, and variations in local

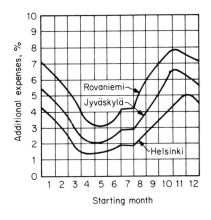

Figure 8.2 Additional expenses for winter construction in three Finnish locations depending on starting time. Construction time for residential building is 9.5 months. *(Tynkkynen and Ojala, 1979.)*

conditions. The additional cost of winter earthworks depends greatly on local conditions. The interior works represent the largest cost fraction and are not interrupted by bad weather. Energy cost is the only costly component in the interior works during winter.

The effect of winter on the different work phases in the example of Helsinki can be seen in Table 8.3. An American survey summarized in Table 8.4 yields similar results. Winter has generally the most severe effect on the cost of earthworks and foundation construction. On the other hand, these phases do not require very large portions of the total construction cost or time.

In most cases year-round construction has proved feasible in developed subarctic areas. The additional cost during the winter period is generally well below 10%. By careful planning and design this additional cost can be further reduced. Scheduling is the most effective way to reduce additional winter construction expenses. By scheduling the start-

TABLE 8.2   Breakdown of Additional Winter Expenses

	Cost portion of actual construction, %	Additional winter expenses, %	Additional winter expenses, % of total
Earthwork	7.9	8–16	0.6–1.3
Foundations	13.6	13–15	1.8–2.0
Framework	35.2	5.5–6.5	1.9–2.3
Interior works	43.3	3.3–3.7	1.4–1.6
Total	100.0		6.3–6.7

SOURCE: Tynkkynen and Ojala (1979).

TABLE 8.3 Percentage of Additional Winter Construction Costs for Different Work Phases, Excluding Electricity, Plumbing, etc., on a Residential Building

	Foundations	Framework	Interior work
Work	2.6–2.9	0.6–0.7	
Material	1.7–3.7	0.6–1.9	
Energy	0.9–1.0	1.2–1.4	2.8–3.2
Machines and equipment	1.8–2.2	1.2–1.4	0.1–0.2
Additional winter work	1.6–1.8	0.7–0.9	0.2–0.4
Additional time	2.0–2.2	1.0–1.2	
Total	13–15	5.5–7.5	3.3–3.7

SOURCE: Tynkkynen and Ojala (1979).

ing date of the example project to May, the cost increase could be reduced to less than half of that with a starting date in late November. The timing of interior work is not as important as the timing of earthwork and foundation construction.

Winter construction should also be considered when construction materials and methods are selected. For example, the use of precast concrete elements may mean savings in the additional winter expenses compared to the cast-in-place alternative. Readiness for winter construction may also cut expenses. It is generally cheaper to have proper storage facilities and adequate means for protection and heating for unusually

TABLE 8.4 Average Percentage Increase in Cost by Building Elements for Worst 3 Months versus Best 3 Months in the United States

Building element	North	Mid south	Deep south	West Mountain	West Desert	West North coast	West South coast
Excavation	16	15	10	15	3	18	16
Foundation	7	7	10	18		15	15
Floors	7	8	5	20		5	11
Rough framework	6	6	5	13	2	8	10
Electrical	0.5		1	7			2
Plumbing	1	1	2	2		2	3
Interior finish	4	8	1	10		4	8
Finish carpentry	1	7		5		3	6
Roofing	7	5	3	7		8	4
Siding	6	7	1	8		8	8
Exterior painting	5	8	2	5		8	6

SOURCE: NAHB Research Foundation (1975).

severe weather conditions than to pay for unexpected damages, repair and the loss of time.

Energy represents a considerable fraction of the total cost. On the average it is about one-third of the additional winter construction expenses. It is the major portion during the interior work. However, it is the rule rather than the exception that a lot of energy is wasted in construction because of improper protection, uneven heat distribution, overheating, and large uncontrolled air leaks. It is important that heating and protection measures be carefully planned and executed.

8.2 Construction Projects in Arctic Environment

8.2.1 Seasonal Restrictions

If seasonal difficulties can be overcome with minor adjustments in developed subarctic areas, this is no longer possible in the arctic or in poorly developed subarctic conditions. This is why seasonal conditions provide a very firm basis for the realization of a construction project.

A typical analysis of arctic conditions and their impact on some essential functions is given in Fig. 8.3. Information on local conditions such as this is very useful for the planning and execution of any arctic construction project. During the winter windchill factors are continuously high enough to reduce work efficiency, and they often become intolerable for any type of effective outside operation. Darkness, poor visibility, and blowing snow are additional difficulties for winter operations. On the other hand, in transportation, which is a major consideration in cold region construction operations, winter conditions with snow, ice, and frozen ground can also be taken as an advantage.

Long-distance transportation in the Arctic occurs mainly through air and water routes (Fig. 8.4). Air transportation, although occasionally interrupted by storm, fog, and whiteout conditions, provides the year-round service that is essential to arctic development. Water transportation is the prime mover of freight. The ice-free season in the Arctic is often only a couple of months long, but the navigation season can be extended by using icebreaker-assisted convoys with ice-reinforced vessels. Some gravel roads and railroads have been constructed in the Arctic, but so far they have only local importance.

Winter roads ranging from ordinary winter trails, snow roads, and roads on the ice cover to high-quality roads made of ice and ice bridges constructed over water represent an excellent seasonal alternative to permanent gravel roads in short- and medium-range transportation

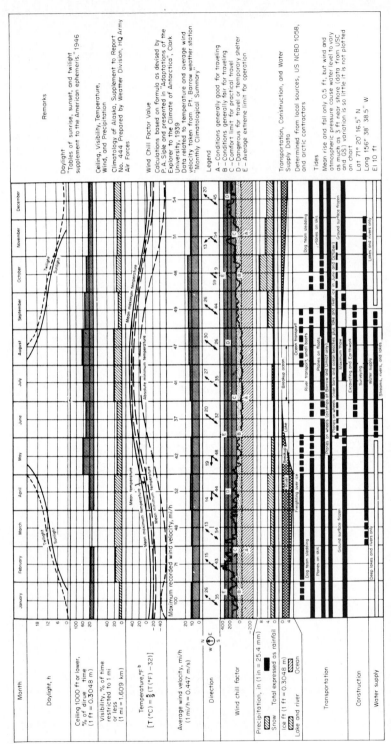

Figure 8.3 Work feasibility chart, Point Barrow, Alaska. (*U.S. Naval Facilities Engineering Command, 1975.*)

(Table 8.5). The annual construction cost of a winter road is typically only 1% of the cost of a permanent gravel road, although high-quality winter roads for heavy transportation purposes may be more expensive. Winter trails are not generally recommended on tundra, because construction and traffic may damage the vegetation and cause environmental disturbance. This hazard may be avoided if rolligons, multiaxial wheeled vehicles with low contact pressures, are used. A good summary of the planning, construction, and maintenance of winter roads is presented in Adam (1978).

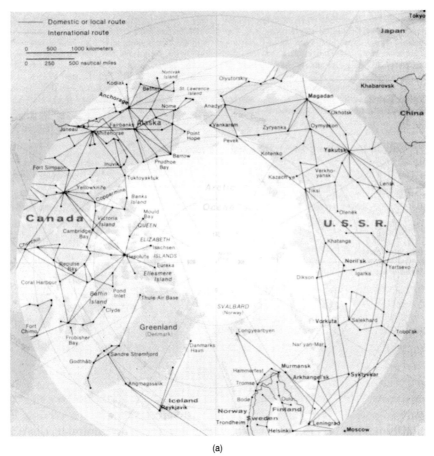

(a)

Figure 8.4 Major transportation networks in the Arctic. (*a*) Scheduled air transportation (representative routes). (*b*) Waterway transportation. (*Central Intelligence Agency, 1978.*)

(*Continued*)

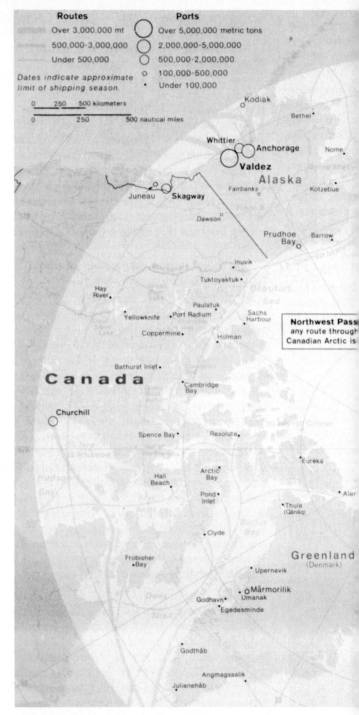

Figure 8.4 *(Continued)* (b)

Japan

Sea of
Okhotsk

China

○Beringovskiy
○ Anadyr'

Magadan ○

○Okhotsk

○ya
vekinot

Arctic Circle

Mys Shmidta

Zelënyy
Mys ○
Pevek ○
○Mikhalkino

○Zyryanka

Yakutsk ○

• Tommot

15 October

East
Siberian
Sea

Verkhoyansk •

○Sangar

Vilyuysk •

○ Nizhneyansk

Tiksi ○

Lensk •

minimum
of sea ice

Laptev
Sea

80

Ocean

5 August
to October

U. S. S. R.

• Khatanga

North
Pole

Lesosibirsk ○

Northern Sea Route
vaya Zemlya–Bering Strait

**Dudinka/
Noril'sk** ○

Dikson ○
Igarka ○

Kara
Sea

○Urengoy

Nadym ○

Nizhnevartovsk ○

Mya
Kharasavey •

Surgut ○

age minimum
nt of sea ice

SVALBARD
(Norway)

NOVAYA
ZEMLYA

•Amderma
Salekhard ○

Khanty-
Mansiysk ○

Longyearbyen
Barentsburg

Barents
Sea

Belush'ya•
Guba

Varandey •
○Abez'

Tobol'sk ○

Nar'yan ○
Mar

Pechora ○

Tyumen' •

Kirkenes
Hammerfest •
Norway
Tromsø

Murmansk

Mezen' •

Solikamsk ○

Arkhangel'sk ○
Syktyvkar ○

Narvik ○

Kandalaksha ○

Kotlas ○

Belomorsk ○

483

TABLE 8.5   Description of Winter Roads

Type of winter road	Description	Remarks
Winter trails		
Temporary winter trails, permanent winter trails	Constructed by clearing vegetation and leveling major bumps with bladed vehicle. Snow is compacted by vehicle traffic.	Construction and operation of winter trails may cause environmental damage. Therefore winter trails are not generally recommended for vulnerable permafrost areas.
Snow roads		
Compacted snow roads	After possible clearing and leveling the route, snow is used as a fill and paving material, which is compacted with rollers, heavy drags, or other means.	Compacted to snow density of about 500–600 kg/m^3 (40–50 lb/ft^3); do not carry heavy traffic, especially if specified densities or hardness values are not reached.
Processed snow roads	Like compacted snow roads, except that snow is agitated or mixed before each compaction in order to obtain harder snow surface.	
Artificial snow roads	Constructed of artificially manufactured snow.	Used only if natural snow is not available in sufficient quantities.
Ice-capped snow roads	Surface of snow road is solidified by distributing typically 2–3 cm (1 in) of water on surface.	Good-quality snow roads; can carry heavy highway traffic without excessive maintenance.
Roads made of ice		
Flooded ice roads	Constructed by surface flooding on ground.	High-quality winter roads capable of carrying very heavy traffic.
Aggregate ice roads	Crushed ice is used as construction material, and water is used to bond the mass.	Road construction rate can be accelerated, but ice aggregate is difficult to produce.
Roads on ice cover		
Ordinary ice roads	Constructed on natural ice cover by clearing snow from route.	Weight restrictions and safety hazards; shorter operation period than with other winter roads; good-quality surface, but vehicle speed and parking may be limited.
Artificially strengthened ice roads	Constructed on natural ice cover by artificial thickening using surface flooding. Can be further strengthened with reinforcement.	

Figure 8.5 Air-cushion vehicles can be used for arctic cargo transport to unbuilt shores and on flat tundra. *(Courtesy of Oy Wärtsilä Ab.)*

During summer the active soil layer usually becomes very moist and soft because drainage is blocked downward by permafrost. Even if tundra would carry the transportation or construction loads, activities could easily damage surface vegetation and cause thermoerosion and other undesirable environmental effects. Only thick gravel pads provide an environmentally acceptable solution for ordinary summer transportation and construction activities.

Other local transportation alternatives include helicopters and air-cushion vehicles. Helicopters are convenient but quite expensive in cargo transportation. The load capacity of a helicopter is generally limited to less than 10 tons. Air-cushion vehicles are even applicable on

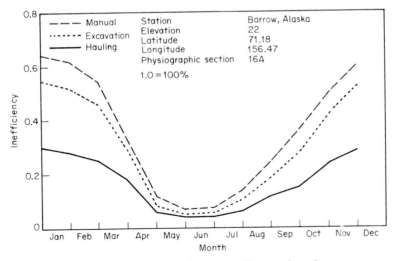

Figure 8.6 Example of arctic earthwork inefficiency based on temperature, lighting, and precipitation. *(Roberts, 1976.)*

tundra surfaces and may provide reliable protection for tundra vegetation and surface layer during the summer thaw. They also provide an attractive alternative to year-round coastal service (Fig. 8.5). Plans have called for air-cushion vehicle transportation with payload capacities of up to 300 tons to support Beaufort Sea oil operations (Walden and Dickens, 1982). In another application air-cushion vehicles serve as short-range transportation between a supply boat and a coast without proper harbor facilities (Thomas, 1981).

Working conditions may be quite uncomfortable or even hazardous in the arctic winter. The extent of operations, therefore, is limited. One example of working inefficiency in arctic conditions based on temperature, lighting, and precipitation is given in Fig. 8.6. However, there are also problems in summer operations. Thawed ground is easy to excavate but if no gravel pads are used, it does not support earthwork machines any better than land transportation equipment. Furthermore, excavation may be quite elaborate, especially in the early summer when the active layer is extremely moist. In late fall or in early spring the frozen ground, although harder to excavate, provides support for the machinery while working conditions are still acceptable.

Seasonal difficulties are also connected to water supply. There is a lot of good-quality surface water available in the spring and summer months, but during winter shallow streams freeze to the bottom and water in shallow lakes becomes contaminated. It is extremely expensive to melt snow or ice in adequate quantities, typically 200 liters (50 gal) per person per day. Treatment of poor-quality water and desalination of seawater are equally expensive. Trying to find good-quality water below permafrost in the high Arctic is seldom economically feasible.

Deep lakes and streams that do not freeze to the bottom during the winter are convenient water sources year-round. The thaw bulbs beneath shallow lakes or even streams that freeze to the bottom (thawed aquifers under the stream bottom may carry considerable flow) are also potential sources. Such water goes through a natural filtration system and may be of good quality (Rice, 1975).

8.2.2 Problems in Project Management

The problems of running a construction project in arctic or undeveloped subarctic conditions differ considerably from those in well-developed areas. In addition to seasonal restrictions, the special features include:

1. Lack of local resources, such as work force, materials, and equipment

2. In many cases lack of local infrastructure

3. Long transportation and service lines

4. Engineering solutions suitable for arctic conditions

5. Social and safety requirements for the work force

The effects of ordinary problems in construction, such as equipment failures, quantity overruns, or design changes, are greatly magnified. The logistics of arctic construction have been most extensively tested in the construction of the trans-Alaska pipeline and are discussed in Krause (1978) and Haugen (1978).

A thorough site survey and a carefully developed design process are essential parts of an arctic construction project, because even minor surprises may cause major delays. Major surprises could delay the project by half a year because of the seasonal difficulties. The design must be based on adequate and unambiguous information. It must be accurate enough to include complete listings of materials, supplies, and equipment.

Because scheduling is a key factor, careful work plans must also be drawn up. The seasonally reduced productivity must be evaluated, and construction strategies and machinery should be suitable to the local seasonal conditions. For example, poor judgment in the selection of earthwork machinery may have very severe effects on the execution of the entire project.

The remoteness, the lack of infrastructure, and the extremely harsh environment require that certain essential items such as power, heat, and communications have backup systems. Separate living shelters and storage spaces for fuel and important materials may be needed to minimize the effects of a possible fire. The availability of well-equipped medical facilities and repair workshops is essential (Harwood, 1973). Finally, it is difficult to recruit enough good permanent workers to the Arctic. Special attention should be given to the social facilities in the camp since productivity depends significantly on the morale and spirit of the work force.

Health inspections and training are also important. The work force should be in good physical and mental condition. Workers should know local conditions well and they should be aware of how to avoid or handle safety hazards such as fire, carbon monoxide, frostbites, animals, and so on. In some cases, structures and machines are preassembled before loading and transportation. This reduces numerous risks at the arctic site and may be especially useful if done by the workers who also do the actual construction.

One major problem in arctic construction is to bring all the materials and equipment to the site on schedule. Shipping is often the primary mode of transportation, but the season is very short. It is therefore most important that all necessary parts arrive on time without any missing items.

The actual execution of an arctic construction project cannot always proceed as planned. Therefore the project organization must have the flexibility to adjust to new situations. The field management should have the capacity to make decisions. A tight cost and scheduling control system should be maintained in order to react better to unsatisfactory developments (Lipinski, 1978). Minor material shortages or equipment needs can be handled with the person and supply transportation system. Approximate backup plans should be prepared in case major surprises, such as exceptionally bad weather conditions, delays in the primary transportation, unsatisfactory performance of a key machinery, major design changes, or severe delays in scheduling. The purpose of this kind of planning is not only to obtain readiness, but also to strengthen the strategy and organization in weak areas.

The cost basis of arctic construction significantly differs from that of construction in well-developed subarctic areas. Arctic construction typically costs two to five times more than construction in more temperate developed regions. The cost components of a subarctic and an arctic construction project are compared in Fig. 8.7. In both cases the cost of materials and equipment has been set equal to 100%. The total cost of maintaining the work force at the construction site is 182% in an arctic project and 47% in a subarctic project. On the other hand, the cost of ocean freight and stevedoring in the arctic project is quite reasonable (31%). The natural conclusion for economic construction is to minimize the amount of work at the site.

The main principles of arctic construction can be summarized as follows:

1. Do the site survey and planning thoroughly and in advance.
2. Adjust the work schedule to the seasonal schedule.
3. Try to use prefabricated elements and units as much as possible.
4. Try to minimize the need for personnel and work phases.
5. Try to predict uncertainties in the planned construction process.
6. Make it simple.

It is to be noted that experience cannot be replaced by quantity in the planning, especially if the time schedule is very tight. Mistakes in the early stages can prove to be very costly later during the construction stage.

When prefabricated structural elements are used, the amount of time and labor can be reduced. The complexity and the degree of uncertainty of the project can also be minimized. However, labor cost should not be reduced by having people work longer hours. Experience has shown that longer work weeks do not mean increased productivity. The practical

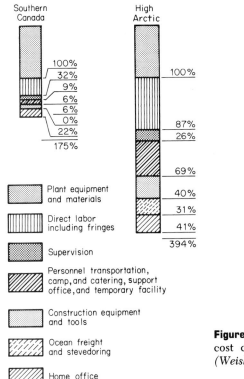

Southern
Canada

High
Arctic

100%
32%
9%
6%
6%
0%
22%
175%

100%

87%
26%

69%

40%

31%

41%

394%

☐ Plant equipment and materials

☐ Direct labor including fringes

☐ Supervision

☐ Personnel transportation, camp, and catering, support office, and temporary facility

☐ Construction equipment and tools

☐ Ocean freight and stevedoring

☐ Home office

Figure 8.7 Comparison of cost components of projects. *(Weishar, 1976.)*

limit for increased productivity may be on the order of 60 hours per week.

The importance of proper project management in arctic construction can be best described by a true story related in Eliason (1978). A major company was involved in a multibillion dollar project. The original estimate was to complete a project at $800 million. At 10% completed, the estimate to complete was $800 million; at 50% completed, the estimate to complete was $800 million; and at 90% completed, the estimate to complete was still $800 million. The lack of early planning and control and of realistic estimating forced the company into refinancing.

8.2.3 Project Execution and Construction Methods

Site investigations provide a sound basis for actual planning and design. A search for existing information such as maps, aerial photographs, previous investigations, and local experience together with the available information on the environmental conditions (Fig. 8.3) is an important first step. The selection of tentative sites and routes will be based on this information and additional ground, air, and water reconnaissance where necessary. In the final phase detailed site investigations are carried out. Items to be studied include:

1. Topography, surface features, and vegetation
2. Local weather conditions, especially winds and locations of snow accumulations
3. Natural features relevant to environmental impact analysis
4. Groundwater, drainage, and icing
5. Special terrain features such as polygons and solifluction
6. Soil and rock formations and conditions with depth
7. Ground temperatures and frost conditions
8. Engineering properties of frozen and thawed soils
9. Sources of water and borrow materials
10. Wave, ice, and sea-bottom conditions for coastal developments

The site investigation can generally be carried out with minor arrangements and easily transportable equipment and facilities (Fig. 8.8).

The schedule of site investigations is also tied to seasonal conditions. For example, air and ground reconnaissance are often pursued in the spring because drainage patterns and areas of aufeis features and snow accumulations are clearly visible. On the other hand, there may be difficulties in subsurface investigations in the spring or early summer because the thawed active layer may not carry the equipment and water tends to penetrate test pits or uncased drill holes, thus disturbing the activities. The philosophy and methodology of site investigations are discussed in detail in Linell and Tedrow (1981).

The actual construction may follow quite rapidly after the site investigations have been concluded, because planning and design work can progress along with the site investigations.

The founding of the construction camp with living quarters, office facilities, power generation, and waste disposal units should be done effectively using combinations of fully equipped space units and advanced prefabricated element systems. The installation can be done over a gravel pad and logs with a specially equipped truck, crane, or helicopter. Temporary warehouses should be light and easy to erect (Fig. 8.9).

Earthwork and foundation construction represent the most laborious and time-consuming working stages in an arctic construction project. Seasonal conditions provide both restrictions and opportunities at these stages. This presents a considerable challenge to project planning and scheduling.

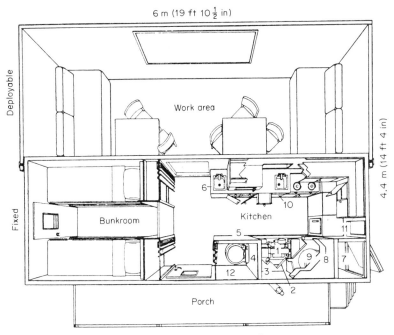

Figure 8.8 Air- and ground-transportable shelter for arctic surveying. 1 — alternator; 2 — auxiliary heater; 3 — battery charger; 4 — electrical distribution panel; 5 — remote start panel; 6 — fan-coil space heater; 7 — snowmelt/holding tank; 8 — gravity tank; 9 — water heater; 10 — wastewater holding tank; 11 — drinking water filter; 12 — safety systems. *(Flanders, 1981.)*

Advanced element systems provide one possibility for the construction of residential buildings and industrial facilities. The construction efficiency is usually sufficient even when the high cost of the work force is considered (Fig. 8.10). Element construction can be continued even in the harshest conditions within shelters. One interesting arrangement is shown in Fig. 8.11. Sections of buildings are constructed within a construction hangar that moves along rail tracks. At the end of the sequence the hangar is anchored permanently and becomes a gymnasium and a bulk storage area.

After the infrastructure has been established, more conventional construction techniques can be applied, as shown in Fig. 8.12. Scheduling is still important so that most of the construction can occur in a sheltered space during the harshest period.

Module construction provides an interesting alternative for element construction, especially in the case of sophisticated applications such as

(a)

(b)

Figure 8.9 Prefabricated units for construction operations. (*a*) Living quarters at construction camp. (*b*) Office buildings. (*c*) Power generation units. (*d*) Temporary storage shelter. (*Courtesy of Finn-Stroi Ltd.*)

industrial facilities (Fig. 8.13). The maximum amount of construction effort occurs within the developed infrastructure when this approach is applied. Module construction started on a large scale when offshore hydrocarbon production facilities were built in the North Sea. It is now widely applied in offshore construction and increasingly also in onshore construction. Modules are generally transported to the construction area on barges. In offshore installations they are lifted directly into place with heavy floating cranes. Onshore they are transported to foundations along gravel or winter roads using special transporters. Typical modules weigh less than 100 tons, but modules weighing more than 2000 tons have been transported and installed on foundations at Prudhoe Bay (Fig. 8.14).

One extreme of module construction is the use of fully prefabricated floating units that are towed and anchored to their operation sites. Pon-

(c)

(d)

toon units that could be towed to the Arctic and used as piers represent an example of potential arctic solutions. Factories and other production units constructed on a barge as well as removable caisson drilling units have already been used in the Arctic (Figs. 8.15 and 8.16).

The philosophy of adjusting to seasonal conditions and using units that are prefabricated in industrialized areas in order to minimize the need for time, personnel, and work phases provides many advantages in the Arctic, including:

A large portion of the available work force can be used in the primary activity, for example, in the search for and production of oil, and not in construction.

Figure 8.10 Construction of arctic villages for the Urengoj-Uzgorod gas pipeline. The roof is erected one week after the first element is installed on the foundation when advanced element techniques are used. *(Courtesy of HT-Group, Huurre Oy and S. A. Tervo Oy.)*

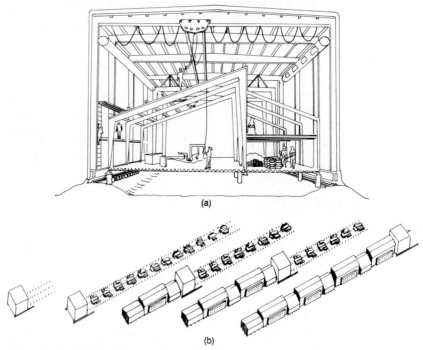

(a)

(b)

Figure 8.11 Polaris accommodation building construction phasing, utilizing mobile hangar system. *(Bent, 1981.)*

(a)

Figure 8.12 Construction of a dairy in Norilsk, U.S.S.R. (*a*) Shipment of prefabricated construction units and equipment. (*b*) Element construction. (*c*) Final equipment deliveries through roof. (*d*) Dairy near completion. *(Courtesy of Finn-Stroi Ltd.)*

(b)

(c)

(d)

Figure 8.13 Module construction in Norman Wells, Northwest Territories.

Figure 8.14 Heavy module transportation using crawlers at Prudhoe Bay, Alaska. *(Courtesy of Neil F. Lampson, Inc.)*

Figure 8.15 The 4200-ton barge-mounted seawater treatment facility was towed to site at end of long causeway at Prudhoe Bay, Alaska. *(Courtesy of ARCO Alaska, Inc.)*

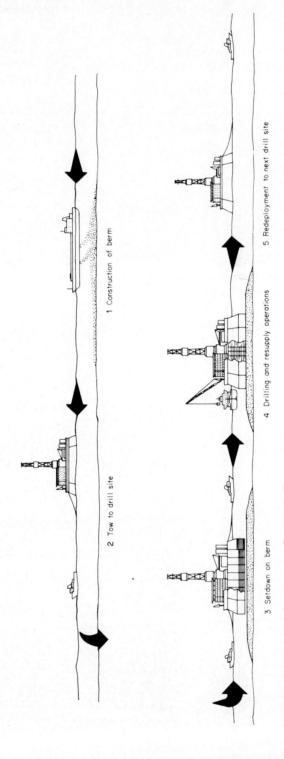

Figure 8.16 Stages of operation of arctic drilling caisson.

1 Construction of berm

2 Tow to drill site

3 Setdown on berm

4 Drilling and resupply operations

5 Redeployment to next drill site

During the construction period capital is tied up inefficiently, and thus rapid construction means savings in interest costs.

Risks in project execution are smaller in number and easier to manage.

Considerable savings are usually achieved when the direct and indirect labor costs can be reduced.

As a matter of fact, this construction philosophy is also rapidly gaining ground in more temperate areas.

BIBLIOGRAPHY

Adam, K. M., 1978: "Winter Road Construction Techniques," *Proc. Conf. Applied Techniques for Cold Environments* (Anchorage, Alaska), pp. 429–440.

Adams, C. M., D. M. French, and W. D. Kingery, 1963: "Field Solidification and Desalination of Sea Ice," in *Ice and Snow — Processes, Properties, and Applications*, W. D. Kingery, Ed., MIT Press, Cambridge, Mass., pp. 277–287.

Afanasev, V. P., 1972: "Ice Pressure on Vertical Structures," Nat. Res. Council of Canada, Ottawa, Ont., Tech. Transl. 1708.

Ahlgren, L., 1972: "Moisture Fixation in Porous Building Materials," Lund Institute of Technology, Div. of Building Technol., Rep. 36.

Aho, O. K., 1977: "Frost Protection of Shallow Foundations" (in Finnish), University of Oulu, Div. of Road and Soil Construction, Oulu, Publ. 24.

Aitken, G. W., 1974: "Reduction of Frost Heave by Surcharge Stress," U.S. Army Cold Regions Res. Eng. Lab., Hanover, N.H., Res. Rep. 184.

Alaluusua, M., 1980: "A Buoy Study at Field" (in Finnish), Finnish Board of Navigation, Int. Rep.

Alasaarela, P., 1979: "Fatigue Strength of Welded Structures" (in Finnish), *Hitsaustekniikka* 29:4–5.

Aldrich, H. P., and H. M. Paynter, 1953: "Analytical Studies of Freezing and Thawing of Soils," U.S. Army ACFEL TR 42.

Algaard, E., 1976: "Frost Protection of Floors, Foundations, and Construction Ground during Construction Period," in *Frost Protection of Cold Structures* (in Norwegian), Roy. Norw. Council Sci. and Ind. Res. and the Public Roads Administration's Committee on Frost Action in Soils, Frost i jord, no. 17, chap. 11, pp. 315–339.

Alkire, B. D., W. M. Haas, and T. J. Kaderabek, 1975: "Improving Low Temperature Compaction of a Granular Soil," *Can. Geotech. J.* 12:527–530.

Allyn, N., and K. Charpentier, 1982: "Modelling Ice Rubble Fields around Arctic Offshore Structures," *Proc. Offshore Technol. Conf.*, OTC 4422.

———, B. R. Wasilewski, 1979: "Some Influences of Ice Rubble Field Formations around Artificial Islands in Deep Water," *Proc. Conf. Port and Ocean Engineering under Arctic Conditions* (Trondheim, Norway), pp. 31–55.

Almazov, V. O., and E. M. Kopaigorodski, 1982: "Bearing Capacity of a Reinforced Concrete Shell in an Arctic Environment," *Cold Regions Sci. Technol.* 6:89–98.

Aluminum Association, 1976: Aluminum standards and data, New York.

American Association of State Highway and Transportation Officials, 1982: "Standard Specifications for Transportation Materials and Methods of Sampling and Testing," AASHTO, Washington, D.C.

American Concrete Institute, 1978: "Cold Weather Concreting," ACI 306R-78.

American National Standards Institute, 1972: "Building Code Requirements for Minimum Design Loads in Buildings and Other Structures," New York, ANSI A58.1-1972.

American Petroleum Institute, 1982: "Bulletin on Planning, Design and Constructing Fixed Offshore Structures in Ice Environments," API, Washington, D.C.

American Society for Testing and Materials, 1977: "Standard Test Method for Resistance of Concrete to Rapid Freezing and Thawing," ASTM C666-77.

———, 1977: "Standard Test Method for Critical Dilation of Concrete Specimens Subjected to Freezing," ASTM C671-77.

———, 1978: "Standard Test Method for Plain Strain Fracture Toughness," ASTM E399-78.

American Society of Civil Engineers, 1981: *Proc. Conf. The Northern Community: A Search for a Quality Environment*, T. S. Vinson, Ed., ASCE, New York.

American Society of Heating, Refrigeration, and Air-Conditioning Engineers, 1980: *Handbook and Product Directory*, vol. *Systems*, chap. 38: "Snow Melting," ASHRAE, New York.

———, 1981: *Handbook of Fundamentals*, ASHRAE, Atlanta, Ga.

Andersland, O. B., and D. M. Anderson, Eds., 1978: *Geotechnical Engineering for Cold Regions*, McGraw-Hill, New York.

———, F. H. Sayles, Jr., and B. Landanyi, 1978: "Mechanical Properties of Frozen Ground," in *Geotechnical Engineering for Cold Regions*, O. B. Andersland and D. M. Anderson, Eds., McGraw-Hill, New York, chap. 5, pp. 216–275.

Anderson, D. L., 1958: "Preliminary Results and Review of Sea Ice Elasticity and Related Studies," *Trans. Eng. Inst. Canada* 2:116–122.

———, 1960: "The Physical Constants of Sea Ice," *Research* 13:310–318.

Anderson, D. M., R. Pusch, and E. Penner, 1978: "Physical and Thermal Properties of Frozen Ground," in *Geotechnical Engineering for Cold Regions*, O. B. Andersland and D. M. Anderson, Eds., McGraw-Hill, New York, chap. 2, pp. 36–102.

———, A. R. Tice, and H. L. McKim, 1973: "The Unfrozen Water and the Apparent Specific Heat Capacity of Frozen Soils," *Proc. 2d Intern. Conf. Permafrost* (Yakutsk, U.S.S.R.), North Amer. Contrib., U.S. Nat. Acad. Sci., Washington, D.C., pp. 289–295.

Anno, Y., 1984: "Requirements for Modeling of a Snowdrift," *Cold Regions Sci. Technol.* 8: 241–252.

Armstrong, M. D., and T. I. Csathy, 1963: "Frost Design Practice in Canada," *Highway Res. Rec.*, no. 33, pp. 170–201.

Arridge, R. G. C., 1975: *Mechanics of Polymers*, Oxford University Press, London.

Ashton, G. D., 1974: "Air Bubbler System to Suppress Ice," U.S. Army Cold Regions Res. Eng. Lab., Hanover, N.H., Spec. Rep. 210.

———, 1978: "Numerical Simulation of Air Bubbler Systems," *Can. J. Civil Eng.* 5:231–238.

———, 1979: "Point Source Bubbler Systems to Suppress Ice," U.S Army Cold Regions Res. Eng. Lab., Hanover, N.H., Rep. 79–12.

Associated General Contractors and House Builders of Sweden, 1963: *Road Construction Year Round* (in Swedish), Byggnadsindustrins Förlags AB.

Assur, A., 1956: "Airfields on Floating Ice Sheets," U.S. Army Corps of Engineers Snow, Ice, and Permafrost Research Establishment (SIPRE), Tech. Rep. 36.

————, 1958: "Composition of Sea Ice and Its Tensile Strength," in *Arctic Sea Ice,* U.S. Nat. Acad. Sci., Nat. Res. Council, Publ. 598, pp. 106–138.

————, 1961: "Traffic over Frozen or Crusted Surfaces," *Proc. 1st Intern. Conf. Mech. Soil Vehicle Systems* (St. Vincent, Torino, Italy), pp. 913–923.

————, 1971: "Forces in Moving Ice Fields," *Proc. Conf. Port and Ocean Engineering under Arctic Conditions* (Trondheim, Norway), pp. 112–118.

Azmi, M. H., 1978: "Steel Construction in the Arctic," *Proc. Conf. Applied Techniques for Cold Environments* (Anchorage, Alaska), pp. 330–341.

Baker, M. C., 1980: "Roofs, Design, Application and Maintenance," Nat. Res. Council of Canada, Multiscience Publ. Ltd., Montreal, Que.

Banke, E. G., and S. D. Smith, 1971: "Wind Stress over Ice and over Water in the Beaufort Sea," *J. Geophys. Res.* 76:7369–7374.

———— and ————, 1973: "Wind Stress on Arctic Sea Ice," *J. Geophys. Res.* 78:7871–7873.

———— and ————, 1976: "Recent Measurements of Wind Stress on Arctic Sea Ice," *J. Fish. Res. Board Canada* 33:2307–2317.

Bates, R. E., and M. A. Bilello, 1966: "Defining the Cold Regions of the Northern Hemisphere," U.S. Army Cold Regions Res. Eng. Lab., Hanover, N.H., Tech. Rep. 178.

Baudais, D. J., D. M. Masterson, and J. S. Watts, 1974: "A System for Offshore Drilling in the Arctic Islands," 25th Ann. Meet., Pet. Soc. Can. Inst. Mining, Calgary, Alta.

Bennett, F. L., 1976: "Temporary Enclosures and Heating in Cold Regions Construction," *Proc. 2d Intern. Symp. Cold Regions Engineering* (Fairbanks, Alaska), pp. 502–516.

Bennett, I., 1959: "Glaze, Its Meteorology and Climatology, Geographical Distribution and Economic Effects," Quartermaster Res. and Eng. Center, Tech. Rep. EP-105.

Bent, C., 1981: "Accommodation Design for Remote Settlements," *Proc. Conf. The Northern Community: A Search for a Quality Environment,* T. S. Vinson, Ed., ASCE, New York, pp. 115–131.

Bercha, F. G., 1979: "The Development and Application of Multimodal Ice Failure Theory," *Proc. IUTAM Symp. Physics and Mechanics of Ice* (Copenhagen), Springer-Verlag, pp. 16–27.

Berg, R. L., G. L. Guymon, and T. C. Johnson, 1980: "Mathematical Model to Correlate Frost Heave of Pavements with Laboratory Predictions," U.S. Army Cold Regions Res. Eng. Lab., Hanover, N.H., CRREL Rep. 80-10.

Bergedahl, L., 1978: "Thermal Ice Pressure in Lake Ice Covers," Chalmers University of Technology, Dept. of Hydraulics, Göteborg, Rep., ser. A:2.

Bilello, M. A., 1960: "Formation, Growth and Decay of Sea Ice in Canadian Arctic Archipelago," U.S. Army Corps of Engineers Ice, Snow and Permafrost Research Establishment (SIPRE), Res. Rep. 65.

Biyanov, G. F., 1973. "Experience on Constructing Dams on Permafrost in Yakutia, U.S.S.R., " *Proc. 2d Intern. Conf. Permafrost* (Yakutsk, U.S.S.R.), U.S.S.R. Contrib., U.S. Nat. Acad. Sci., Washington, D.C., pp. 594–598.

Bogorodsky, V. V., and V. P. Gavrilo, 1980: *Ice, Physical Properties, Modern Methods of Glaciology* (in Russian), Gidrometeoizdat, Leningrad.

Botz, J. J., and W. M. Haas, 1980: "The Construction of an Embankment with Frozen Soil," U.S. Army Cold Regions Res. Eng. Lab., Hanover, N.H., CRREL Spec. Rep. 80-21.

Bragg, R. A., and O. B. Andersland, 1980: "Strain Rate, Temperature, and Sample Size Effects on Compression and Tensile Properties of Frozen Sand," *Proc. 2d Intern. Symp. Ground Freezing* (Trondheim, Norway), pp. 34–47.

British Standards Institute, 1977: "Methods for Crack Opening Displacement (COD) Testing," BS5762.

Brown, J. H., 1963: "Elasticity and Strength of Sea Ice," in *Ice and Snow—Processes, Properties and Applications,* W. D. Kingery, Ed., MIT Press, Cambridge, Mass., pp. 76–106.

Brown, W. G., 1963: "Graphical Determination of Temperature under Heated or Cooled Areas on the Ground Surface," Nat. Res. Council of Canada, Ottawa, Ont., Div. of Building Res., Tech. Pap. 163.

Browne, R. D., and P. B. Bamforth, 1981: "The Use of Concrete for Cryogenic Storage; A Summary of Research Past and Present," presented at the 1st Intern. Conf. Cryogenic Concrete (Newcastle).

Bruun, P. M., and P. Johannesson, 1971: "The Interaction between Ice and Coastal Structures," *Proc. Conf. Port and Ocean Engineering under Arctic Conditions* (Trondheim, Norway), pp. 683–712.

Burdick, J. L., E. F. Rice, and A. Phukan, 1978: "Cold Regions: Descriptive and Geotechnical Aspects," in *Geotechnical Engineering for Cold Regions*, O. B. Andersland and D. M. Anderson, Eds., McGraw-Hill, New York, chap. 1, pp. 1–36.

Butkovich, T. R., 1954: "Ultimate Strength of Ice," U.S. Army Corps of Engineers Snow, Ice and Permafrost Research Establishment (SIPRE), Res. Pap. 11.

————, 1956: "Strength Studies of Sea Ice," U.S. Army Corps of Engineers Snow, Ice and Permafrost Research Establishment (SIPRE), Res. Pap. 20.

Calkins, D. J., 1983: "Ice Jams in Shallow Rivers with Floodplain Flow," *Can. J. Civil Eng.* 10:538–548.

———— and G. D. Ashton, 1976: "Passage of Ice at Hydraulic Structures," in *Rivers '76, Symp. Inland Waterways for Navigation, Flood Control and Water Diversion* (Colorado State University, Fort Collins), Reprint, pp. 1726–1736.

———— and M. Mellor, 1975: "Cost Comparison for Lock Wall Deicing," *Proc. Intern. Ass. Hydr. Res. (IAHR), Intern. Symp. Ice* (Hanover, N.H.), pp. 51–66.

Cammaert, A. B., D. R. Miller, and R. J. Gill, 1979: "Concepts for Ice Management at Arctic LNG Terminal," *Proc. Conf. Port and Ocean Engineering under Arctic Conditions* (Trondheim, Norway), pp. 1257–1268.

———— and G. P. Tsinker, 1981: "Impact of Large Ice Floes and Icebergs on Marine Structures," *Proc. Conf. Port and Ocean Engineering under Arctic Conditions* (University of Laval, Quebec), pp. 653–662.

Canadian Standards Association, 1974: "Design of Highway Bridges," Std. S-6.

Carey, K. L., 1973: "Icings Developed from Surface Water and Ground Water," U.S. Army Cold Regions Res. Eng. Lab., Hanover, N.H., Monograph M III-D3.

————, 1977: "Solving Problems of Ice-Blocked Drainage Facilities," U.S. Army Cold Regions Res. Eng. Lab., Hanover, N.H., CRREL SR 77-25.

Carlsson, J., 1976: *Fracture Mechanics* (in Swedish), Ingenjörsförslaget, Stockholm.

Casagrande, A., 1931: "Discussion of Frost Heaving," *Highway Res. Board Proc.* 11:163–172.

Central Intelligence Agency, 1978: *Polar Regions Atlas*, Nat. Foreign Assessment Center, C.I.A., Washington, D.C.

Chamberlain, E. J., 1981. "Frost Susceptibility of Soil, Review of Index Tests," U.S. Army Cold Regions Res. Eng. Lab., Hanover, N.H., CRREL Monograph 81-2.

Chari, T. R., and K. Muthukrishnaiah, 1978: "Iceberg Threat to Ocean Floor Structures," *Proc. Intern. Ass. Hydr. Res. (IAHR), Intern. Symp. Ice* (Luleå, Sweden), pp. 421–435.

Colbeck, S. C., Ed., 1980: *Dynamics of Snow and Ice Masses*, Academic Press, New York.

Cotton, H. C., and I. M. Macaulay, 1976: "Using Steel in Arctic Construction," *Proc. Conf. Materials Engineering in the Arctic* (St. Jovite, Que.), American Society for Metals, pp. 50–58.

Cox, G. F. N., 1979: "Artificial Ice Islands for Exploratory Drilling," *Proc. Conf. Port and Ocean Engineering under Arctic Conditions* (Trondheim, Norway), pp. 147–162.

————, J. A. Richter-Menge, W. F. Weeks, M. Mellor, and H. W. Bosworth, 1984: "The Mechanical Properties of Multi-Year Sea Ice, Phase I: Test Results," U.S. Army Cold Regions Res. Eng. Lab., Hanover, N.H., CRREL Rep. 84-9.

Croasdale, K. R., 1974: "Crushing Strength of Arctic Ice," *Proc. Symp. Beaufort Sea Coast and Shelf Research*, Arctic Inst. of North America, pp. 377–399.

————, 1980: *Ice Forces on Fixed, Rigid Structures*, T. Carstens, Ed., IAHR Working Group on Ice Forces on Structures, U.S. Army Cold Regions Res. Eng. Lab., Spec. Rep. 80–26, pp. 34–106.

———— and R. W. Marcellus, 1981: "Ice Force on Large Marine Structures," *Proc. Intern. Ass. Hydr. Res. (IAHR), Intern. Symp. Ice* (Quebec, Que.), pp. 755–764.

————, N. R. Morgenstern, and J. B. Nuttal, 1976: "Indentation Tests to Investigate Ice Pressures on Vertical Piers," *J. Glaciology* **19**(81):301–312.

Crory, F. E., 1963: "Pile Foundations in Permafrost," *Proc. 1st Intern. Conf. Permafrost* (Lafayette, Ind.), pp. 467–476.

————, 1968: "Bridge Foundations in Permafrost Areas: Goldstream Creek, Fairbanks, Alaska," U.S. Army Cold Regions Res. Eng. Lab., Hanover, N.H., Tech. Rep. 180.

————, 1973: "Settlement Associated with the Thawing of Permafrost," *Proc. 2d Intern. Conf. Permafrost* (Yakutsk, U.S.S.R.), North Amer. Contrib., U.S. Nat. Acad. Sci., Washington, D.C., pp. 599–607.

————, 1978: "Design and Construction of Temporary Air Fields in the National Petroleum Reserve—Alaska," *The Northern Eng.* **10**(3), Fall.

————, 1980: "Use of Piling in Frozen Ground," presented at the ASCE Nat. Conv., Session 3, Cold Regions Eng. (Portland, Ore.).

Danys, J. V., 1978: "Ice Management of Lac St. Pierre, Quebec," *Can. J. Civil Eng.* **5**:374–390.

Davison, B. E., J. W. Rooney, and D. E. Bruggers, 1978: "Design Variables Influencing Piles Driven in Permafrost," *Proc. Conf. Applied Techniques for Cold Environments* (Anchorage, Alaska), pp. 307–318.

de Jong, J. J. A., C. Stigter, and B. Steyn, 1975: "Design and Building of Temporary Artificial Islands in the Beaufort Sea," *Proc. Conf. Port and Ocean Engineering under Arctic Conditions* (Fairbanks, Alaska), pp. 753–789.

Dempsey, B. J., J. Ingersoll, T. C. Johnson, and M. Y. Shahin, 1980: "Asphalt Concrete for Cold Regions," U.S. Army Cold Regions Res. Eng. Lab., Hanover, N.H., CRREL Rep. 80-5.

Dolgopolov, Y. V., V. P. Afanasev, V. A. Korenkov, and D. F. Panfilov, 1975: "Effect of Hummocked Ice on Piers of Marine Hydraulic Structures," *Proc. Intern. Ass. Hydr. Res. Symp. Ice Problems* (Hanover, N.H.), pp. 469–477.

Donelly, P., 1966: "An Outline of the Design and Operation of the Montreal Ice Control Structure," *Proc. Conf. Ice Pressures against Structures* (University of Laval, Quebec), Nat. Res. Council of Canada, Ottawa, Ont., TM 92, pp. 171–184.

Drouin, M., and B. Michel, 1971: "Pressure of Thermal Origin Exerted by Ice Sheets upon Hydraulic Structures" (in French), University of Laval, Quebec, Rep. S-22; also U.S. Army Cold Regions Res. Eng. Lab., Hanover, N.H., Draft Transl. 427.

Eaton, R., and R. Berg, 1978: "Temperature Effects in Compacting an Asphalt Concrete Overlay," *Proc. Conf. Applied Techniques for Cold Environment* (Anchorage, Alaska), pp. 146–158.

Eliason, K., 1978: "Materials, Manpower, Equipment and Methods," *Proc. Conf. Applied Techniques for Cold Environments* (Anchorage, Alaska), pp. 674–683.

Eranti, E., 1978: "Ice Forces on Structures and Bearing Capacity of Ice Cover" (in Finnish), Finnish Roads and Waterways Administration, Rep. 753320.

————, 1979: "Indentation Experiments with Natural Ice Plates," Helsinki University of Technology, Div. of Struct. Eng., Publ. 21.

————, F. D. Hayes, M. Määttänen, and T. T. Song, 1981: "Dynamic Ice-Structure Interaction Analysis for Narrow Vertical Structures," *Proc. Conf. Port and Ocean Engineering under Arctic Conditions* (University of Laval, Quebec), pp. 472–479.

———— and G. C. Lee, 1981: "Introduction to Ice Problems in Civil Engineering," State University of New York at Buffalo, Center for Cold Regions Eng., Sci. and Technol., 81-1.

———— and G. C. Lee, 1983: "Introduction to Cold Regions Structural Design and Construction," State University of New York at Buffalo, Center for Cold Regions Eng., Sci. and Technol., 83-1A.

————, E. Leppänen, and M. Penttinen, 1983: "Ice Control in Finnish Harbours," *Proc. Conf. Port and Ocean Engineering under Arctic Conditions* (Helsinki, Finland), pp. 370–380.

————, M. Penttinen, and T. Rekonen, 1983: "Extending the Ice Navigation Season in the Saimaa Canal," *Proc. Conf. Port and Ocean Engineering under Arctic Conditions* (Helsinki, Finland), pp. 381–391.

Esch, D. C., 1983: "Evaluation of Experimental Design Features for Roadway Construction over Permafrost," *Proc. 4th Intern. Conf. Permafrost* (Fairbanks, Alaska), University of Alaska and Nat. Acad. Sci., Washington, D.C., pp. 283–288.

Everett, D. H., and J. M. Haynes, 1965: "Capillary Properties of Some Model Pore Systems with Reference to Frost Damage," *RILEM Bull.*, new ser. 27:31–38.

Fagerlund, G., 1977: "The Critical Degree of Saturation Method of Assessing the Freeze/Thaw Resistance of Concrete," RILEM Committee 4 CDC, *Matér. Constr.* 10(58): 217–229.

———, 1981: "Principles for Frost-Susceptibility of Concrete" (in Swedish), *Nordisk betong*, no. 2, pp. 5–13.

Farouki, O. T., 1981: "Thermal Properties of Soils," U.S. Army Cold Regions Res. Eng. Lab., Hanover, N.H., Monograph 81-1.

Finnish Ministry of the Interior, 1978: Finnish collection of building regulations. C3 "Insulation—Regulations;" C4 "Insulation—Instructions" (in Finnish), Helsinki.

Finnish Roads and Waterways Administration, 1980: "Improvement of the Pavement Structure, Design Instructions" (in Finnish), TVH 722336.

Flanders, S. N., 1981: "Cold Regions Testing of an Air-Transportable Shelter," U.S. Army Cold Regions Res. Eng. Lab., Hanover, N.H., CRREL Rep. 81-16.

Foster-Miller Associates, Inc., 1973: "Fundamental Concepts for the Rapid Disengagement of Frozen Soil, Phase 1," U.S. Army Cold Regions Res. Eng. Lab., Hanover, N.H., CRREL Tech. Rep. 233.

Frankenstein, G. E., and N. Smith, 1970: "The Use of Explosives in Removing Ice Jams," *Proc. Intern. Ass. Hydr. Res. (IAHR), Intern. Symp. Ice* (Reykjavik, Iceland).

Frederking, R. N. W., 1974: "Downdrag Loads Developed by a Floating Ice Cover," *Can. Geotech. J.* 11:339–347.

———, 1977: "Plain Strain Compressive Strength of Columnar-Grained and Granular-Snow Ice," *J. Glaciology* 18(89):506–516.

———, 1979: "Laboratory Tests on Downdrag Loads Developed by Floating Ice Covers on Vertical Piles," *Proc. Conf. Port and Ocean Engineering under Arctic Conditions* (Trondheim, Norway), pp. 1097–1110.

——— and L. W. Gold, 1975: "Experimental Study of Edge Loading of Ice Plates," *Can. Geotech. J.* 12:456–463.

——— and ———, 1976: "The Bearing Capacity of Ice Covers under Static Loads," *Can. J. Civil Eng.* 3:288–293.

Fullwider, C. W., 1973: "Thermal Regime in an Arctic Earthfill Dam," *Proc. 2d Intern. Conf. Permafrost* (Yakutsk, U.S.S.R.), North Amer. Contrib., U.S. Nat. Acad. Sci., Washington, D.C., pp. 622–628.

Gardner, W. R., 1958: "Some Steady State Solutions of the Unsaturated Flow Equation with Application to Evaporation from a Water Table," *Soil Sci.* 88:228–232.

Gevay, B. J., and H. A. Erith, 1979: "Electric Heating of Intake Trash Racks at Twin Falls, Labrador," *Can. J. Civil Eng.* 6:319–324.

Gevirtz, G., and V. Mostkov, 1978: "Specific Features of Constructing Underground Hydraulic Structures in Permafrost Conditions," *Proc. Conf. Applied Techniques for Cold Environments* (Anchorage, Alaska), pp. 245–262.

Glukhov, V., 1971: "Evaluation of Ice Loads on High Structures from Aerological Observations" (in Russian), *Soviet Hydrology: Selected Papers*, no. 3/1971.

Gold, L. W., 1966: "Observations on the Movement of Ice at a Bridge Pier," *Proc. Conf. Ice Pressures against Structures* (University of Laval, Quebec), Nat. Res. Council of Canada, Ottawa, Ont., TM 92, pp. 135–141.

———, 1971: "Use of Ice Covers for Transportation," *Can. Geotech. J.* 8:170–181.

———, 1973: "Ice—A Challenge to the Engineer," *Proc. 4th Can. Conf. Applied Mechanics* (Montreal, Que.), pp. G19–G36; Nat. Res. Council of Canada, Ottawa, Ont., Tech. Pap. 395, NRC 13436.

Gray, D. M., and D. H. Male, 1981: *Handbook of Snow*, Pergamon Press, Toronto, Ont., Canada.

Gundersen, P., 1976: "Frost Protection of Utility Lines" (in Norwegian), Roy. Norw. Council Sci. and Ind. Res. and the Public Roads Administration's Committee on Frost Action in Soils, Frost i jord, no. 17, chap. 9, pp. 233–286.

Gurtin, M. E., 1964: "Variational Principles for Linear Initial-Value Problems," *Quart. Appl. Math.* 23:252–256.

Guymon, G. L., and T. V. Hromadka, 1977: "Finite Element Model of Transient Heat Conduction with Isothermal Phase Change (Two and Three-Dimensional)," U.S. Army Cold Regions Res. Eng. Lab., Hanover, N.H., CRREL Spec. Rep. 77-38.

———, ———, and R. L. Berg, 1984: "Two-Dimensional Model of Coupled Heat and Moisture Transport in Frost Heaving Soils," *Proc. 3d Intern. Offshore Mechanics and Arctic Engineering Symp.*, pp. 91–98.

Hansen, J. B., 1961: "A General Formula for Bearing Capacity," Danish Geotech. Inst., Copenhagen, Bull. 11.

Harlan, R. L., and J. F. Nixon, 1978: "Ground Thermal Regime, in *Geotechnical Engineering for Cold Regions*, O. B. Andersland and D. M. Anderson, Eds., McGraw-Hill, New York, chap. 3, pp. 103–163.

Harrison, J. D., M. G. Dawes, G. L. Archer, and M. S. Kamath, 1979: "The COD Approach and Its Application to Welded Structures, Elastic-Plastic Fracture," ASTM STP 668, pp. 606–631.

Harwood, T. A., 1973: "A Short Summary of Safety Problems in the Arctic," *Proc. Arctic Oil and Gas, Problems and Possibilities*, 5th Cong. Intern. Fondation Française d'Etudes Nordiques (Le Havre), pp. 474–484.

Haugen, D. W., 1978: "Project Management and the Construction of the Trans- Alaska Pipeline System," *Proc. Conf. Applied Techniques for Cold Environments* (Anchorage, Alaska), pp. 922–936.

Hawkes, J., and M. Mellor, 1972: "Deformation and Fracture of Ice under Uniaxial Stress," *J. Glaciology* 11(61):103–131.

Hayley, D. W., 1981: "Application of Heat Pipes to Design of Shallow Foundations on Permafrost," *Proc. 4th Can. Permafrost Conf.* (Calgary, Alta.), Nat. Res. Council of Canada, Ottawa, Ont., pp. 535–544.

Hedberg, B., L. Berntsson, and O. Berge, 1979: "The Hydrophobe 3L-Concrete, A State-of-Art Report," Chalmers University of Technology, Div. of Concrete Structures, Göteborg, Rep. 79:2.

Heger, F. J., 1978: "Thermal Gradient Deflections and Stresses in Structural Sandwich Insulating Panels," *Proc. Conf. Applied Techniques for Cold Environments* (Anchorage, Alaska), pp. 385–402.

Heiner, A., 1972: "Strength and Compaction Properties of Frozen Soil," Nat. Swedish Building Res., D11.

Heuer, C. E., J. B. Caldwell, and B. Zamsky, 1983: "Design of Buried Seafloor Pipelines for Permafrost Thaw Settlement," *Proc. 4th Intern. Conf. Permafrost* (Fairbanks, Alaska), University of Alaska and U.S. Nat. Acad. Sci., Washington, D.C., pp. 486–491.

Hoekstra, P., E. Chamberlain, and A. Frate, 1965: "Frost Heaving Pressures," U.S. Army Cold Regions Res. Eng. Lab., Hanover, N.H., CRREL Res. Rep. 176.

Hopper, H. R., C. P. S. Simonsen, and W. J. S. Poulier, 1978: "Churchill River Diversion, Burntwood River Waterway, Studies to Evaluate Winter Regime," *Can. J. Civil Eng.* 5:586–594.

Huck, R. W., 1967: "Interim Report on the Behavior of Timber Piles under Static Load in Permafrost," U.S. Army Cold Regions Res. Eng. Lab. Hanover, N.H., Tech. Note.

Hudson, R. D., 1983: "Observations on the Extrusion of Sea Ice Rubble," *Intern. Conf. Port and Ocean Engineering under Arctic Conditions* (Tech. Res. Centre of Finland, Espoo), pp. 99–108.

Hwang, C. T., R. Seshadri, and V. G. Krishnayya, 1980: "Thermal Design for Insulated Pipes," *Can. Geotech. J.* 17:613–622.

Hyvärinen, A., 1980: "Electrical Heating of Concrete with Embedded Wires" (in Finnish), in *Rakentajan Kalenteri*, Rakentajain Kustannus Oy, Helsinki, pp. 585–604.

Jahns, H. O., and C. E. Heuer, 1983: "Frost Heave Mitigation and Permafrost Protection for a Buried Chilled-Gas Pipeline," *Proc. 4th Intern. Conf. Permafrost* (Fairbanks, Alaska), University of Alaska and U.S Nat. Acad. Sci., Washington, D.C., pp. 531–536.

———, T. W. Miller, L. D. Power, W. P. Rickey, T. P. Taylor, and J. A. Wheeler, 1973: "Permafrost Protection for Pipelines," *Proc. 2d Intern. Conf. Permafrost* (Yakutsk, U.S.S.R.), North Amer. Contrib., U.S. Nat. Acad. Sci., Washington, D.C., pp. 673–684.

Jessberger, H. L., 1975: "Bearing Strength of Frost-Sensitive Soils after Thawing as a Parameter for Dimensioning Roads and as a Measure for Evaluating Frost Criteria" (in German), *Strasse u. Autobahn* 24:511–519.

Jobson, H. E., 1973: "The Dissipation of Excess Heat from the Water Systems," *ASCE J. Power Div.* 99:89–103.

Johannessen, O. M., 1970: "Note on Some Vertical Profiles below Ice Floes in the Gulf of St. Lawrence Near the North Pole," *J. Geophys. Res.* 75:2857–2862.

Johnson, R., 1980: "Resins and Non-Portland Cements for Construction in the Cold," U.S. Army Cold Regions Res. Eng. Lab., Hanover, N.H., CRREL Spec. Rep. 80-35.

Johnson, T. C., R. L. Berg, K. L. Carey, and C. W. Kaplar, 1975: "Roadway Design in Seasonal Frost Areas," U.S. Army Cold Regions Res. Eng. Lab., Hanover, N.H., CRREL Tech. Rep. 259.

———, D. M. Cole, and E. J. Chamberlain, 1978: "Influence of Freezing and Thawing on the Resilient Properties of a Silt Soil beneath an Asphalt Concrete Pavement," U.S. Army Cold Regions Res. Eng. Lab., Hanover, N.H., Rep. 78-23.

——— and F. H. Sayles, 1980: "Embankment Dams on Permafrost in the USSR," U.S. Army Cold Regions Res. Eng. Lab., Hanover, N.H., CRREL Spec. Rep. 80-40.

Johnston, G. H., 1980: "Permafrost and the Eagle River Bridge, Yukon Territory, Canada," *Permafrost Eng. Workshop Proc.* (Sept. 27–28, 1979), Nat. Res. Council of Canada (Associate Committee in Geotech. Res.), Tech. Memo. 130, pp. 12–28.

———, Ed., 1981: *Permafrost, Engineering Design and Construction*, Nat. Res. Council of Canada, Wiley, New York.

——— and B. Ladanyi, 1972: "Field Tests of Grouted Rod Anchors in Permafrost," *Can. Geotech. J.* 9:176–194.

Jonson, J.-Å., 1977: "Temperatures in Masonry Works during Winter" (in Swedish), Byggforskning, Stockholm, Rep. R-38.

Jumikis, A. R., 1966: *Thermal Soil Mechanics*, Rutgers University Press, New Brunswick, N.J.

Kane, D. L., 1981: "Physical Mechanics of Aufeis Growth," *Can. J. Civil Eng.* 8:186–195.

Kaplar, C. W., 1954: "Investigation of the Strength Properties of Frozen Soils," U.S. Army ACFEL TR 48/1.

———, 1969: "Laboratory Determination of Dynamic Moduli of Frozen Soils and of Ice," U.S. Army Cold Regions Res. Eng. Lab., Hanover, N.H., CRREL Res. Rep. 163.

———, 1971: "Some Strength Properties of Frozen Soil and Effects of Loading Rate," U.S. Army Cold Regions Res. Eng. Lab., Hanover, N.H., CRREL Spec. Rep. 159.

———, 1974 *a*: "Moisture and Freeze-Thaw Effects on Rigid Thermal Insulations," U.S. Army Cold Regions Res. Eng. Lab., Hanover, N.H., CRREL Tech. Rep. 249.

———, 1974 *b*: "A Laboratory Freezing Test to Determine the Relative Frost Susceptibility of Soils," U.S. Army Cold Regions Res. Eng. Lab., Hanover, N.H., Tech. Rep. 250.

Karri, J., 1979: "Reinforcement of Ice" (in Finnish), *J. Struct. Mech.* 12:4, Helsinki University of Technology.

Kaufman, J. G., D. J. Lege, and R. A. Kelsey, 1978: "Aluminum Alloys for Cold Environments," *Proc. Conf. Applied Techniques for Cold Environments* (Anchorage, Alaska), pp. 186–200.

Kerr, A. D., 1975: "Ice Forces on Structures Due to Change in Water Level," *Proc. Intern. Ass. Hydr. Res. (IAHR), Intern. Symp. Ice* (Hanover, N.H.), pp. 419–427.

———, 1978: "Forces an Ice Cover Exerts on Rows or Clusters of Piles Due to a Change of the Water Level," *Proc. Intern. Ass. Hydr. Res. (IAHR), Intern. Symp. Ice* (Luleå, Sweden), pp. 511–525.

Kersten, M. S., 1949: "Thermal Properties of Soils," University of Minnesota, Eng. Experiment Station, Bull. 28.

Kilpi, E., and A. Sarja, 1981: "Winter Concreting Guide for Constructors, Safe Winter-Concreting" (in Finnish), Tech. Res. Centre of Finland, VTT, Espoo, Res. Note 62/1981.

Kingery, W. D., 1962: "Sea Ice Engineering Summary Report, Project ICE WAY," U.S. Naval Civil Eng. Lab., Port Hueneme, Calif., Tech. Rep. R189.

Kivisild, H., G. Rose, and D. M. Masterson, 1975: "Salvage of Heavy Construction Equipment by a Floating Sea Bridge," *Can. Geotech. J.* 12:58–69.

Koivupalo, A., 1981: "The Properties and Use of Hot Concrete in Winter Concreting" (in Finnish), Ass. of the Concrete Industry in Finland, *Betonituote* 4/81.

Kokki, P., and H. Mäkelä, 1980: *Winter Construction, Thermal Protection and Use of Energy* (in Finnish), Rakentajain Kustannus Oy, Helsinki.

Kollmann, F., 1951: *Technology of Wood and Wood Materials* (in German), vol. 1, 2d ed., Springer-Verlag, Berlin.

Konrad, J.-M., and N. R. Morgenstern, 1984: "Frost Heave Prediction of Chilled Pipelines Buried in Unfrozen Soils," *Can. Geotech. J.* 21:100–115.

Korzhavin, K. N., 1962: "Action of Ice on Engineering Structures," U.S. Army Cold Regions Res. Eng. Lab., Hanover, N. H., Draft Transl. 260.

Kovacs, A., 1976: "Study of Piles Installed in Polar Snow," U.S. Army Cold Regions Res. Eng. Lab., Hanover, N.H., CRREL Rep. 76-23.

———, 1983: "Sea Ice on the Norton Sound and Adjacent Bering Sea Coasts," *Proc. Conf. Port and Ocean Engineering under Arctic Conditions* (Tech. Res. Centre of Finland, Espoo), pp. 654–666.

——— and D. S. Sodhi, 1979: "Ice Pile-up and Ride-up on Arctic and Subarctic Beaches," *Proc. Conf. Port and Ocean Engineering under Arctic Conditions* (Trondheim, Norway), pp. 127–146.

Krause, G. W., 1978: "Impact of Cold Environments on Contractors," *Proc. Conf. Applied Techniques for Cold Environments* (Anchorage, Alaska), pp. 896–902.

Kreig, R. A., and R. D. Reger, 1982: "Air-Photo Analysis and Summary of Landform Soil Properties along the Route of the Trans-Alaska Pipeline System," State of Alaska, Div. of Geol. and Geophys. Surveys, Geol. Rep. 66.

Kry, P. R., 1977: "Ice Rubble Fields in the Vicinity of Artificial Islands," *Proc. Conf. Port and Ocean Engineering under Arctic Conditions* (St. John's, Nfld.), pp. 200–211.

———, 1978: "Statistical Prediction of Effective Ice Crushing Stresses on Wide Structures," *Proc. Intern. Ass. Hydr. Res. (IAHR), Intern. Symp. Ice* (Luleå, Sweden), pp. 33–47.

———, 1979: "Implications of Structure Width for Design Ice Forces," *Proc. IUTAM Symp. Physics and Mechanics of Ice* (Copenhagen), Springer-Verlag, pp. 189–191.

———, 1980: "Ice Forces on Wide Structures," *Can. Geotech. J.* 17:97–113.

Krzewinski, T. G., E. S. Clarke, and M. C. Metz, 1981: "Present Condition (1980) of the TAPS Gravel Workpad," *Proc. Conf. The Northern Community: A Search for a Quality Environment*, T. S. Vinson, Ed., ASCE, New York, pp. 678–692.

Kudoyarov, V. I., M. P. Pavchich, and V. G. Radchenko, 1978: "Earth and Rock Excavation Technique for Construction of Hydraulic Structures in Regions of Permafrost," *Proc. Conf. Applied Techniques for Cold Environments* (Anchorage, Alaska), pp. 201–215.

Kuroiwa, D., 1965: "Icing and Snow Accretion on Electric Wires," U.S. Army Cold Regions Res. Eng. Lab., Hanover, N.H., CRREL Res. Rep. 123.

Lachenbruch, A. H., 1959: "Periodic Heat Flow in a Stratified Medium with Application in Permafrost Problems," U.S. Geol. Survey, Bull. 1052, 51 pp.

———, 1962: "Mechanics of Thermal Contraction Cracks and Ice-Wedge Polygons in Permafrost," Geol. Soc. America, Spec. Pap. 70.

———, 1970: "Some Estimates of the Thermal Effects of a Heated Pipeline in Permafrost," U.S. Geol. Survey, Circular 632.

Ladanyi, B., 1976: "Use of the Static Penetration Test in Frozen Soils," *Can. Geotech. J.* 13:95–110.

———, 1983: "Shallow Foundations on Frozen Soil: Creep Settlement," *ASCE J. Geotech. Eng.* 109(11).

——— and G. H. Johnston, 1974: "Behaviour of Circular Footings and Plate Anchors Embedded in Permafrost," *Can. Geotech. J.* 11:531–553.

——— and ———, 1978: "Field Investigations of Frozen Ground," in *Geotechnical Engineering for Cold Regions*, O. B. Andersland and D. M. Anderson, Eds., McGraw-Hill, New York, chap. 9, pp. 459–504.

Lang, T. E., and R. L. Brown, 1980: "Snow-Avalanche Impact on Structures," *J. Glaciology* 25(93).

Langmuir, I., and K. Blodgett, 1946: "Mathematical Investigation of Water Droplet Trajectories," U.S. Army Air Force, Tech. Rep. 5418.

Larsson, N. O., 1974: "Ice Pressure at Bridges," *Symp. Ice Problems in Stockholm* (in Swedish), IVA Rep. 190.

Lavonie, N. Y., 1966: "Ice Effects on Structures in the Northumberland Strait Crossing," *Proc. Conf. Ice Pressures against Structures* (University of Laval, Quebec), Nat. Res. Council of Canada, Ottawa, Ont., TM 92.

Länsiluoto, J., 1977: "Fabrication and Qualities of Steel and Steel Products," in *Steel Structures* (in Finnish), Finnish Asso. for Civil Eng., RIL 113, chap. 2.

Lehtinen, E., E. Törmi, H. Miettinen, and A. Sarja, 1979: "Infrared Heating of Concrete in Angle Forms" (in Finnish), Tech. Res. Centre of Finland, VTT, Concrete and Silicate Lab., Espoo, Note 59.

Leppävuori, E. K. M., 1977: "Creep of Ice, II" (in Finnish), *J. Struct. Mech.* 10:4, Helsinki University of Technology.

Linell, K. A, and C. W. Kaplar, 1966: "Description and Classification of Frozen Soils," *Proc. Intern. Conf. Permafrost* (Lafayette, Ind., 1963), U.S. Nat. Acad. of Sci., Washington, D.C., Publ. 1287, pp. 481–487.

———— and E. F. Lobacz, 1980: "Design and Construction of Foundations in Areas of Deep Seasonal Frost and Permafrost," U.S. Army Cold Regions Res. Eng. Lab., Hanover, N.H., CRREL Spec. Rep. 80-34.

———— and J. C. F. Tedrow, 1981: *Soil and Permafrost Surveys in the Arctic*, Oxford University Press, New York.

Lipinski, R., 1978: "The Use of Cost and Schedule Control during Construction of the Trans-Alaska Pipeline," *Proc. Conf. Applied Techniques for Cold Environments* (Anchorage, Alaska), pp. 951–959.

Livingston, H., and E. Johnson, 1978: "Insulated Roadway Subdrains in the Subarctic for the Prevention of Spring Icings," *Proc. Conf. Applied Techniques for Cold Environments* (Anchorage, Alaska), ASCE, pp. 513–521.

Lobacz, E. F., G. D. Gilman, and F. B. Hennion, 1973: "Corps of Engineers Design of Highway Pavements in Areas of Seasonal Frost," *Proc. Symp. Frost Action Roads* (Oslo), pp. 142–152.

Logan, T. H., 1974: "Prevention of Frazil Ice Clogging of Water Intakes by Application of Heat," prepared for Ice Research Committee, Eng. Res. Center, Bureau of Reclamation, Rec-ERC-74-15.

Long, E. L., 1973: "Designing Friction Piles for Increasing Stability at Lower Installed Cost in Permafrost," *Proc. 2d Intern. Conf. Permafrost* (Yakutsk, U.S.S.R.), North Amer. Contrib., U.S. Nat. Acad. Sci., Washington, D.C., pp. 693–699.

————, 1978: "Permafrost Foundation Designs," *Proc. Conf. Applied Techniques for Cold Environments* (Anchorage, Alaska), pp. 973–987.

Lottman, R. P., 1978: "Predicting Moisture-Induced Damage to Asphaltic Concrete," Nat. Cooperat. Highway Res. Program, Rep. 192.

Löfquist, B., 1954: "Ice Pressure: Studies of the Effects of Temperature Variations," *Trans. ASCE* 119, Pap. 2656.

Lunardini, V. J., 1978: "Theory of *n*-Factors and Correlation of Data," *Proc. 3d Intern. Conf. Permafrost*, Nat. Res. Council of Canada, Ottawa, Ont., pp. 41–46.

————, 1982: "Conduction Phase Change beneath Insulated Heated or Cooled Structures," U.S. Army Cold Regions Res. Eng. Lab., Hanover, N.H., CRREL Rep. 82-22.

Maclean, C., W. Semotiuk, A. Strandberg, and D. M. Masterson, 1981: "Ice Platforms with Urethane Foam Cells in the Neutral Axis Zone and Their Application in Arctic Offshore Drilling," *Proc. Intern. Conf. Port and Ocean Engineering under Arctic Conditions* (University of Laval, Quebec), pp. 49–59.

Manikian, V., 1983: "Pile Driving and Load Tests in Permafrost for the Kuparuk Pipeline System," *Proc. 4th Intern. Conf. Permafrost* (Fairbanks, Alaska), University of Alaska and U.S. Nat. Acad. Sci., Washington, D.C., pp. 804–810.

Mariusson, J. M., S. Freysteisson, and E. B. Eliasson, 1975: "Ice Jam Control, Experience from the Burfell Power Plant, Iceland," *Proc. Intern. Ass. Hydr. Res. (IAHR), Intern. Symp. Ice* (Hanover, N.H.).

Markin, K. F., and Y. O. Targulyan, 1973: "Bearing Capacity of Piles in Permafrost," in *Principles of the Control of Cryogenic Processes during the Development of Permafrost Regions*, S. S. Vyalov, Ed., U.S. Army Cold Regions Res. Eng. Lab., Hanover, N.H., Transl. TL 438.

Masterson, D. M., K. G. Anderson, and A. G. Strandberg, 1979: "Strain Measurements in Floating Ice Platforms and Their Application to Platform Design," *Can. J. Civil Eng.* 6:394–400.

Matilainen, V., and P. Suontausta, 1970: "Artificial Thawing of Snow" (in Finnish), *Rakennustekniikka* 10, Helsinki.

Määttänen, M., 1975: "Ice Forces and Vibrational Behavior of Bottom-Founded Steel Lighthouses," *Proc. Intern. Ass. Hydr. Res. (IAHR), Intern. Symp. Ice* (Hanover, N.H.), pp. 345–355.

———, 1977: "Stability of Self-Excited Ice-Induced Structural Vibrations," *Proc. Conf. Port and Ocean Engineering under Arctic Conditions* (St. John's, Canada), pp. 684–694.

———, 1979: "Laboratory Tests for Dynamic Ice Structure Interaction," *Proc. Intern. Conf. Port and Ocean Engineering under Arctic Conditions* (Trondheim, Norway), pp. 1139–1154.

———, T. Krankala, J. Hoikkanen, and E. Pulkkinen, 1984: "Calculation Methods for Ice Loads against Offshore Structures," University of Oulu, Dept. of Mech. Eng., Rep. 43.

Mäkelä, H., 1982: "Frost Protection and Insulation of Shallow-Laid Pipelines" (in Finnish), Tech. Res. Centre of Finland, VTT, Espoo, Res. Notes 113.

——— and S. Saarelainen, 1978: "Working Methods for House Foundation Construction in Winter Conditions" (in Finnish), Tech. Res. Centre of Finland, VTT, Geotech. Lab., Espoo, Note 33.

——— and M. Tammirinne, 1979: "Instructions for Frost Protection of Building Foundations" (in Finnish), Tech. Res. Centre of Finland, VTT, Geotech. Lab., Espoo, Rep. 37.

McDougall, J. C., 1977: "The Beaufort Gas Project Surface Facilities," *Proc. 2d Intern. Symp. Cold Regions Engineering* (University of Alaska), pp. 383–400.

McGonical, D., and B. D. Wright, 1982: "First-Year Pressure Ridges in the Beaufort Sea," *Proc. Intermaritec* (Hamburg), pp. 444–459.

McLeod, W. R., 1977: "Atmospheric Superstructure Ice Accumulation Measurements," *Proc. Offshore Technol. Conf.* (Houston, Tex.), OTC 2950.

McRoberts, E. C., 1973: "Stability of Slopes in Permafrost," Ph.D. thesis, University of Alberta, Dept. of Civil Eng., Edmonton.

———, T. C. Law, and T. K. Murray, 1978: "Creep Tests on Undisturbed Ice-Rich Silt," *Proc. 3d Intern. Conf. Permafrost* (Edmonton, Alta.), Nat. Res. Council of Canada, Ottawa, Ont., vol. 1, pp. 539–545.

Mellor, M., 1964: "Snow and Ice on the Earth's Surface," U.S. Army Cold Regions Res. Eng. Lab., Hanover, N.H., Monograph II-C1.

———, 1965: "Blowing Snow," U.S. Army Cold Regions Res. Eng. Lab., Hanover, N.H., Monograph III-A3c.

———, 1968: "Avalanches," U.S. Army Cold Regions Res. Eng. Lab., Hanover, N.H., Monograph III-A3d.

———, 1969: "Foundations and Subsurface Structures in Snow," U.S. Army Cold Regions Res. Eng. Lab., Hanover, N.H., Monograph III-A2c.

——— and S. Reed, 1967: "Ice Cap Strains and Some Effects on Engineering Structures," U.S. Army Cold Regions Res. Eng. Lab., Hanover, N.H., CRREL Tech. Rep. 202.

——— and P. V. Snellman, 1970: "Experimental Blasting in Frozen Ground," U.S. Army Cold Regions Res. Eng. Lab., Hanover, N.H., CRREL Spec. Rep. 153.

Meyerhof, G. G., 1962: "Bearing Capacity of Floating Ice Sheets," *Trans. ASCE* 127:524–581.

Michel, B., 1966: "Thrust Exerted by an Unconsolidated Ice Cover on a Boom," *Proc. Conf. Ice Pressures against Structures* (University of Laval, Quebec), Nat. Res. Council of Canada, Ottawa, Ont., TM 92, pp. 163–170.

———, 1971: "Winter Regime in Rivers and Lakes," U.S. Army Cold Regions Res. Eng. Lab., Hanover, N.H., Monograph III-B1a.

———, 1978: *Ice Mechanics*, University of Laval Press, Quebec.

——— and D. Berenger, 1975: "Algorithm for Accelerated Growth of Ice in a Ship's Track," *Proc. Intern. Ass. Hydr. Res. (IAHR), Intern. Symp. Ice* (Hanover, N.H.).

——— and R. O. Ramseier, 1971: "Classification of River and Lake Ice," *Can. Geotech. J.* 8: pp. 36–45.

———— and N. Toussaint, 1976: "Mechanism and Theory of Indentation of Ice Plates," *J. Glaciology* 18(81):285–300.

Miettinen, H., J. Vuorinen, and H. Kukko, 1981: "Comparison of Codes Concerning Winter Concreting in Finland and in the Soviet Union," Tech. Res. Centre of Finland, VTT, Espoo, Res. Notes 11/1981.

Minsk, L. D., 1977: "Ice Accumulation on Ocean Structures," U.S. Army Cold Regions Res. Eng. Lab., Hanover, N.H., CRREL Rep. 77-17.

————, 1980: "Icing on Structures," U.S. Army Cold Regions Res. Eng. Lab., Hanover, N.H., CRREL Rep. 80-31.

Mironov, S. A., 1956: *Theory and Methods of Winter Concreting* (in Russian), Moscow; also U.S. Army Cold Regions Res. Eng. Lab., Draft Transl. 636.

Mohan, A., 1975: "Heat Transfer in Soil-Water-Ice Systems," *ASCE J. Geotech. Div.* 101(GT2):97–113.

Monfore, G. E., 1954: "Ice Pressure against Dams. Experimental Investigation by the Bureau of Reclamation," *Trans. ASCE* 119:26–38.

Moore, H. E., and F. H. Sayles, 1980: "Excavation of Frozen Materials," collection of papers from a U.S.–Soviet Union joint seminar: *Building under Cold Climates and on Permafrost,* U.S. Dept. of Housing and Urban Develop. and U.S. Army Corps of Engineers, pp. 323–345.

Morgenstern, N. R., and J. F. Nixon, 1971: "One-Dimensional Consolidation of Thawing Soils," *Can. Geotech. J.* 8:558–565.

NAHB Research Foundation, Inc., 1975: "All-Weather Home Building Manual," U.S. Dept. of Housing and Urban Develop. and U.S. Dept. of Housing, Rockville, Md.

Nakano, Y., and N. H. Froula, 1973: "Sound and Shock Transmission in Frozen Soils," *Proc. 2d Intern. Conf. Permafrost* (Yakutsk, U.S.S.R.), North Amer. Contrib., U.S. Nat. Acad. Sci., Washington, D.C., pp. 359–369.

Nevel, D. E., 1961: "The Narrow Free Infinite Wedge on an Elastic Foundation," U.S. Army Corps of Engineers Snow, Ice and Permafrost Research Establishment (SIPRE), Res. Rep. 79.

————, 1970: "Moving Loads on a Floating Ice Sheet," U.S. Army Cold Regions Res. Eng. Lab., Hanover, N.H., Res. Rep. 261.

————, 1978: "Bearing Capacity of River Ice for Vehicles," U.S. Army Cold Regions Res. Eng. Lab., Hanover, N.H., CRREL Rep. 78-3.

Nilson, L. O., 1976: "Moisture Problems at Concrete Floors" (in Swedish), Lund Institute of Technology, Rep. TVBM-3002.

Nixon, J. F., 1978: "First Canadian Geotechnical Colloquium: Foundation Design Approaches in Permafrost Areas," *Can. Geotech. J.* 15:96–112.

————, 1979: "Some Aspects of Road and Airstrip Pad Design in Permafrost Areas," *Can. Geotech. J.* 16:222–225.

————, 1983: "Geothermal Design of Insulated Foundations for Thaw Prevention," *Proc. 4th Intern. Conf. Permafrost* (Fairbanks, Alaska), University of Alaska and U.S. Nat. Acad. Sci., Washington, D.C., pp. 924–927.

————, 1984: "Laterally Loaded Piles in Permafrost," *Can. Geotech. J.* 21:431–438.

———— and B. Ladanyi, 1978: "Thaw Consolidation," in *Geotechnical Engineering for Cold Regions,* O. B. Andersland and D. M. Anderson, Eds., McGraw-Hill, New York, chap. 4, pp. 164–215.

———— and G. Lem, 1984: "Creep and Strength Testing of Frozen Saline Fine-Grained Soils," *Can. Geotech. J.* 21:518–529.

———— and E. C. McRoberts, 1976: "A Design Approach for Pile Foundations in Permafrost," *Can. Geotech. J.* 13:40–57.

————, J. Stuchly, and A. R. Pick, 1984: "Design of Norman Wells Pipeline for Frost Heave and Thaw Settlement," *Proc. 3d Intern. Offshore Mechanics and Arctic Engineering Symp.*, vol. 3, V. J. Lunardini, Ed., U.S. Army Cold Regions Res. Eng. Lab., Hanover, N.H., ASME, pp. 69–76.

Nottingham, D., and A. B. Christopherson, 1983: "Driven Piles in Permafrost: State of the Art," *Proc. 4th Intern. Conf. Permafrost* (Fairbanks, Alaska), University of Alaska and U.S. Nat. Acad. Sci., Washington, D.C., pp. 928–933.

Nykänen, A., and P. Ahtola, 1967: "Winter Concreting" (in Finnish), Finnish Society of Civil Engineers, Helsinki, RIL 51.

Oberbach, K., 1975: *The Properties of Polymers for Structures* (in German), Carl Hanser Verlag, Munich.

Odar, F., 1965: "Simulation of Drifting Snow," U.S. Army Cold Regions Res. Eng. Lab., Hanover, N.H., CRREL Res. Rep. 174.

Ohstrom, E. G., and S. L. DenHartog, 1976: "Cantilever Beam Tests on Reinforced Ice," U.S. Army Cold Regions Res. Eng. Lab., Hanover, N.H., CRREL Rep. 76-7.

Oksanen, P., 1980: "Adhesion Strength Measurements between Ice and Some Materials" (in Finnish), Tech. Res. Centre of Finland, VTT, Lab. Struct. Eng., Espoo.

O'Neill, K., and R. D. Miller, 1982: "Numerical Solutions for a Rigid-Ice Model of Secondary Frost Heave," U.S. Army Cold Regions Res. Eng. Lab., Hanover, N.H., CRREL Rep. 82-13.

Paige, R. A., and C. W. Lee, 1966: "Sea Ice Strength Studies on the McMurdo Sound during the Austral Summer 1964–65," U.S. Naval Civil Eng. Lab., Tech. Rep. R-437.

Panfilov, D. F., 1960: "Experimental Investigation of the Carrying Capacity of an Ice Cover," Izvestiya Vsesoyuznogo Nauchno-Issledovatel skogo, Instituta Gidrotekhniki, vol. 64, pp. 101–115.

Pariset, E., R. Hausser, and A. Gagnon, 1966: "Formation of Ice Covers and Ice Jams in Rivers," *ASCE J. Hydr. Div.* 92(6):1–24.

Parmerter, R. R., and M. D. Coon, 1973: "Model of Pressure Ridge Formation in Sea Ice," *J. Geophys. Res.* 77(33).

Pedersen, K. B., 1977: "Frost Protection of Norwegian Road Tunnels" (in Norwegian), Frost i jord, Oslo, no. 19, pp. 27–32.

Pellini, W. S., 1971. AWS Adams Lecture, "Principles of Fracture Safe Design, Part I," *Welding J.*, Res. Suppl., Mar.

Penner, E., and L. E. Goodrich, 1983: "Adfreezing Stresses on Steel Pipe Piles, Thompson, Manitoba," *Proc. 4th Intern. Conf. Permafrost* (Fairbanks, Alaska), University of Alaska and U.S. Nat. Acad. Sci., Washington, D.C., pp. 979–983.

Penttala, V., and E. Törmi, 1977: "Infrared Heating of Concrete Structures in Winter Construction Circumstances" (in Finnish), Tech. Res. Centre of Finland, VTT, Concrete and Silicate Lab., Espoo, Note 45.

Perdichizzi, P., and T. Yasuda, 1978: "Port of Anchorage Marine Terminal Design," *Proc. Conf. Applied Techniques for Cold Environments* (Anchorage, Alaska), pp. 875–886.

Perham, R. E., 1976: "Some Economic Benefits of Ice Booms," *Proc. 2d Intern. Symp. Cold Regions Engineering* (Fairbanks, Alaska), pp. 570–591.

———, 1977: "St. Mary River Ice Booms, Design Force Estimate and Field Measurements," U.S. Army Cold Regions Res. Eng. Lab., Hanover, N.H., CRREL Rep. 77-4.

———, 1978: "Righting Moment in a Rectangular Ice Boom Timber or Pontoon," *Proc. Intern. Ass. Hydr. Res. (IAHR), Intern. Symp. Ice* (Luleå, Sweden), pp. 273–289.

———, 1980: "Harnessing Frazil Ice," presented at the Workshop of Hydraulic Resistance of River Ice, Canada Center for Inland Waters, Burlington, Ont.

Peterson, N., 1966: "Concrete Quality Control and Authorization of Ready Mixed Concrete Factories in Sweden," Swedish Cement and Concrete Res. Inst. of Technol., Stockholm.

Péwé, T. L., 1982: "Geologic Hazard of the Fairbanks Area, Alaska," Alaska Geol. and Geophys. Surveys, Spec. Rep. 15.

Peyton, H. R., 1966: "Sea Ice Strength," University of Alaska, Geophys. Inst., Rep. UAG R-182.

———, 1968: "Ice and Marine Structures, Parts I, II, and III," *Ocean Industry*, Mar., Sept., Dec.

Pihlajavaara, S. E., and H. Syrjälä, 1960: "Temperature Changes of Fresh Concrete in Winter Concreting, Theory and Applications" (in Finnish), Tech. Res. Centre of Finland, VTT, Concrete and Silicate Lab., Espoo, Publ. 50.

Pilkington, G. R., and R. W. Marcellus, 1981: "Methods of Determining Pipeline Trench Depths in the Canadian Beaufort Sea," *Proc. Conf. Port and Ocean Engineering under Arctic Conditions* (University of Laval, Quebec), pp. 674–687.

Portland Cement Association, 1979: *Design and Control of Concrete Mixtures*, 12th ed.

Potter, R. E., D. L. Reid, J. C. Bruce, and P. G. Noble, 1982: "Development and Field Testing of a Beaufort Sea Ice Boom," *Proc. Offshore Technol. Conf.*, OTC 4381.

Price, W. I. J., 1961: "The Effect of the Characteristics of Snow Fences on the Quality and

Shape of the Deposited Snow," General Assembly of International Union of Geodesy and Geophysics, Publ. 54.

Prodanovic, A., 1981: "Upper Bounds of Ridge Pressure on Structures," *Proc. Conf. Port and Ocean Engineering under Arctic Conditions* (University of Laval, Quebec), pp. 1288–1302.

Pufahl, D. E., and N. R. Morgenstern, 1980: "Remedial Measures for Slope Instability in Thawing Permafrost," *Proc. 2nd Intern. Symp. Ground Freezing* (Trondheim, Norway), Trondheim University, Norwegian Inst. Technol., pp. 1089–1101.

Pugh, H. L. D., and W. I. J. Price, 1954: "Snow Drifting and the Use of Snow Fences," *Polar Record* 7(47).

Ralston, T. D., 1977: "Ice Force Design Consideration for Conical Offshore Structures," *Proc. Conf. Port and Ocean Engineering under Arctic Conditions* (St. John's, Nfld.), pp. 741–752.

———, 1978: "An Analysis of Ice Sheet Indentation," *Proc. Intern. Ass. Hydr. Res. (IAHR), Intern. Symp. Ice* (Luleå, Sweden), pp. 5–31.

Rantamäki, J., 1972: "Ice Indentation Tests during Winter 1971–1972 for Oulu-Kemi Navigation Channel" (in Finnish), Finnish Board of Navigation.

Räsänen, E., 1980: "Steel Structures for Low Application Temperatures" (in Finnish), Insinöörijärjestöjen Koulutuskeskus, INSKO 124-80 I.

———, 1982: "Use of High Tensile Steels in Welded Constructions," *Scand. J. Metallurgy* 11:115–121.

Reike, J. K., and G. E. Clock, 1966: "A New Adherent Thermoplastic Polyethylenic Material — High Density Polyethylene Acrylic Acid Graft Copolymers," *Proc. 22d Ann. Tech. Conf. of the Society of Plastics Engineers* (Montreal, Que.), sec. VI-6.

Reinicke, K. M., 1979: "Analytical Approach for the Determination of Ice Forces Using Plasticity Theory," *IUTAM Symp. Physics and Mechanics of Ice* (Copenhagen), Springer-Verlag, pp. 325–341.

Rekonen, T., 1973: "Effects of Ice on Navigation and Navigational Aids," *PIANC 23, Intern. Navigation Conf.* (Ottawa), Pap. S II-4.

Réunion Internationale des Laboratoires d'Essais et de Recherches sur les Matériaux et les Constructions, 1980: "International Recommendations on Winter Concreting" (first draft), RILEM Tech. Comm. on Winter Concreting (BH-39), B. A. Krylov, Chairman, Moscow.

Rice, E., 1975: "Building in the North," *The Northern Eng.*, University of Alaska, Geophys. Inst., Fairbanks.

Riska, K., P. Kujala, and J. Vuorio, 1983: "Ice Load and Pressure Measurements on Board I. B. Sisu," *Proc. Conf. Port and Ocean Engineering under Arctic Conditions*, Tech. Res. Centre of Finland, VTT, Espoo, pp. 1055–1069.

Roberts, W. S., 1976: "Regionalized Feasibility Study of Cold Weather Earth Work," U.S. Army Cold Regions Res. Eng. Lab., Hanover, N.H., CRREL Spec. Rep. 76-2.

Rolfe, S. T., and J. M. Barsom, 1977: *Fracture and Fatigue Control in Structures*, Prentice-Hall, Englewood Cliffs, N.J.

Rowley, R. K., G. H. Watson, and B. Ladanyi, 1973: "Vertical and Lateral Pile Load Tests in Permafrost," *Proc. 2d Intern. Conf. Permafrost* (Yakutsk, U.S.S.R.), U.S. Nat. Acad. Sci., Washington, D.C., pp. 712–721.

———, ———, and ———, 1975: "Prediction of Pile Performance in Permafrost under Lateral Load," *Can. Geotech. J.* 12:510–523.

Royal Norwegian Council for Scientific and Industrial Research and the Public Roads Administration's Committee, 1976: "Frost Action in Soils" (in Norwegian), Vegdirektoratet/NTNF.

Rumer, R. R., R. Crissman, and A. Wake, 1979: "Ice Transport in Great Lakes," State University of New York at Buffalo, Dept. of Civil Eng., Water Resources and Environmental Eng. Res. Rep. 79-3, and Center for Cold Regions Eng. and Sci. Technol., Contrib. 79-1.

Sadovsky, S. I., N. A. Zinchenko, and S. I. Korol, 1978: "Winter Concreting at the Sayano-Shushenskaya Hydro Project," *Proc. Conf. Applied Techniques for Cold Environments* (Anchorage, Alaska), pp. 275–286.

Sanger, F. J., 1963: "Degree Days and Heat Conduction in Soils," *Proc. Intern. Conf. Permafrost*, NAS-NRC, pp. 253–263.

————, 1969: "Foundations of Structures in Cold Regions," U.S. Army Cold Regions Res. Eng. Lab., Hanover, N.H., CRREL Monograph MIII-C4.

Saunders, T. F., and M. Timascheff, 1973: "Ice Effects on Planning, Design and Operation of Major Oil Terminal," *Proc. 23d Intern. Navigation Cong.* (Ottawa, Ont.), General Secreteriat of P.I.A.N.C., Brussels, Pap. S.II-4.

Save, S., 1975: "Properties of Wood and Strength of Glued Joints at Low Temperatures" (in Finnish), Technical University of Helsinki, Dept. of Wood Processing, Otaniemi.

Sayles, F. H., 1968: "Creep of Frozen Sands," U.S. Army Cold Regions Res. Eng. Lab., Hanover, N.H., Tech. Rep. 190.

————, 1973: "Triaxial and Creep Tests on Frozen Ottawa Sand," *Proc. 2d Intern. Conf. Permafrost* (Yakutsk, U.S.S.R.), North Amer. Contrib., U.S. Nat. Acad. Sci., Washington, D.C., pp. 384–391.

———— and D. Haines, 1974: "Creep of Frozen Silt and Clay," U.S. Army Cold Regions Res. Eng. Lab., Hanover, N.H., Tech. Rep. 252.

Shira, D. L., 1978: "Hydroelectric Power Plant Siting in Glacial Areas," *Proc. Conf. Applied Techniques for Cold Environments* (Anchorage, Alaska), pp. 59–76.

Siple, P. A., 1945: "Measurements of Dry Atmospheric Cooling in Subfreezing Temperatures," *Proc. Amer. Philos. Soc.* 89:177–199.

Sodhi, D. S., 1979: "Buckling Analysis of Wedge Shaped Floating Ice Sheet," *Proc. Intern. Conf. Port and Ocean Engineering under Arctic Conditions* (Trondheim, Norway), pp. 797–810.

Sundberg-Falkemark, N., 1963: "Load Bearing Capacity of Ice," U.S. Army Cold Regions Res. Eng. Lab., Hanover, N.H., Draft Transl. 684.

Svenska Byggnadsentreprenörföreningens Produktionsråd (Production Council of Swedish Contractors), 1963: "Year-Round Road Construction" (in Swedish), Rep. 1.

Swinzow, G. K., 1970: "Permafrost Tunnelling by a Continuous Mechanical Method," U.S. Army Cold Regions Res. Eng. Lab., Hanover, N.H., Tech. Rep. TR 221.

Sykes, J. F., 1973: "The Thaw and Consolidation of Permafrost," Thesis, University of Waterloo, Ont., Canada.

————, W. C. Lennox, and R. G. Charlwood, 1974a: "Finite Element Permafrost Thaw Settlement Model," *ASCE J. Geotech. Div.* 100(GT11):1185–1201.

————, and T. E. Unny, 1974b: "Two-Dimensional Heated Pipeline in Permafrost," *ASCE J. Geotech. Div.* 100(GT11):1203–1214.

Tada, H., P. Paris, and G. Irwin, 1973: "The Stress Analysis of Cracks Handbook," DEL Research Corp., Hellertown, Penn.

Tart, R. G., Jr., 1983: "Winter Constructed Gravel Islands," *Proc. 4th Intern. Conf. Permafrost,* University of Alaska and U.S. Nat. Acad. Sci., Washington, D.C., pp. 1233–1238.

———— and M. Luscher, 1981: "Construction and Performance of Frozen Gravel Fills," *Proc. Conf. The Northern Community: A Search for a Quality Environment,* T. S. Vinson, Ed., ASCE, New York, pp. 693–704.

Tattelman, P., and I. Gringorten, 1973: "Estimated Glaze Ice and Wind Loads at the Earth's Surface for the Contiguous United States," U.S. Air Force Cambridge Res. Lab., Bedford, Mass., Rep. AFCRL-TR-73-0646.

Taylor, D. A., 1981: "Snow Loads for the Design of Cylindrical Curved Roofs in Canada, 1953–1980," *Can. J. Civil Eng.* 8:63–76.

Templin, J. T., and W. R. Schriever, 1982: "Loads Due to Drifted Snow," *ASCE J. Struct. Div.,* 108(ST8):1916–1925.

Tesaker, E., 1975: "Accumulation of Frazil Ice in an Intake Reservoir," *Proc. Intern. Ass. Hydr. Res. (IAHR), Intern. Symp. Ice* (Hanover, N.H.).

Thomas, B. C., 1981: "Lightering with Air Cushion Vehicles in Alaska," *Proc. Conf. The Northern Community: A Search for a Quality Environment,* T. S. Vinson, Ed., ASCE, New York, pp. 249–263.

Thompson, E. G., and F. H. Sayles, 1972: "In Situ Creep Analysis of Room in Frozen Soil," *ASCE J. Soil Mech. Foundations Div.* 98:899–915.

Thunell, B., 1942: "Quality and Strength in Practice" (in Swedish), Statens Provningsanst., Stockholm, Medd. 89.

Titus, J. B., 1967: "Effect of Low Temperature (0 to −65°F) on the Properties of Plastics," Plastics Tech. Evaluation Center, Picatinny Arsenal, Dover, N.J., Plastec Rep. 30.

Tobiasson, W., 1971: "Deterioration of Structures in Cold Regions," *Proc. Symp. Cold Regions Engineering* (University of Alaska College), pp. 425–448.

————, 1980: "Roofs in Cold Regions," ASCE, Preprint 80-172A.

————, H. Ueda, and G. Hine, 1974: "Measurement of Forces within the Structural Frame of DEW Line Ice Cap Stations DYE-2 and DYE-3," U.S. Army Cold Regions Res. Eng. Lab., Hanover, N.H., CRREL Spec. Rep. 205.

Torgersen, S. E., 1976: "Design of Ground Supported Floors against Frost Action" (in Norwegian), Roy. Norw. Council for Sci. and Ind. Res. and the Public Roads Administration's Committee on Frost Action in Soils, Frost i jord, no. 17, chap. 10, pp. 287–314.

Tsytovich, N. A., 1975. *The Mechanics of Frozen Ground,* McGraw-Hill, New York.

Tynkkynen, S., and I. Ojala, 1979: "Development of Energy Saving Winter Construction Techniques, Additional Winter Expenses, State of the Art" (in Finnish), Tech. Res. Centre of Finland, VTT, Lab. of Construction Economy, Espoo, Note 56/79.

UNESCO, 1970: *Proc. Symp. Ecology of Subarctic Regions* (Helsinki, Finland), UNESCO, Paris.

Urabe, N., and A. Yoshitake, 1981: "Steel Selection System and Reliability Analysis of Structures in Cold Regions," *Proc. Conf. Port and Ocean Engineering under Arctic Conditions* (University of Laval, Quebec), pp. 462–471.

U.S. Air Force, Geophys. Res. Directorate, 1960: *Handbook of Geophysics,* Macmillan, New York.

U.S. Forest Service, 1961: "Snow and Avalanches: A Handbook of Forecasting and Control Measures," Dept. of Agriculture, Agriculture Handbook 194.

U.S. Naval Facilities Engineering Command, 1975: "Design Manual, Cold Regions Engineering," Alexandria, Va., NAVFAC DM-9.

U.S.S.R. design code SNiP II-18-76, 1977: "Bases and Foundations on Permafrost" (in Russian), Gosstroi, Strojizdat, Moscow.

U.S.S.R. design code SNiP III-15-76, 1977: "Cast in Place Concrete and Reinforced Concrete Structures" (in Russian), Gosstroi, Strojizdat, Moscow.

Uzuner, M. S., 1975: "Stability of Ice Blocks beneath the Ice Cover," *Proc. Intern. Ass. Hydr. Res. (IAHR), Intern. Symp. Ice* (Hanover, N.H.)

Vance, G. P., 1980: "Clearing Ice-Clogged Shipping Channels," U.S. Army Cold Regions Res. Eng. Lab., Hanover, N.H., CRREL Rep. 80-28.

Van Ginkel Associates, Ltd., 1976: "Building in the North," vols. 1 and 2, prepared for Canadian Arctic Pipeline, Ltd., Gulf Oil Canada, Ltd., Imperial Oil, Ltd., and Shell Canada, Ltd.

Varsta, P., 1983: "On the Mechanics of Ice Load on Ships in Level Ice in the Baltic Sea," Techn. Res. Centre of Finland, VTT, Espoo, Publ. 11.

Veldman, W. M., and E. K. Yaremko, 1978: "Design and Construction of River Training Structures." *Proc. Conf. Applied Techniques for Cold Environments* (Anchorage, Alaska), pp. 852–863.

Velli, J. J., V. V. Dokutsajeva, and N. F. Fjedorova, 1977: *Handbook of Building in Permafrost Area* (in Russian), Strojizdat, Leningradskoje otdelenije, Leningrad.

Ven Te Chow, 1959: *Open Channel Hydraulics,* McGraw-Hill, New York.

Vinson, T. S., 1978: "Response of Frozen Ground to Dynamic Loading," in *Geotechnical Engineering for Cold Regions,* O. B. Andersland and D. M. Anderson, Eds., McGraw-Hill, New York, chap. 8, pp. 405–458.

Virtanen, R., 1978: "Breaking and Thawing Frost" (in Finnish), *Maansiirto* 1:42–47.

Vita, C. L., and J. W. Rooney, 1978: "Seepage-Induced Erosion along Buried Pipelines," *Proc. Conf. Applied Techniques for Cold Environments* (Anchorage, Alaska), pp. 864–874.

Vivitrat, V., and J. R. Kreider, 1981: "Ice Force Prediction Using a Limited Driving Force Approach," *Proc. Offshore Technol. Conf.,* OTC 4115, pp. 471–485.

Vosikovsky, O., 1968: "Critical Stress Model for Determining the Fracture Stress Transition Temperature in Mild Steel," *Techn. Dig.* 10:271–277.

Vyalov, S. S., 1959: *Rheological Properties and Bearing Capacity of Frozen Soils* (in Russian), U.S.S.R. Acad. Sci. Press; also U.S. Army Cold Regions Res. Eng. Lab., Hanover, N.H., Transl. 74 (1965).

————, Ed., 1962: "The Strength and Creep of Frozen Soils and Calculations for Ice-Soil Retaining Structures," U.S. Army Cold Regions Res. Eng. Lab., Hanover, N.H., SIPRE Transl. 76(1965).

————, 1963: "Rheology of Frozen Soils," *Proc. Intern. Conf. Permafrost* (Lafayette, Ind.), U.S. Nat. Acad. Sci., Washington, D.C., Publ. 1287, pp. 332–339.

————, V. V. Dokuchayev, D. A. Sheynkmen, Y. I. Gaydayenko, and Y. M. Goncharov, 1973: "Ground Ice as the Bearing Surface for Construction," in *Principles of the Control of Cryogenic Processes during the Development of Permafrost Regions*, S. S. Vyalov, Ed., U.S. Army Cold Regions Res. Eng. Lab., Hanover, N.H., Transl. TL 438.

Wahanik, R. J., 1978: "Influence of Ice Formations in the Design of Intakes," *Proc. Conf. Applied Techniques for Cold Environments* (Anchorage, Alaska), pp. 582–597.

Walden, J. T., and D. F. Dickens, 1982: "Design of an Air Cushion Transporter for Arctic Operations," *Offshore Technol. Conf.* (Houston, Tex.), OTC 4379.

Walker, D. B. L., D. W. Hayley, and A. C. Palmer, 1983: "The Influence of Subsea Permafrost on Offshore Pipeline Design," *Proc. 4th Intern. Conf. Permafrost* (Fairbanks, Alaska), University of Alaska and U.S. Nat. Acad. Sci., Washington, D.C., pp. 1338–1343.

Wang, Y. S., 1978: "Buckling Analysis of a Semi-Infinite Ice Sheet Moving against Cylindrical Structures," *Proc. Intern. Ass. Hydr. Res. (IAHR), Intern. Symp. Ice* (Luleå, Sweden), pp. 117–133.

Watson, G. H., W. A. Slusarchuk, and R. K. Rowley, 1973: "Determination of Some Frozen and Thawed Properties of Permafrost Soils," *Can. Geotech. J.* 10:592–606.

Weaver, J. S., and N. R. Morgenstern, 1981: "Pile Design in Permafrost," Nat. Res. Council of Canada, *Can. Geotech. J.* 18: pp. 357–370.

Webb, W. E., and W. F. Blair, 1975: "Ice Problems in Locks and Channels on the St. Lawrence River," *Proc. Intern. Ass. Hydr. Res. (IAHR), Intern. Symp. Ice* (Hanover, N.H.), pp. 15–24.

Weeks, W. F., and D. L. Anderson, 1958: "An Experimental Study of the Strength of Young Sea Ice," *Trans. Amer. Geophys. Union* 39:641–647.

———— and A. Assur, 1967: "The Mechanical Properties of Sea Ice," U.S. Army Cold Regions Res. Eng. Lab., Hanover, N.H., Monograph II-C3.

Weishar, A. L., 1976: "Special Problems of Construction in the Arctic," *Proc. Intern. Conf. Materials Engineering in the Arctic* (St. Jovite, Que.), Amer. Soc. for Metals, pp. 16–20.

Wellman, J. H., E. S. Clarke, and A. C. Condo, 1977: "Design and Construction of Synthetically Insulated Gravel Pads in the Alaskan Arctic," *Proc. 2d Intern. Symp. Cold Regions Engineering* (University of Alaska), pp. 62–85.

Williams, C. J., 1978: "Snow Road Construction," *Proc. Conf. Applied Techniques for Cold Environments* (Anchorage, Alaska), ASCE, New York, pp. 598–609.

Wilson, C., 1967: "Introduction to Northern Hemisphere I," U.S. Army Cold Regions Res. Eng. Lab., Hanover, N.H., CRREL Monograph I-A3a.

————, 1969: "Climatology of the Cold Regions, Northern Hemisphere II," U.S. Army Cold Regions Res. Eng. Lab., Hanover, N.H., CRREL Monograph I-A3b.

Wise, J. L., and A. L. Comiskey, 1980: "Superstructure Icing in Alaskan Water," U.S. Nat. Oceanic and Atmospheric Administration, Pacific Marine Environmental Lab., Seattle, Wash., Spec. Rep.

Wortley, C. A., 1978: "Ice Engineering Guide for Design and Construction of Small Craft Harbors," University of Wisconsin, Advisory Rep. SG-78-417.

Wu, R. C., K. J. Chang, and J. Schwartz, 1976: "Fracture in the Compression of Columnar-Grained Ice," *Eng. Fracture Mech.* 8:365–372.

Wyman, M., 1950: "Deflections of an Infinite Plate," *Can. J. Res.* 28: pp. 293–302.

Yarmak, E., Jr., and E. L. Long, 1983: "Some Considerations Regarding the Design of Two Phase Liquid/Vapor Convection Type Passive Refrigeration Systems," Arctic Foundations, Inc., Anchorage, Alaska, unpublished.

Yoder, E. J., and M. W. Witczak, 1975: *Principles of Pavement Design*, 2d ed., Wiley, New York.

Zirjacks, W. L., and C. T. Hwang, 1983: "Underground Utilidors at Barrow, Alaska: A Two-Year History," *Proc. 4th Intern. Conf. Permafrost* (Fairbanks, Alaska), University of Alaska and U.S. Nat. Acad. Sci., Washington, D.C., pp. 1513–1517.

UNIT CONVERSION TABLE

Length

Metric units	U.S. customary units	U.S. customary units	Metric units
1 km = 1000 m	= 0.6214 mi	1 mi = 1760 yd	= 1.60934 km
1 m = 1000 mm	= 1.0936 yd	1 yd = 3 ft	= 0.91440 m
	= 3.2808 ft	1 ft = 12 in	= 0.3048 m
1 mm	= 0.0394 in	1 in	= 25.4 mm

Area

Metric units	U.S. customary units	U.S. customary units	Metric units
1 km^2 = 100 ha	= 0.3861 mi^2	1 mi^2 = 640 acres	= 2.58999 km^2
1 ha = 10,000 m^2	= 2.47105 acre	1 acre = 4840 yd^2	= 0.40469 ha
1 m^2 = 1,000,000 mm^2	= 1.1960 yd^2	1 yd^2	= 0.83613 m^2
	= 10.7639 ft^2	1 ft^2	= 0.09290 m^2
1 mm^2	= 0.00155 in^2	1 in^2	= 645.160 mm^2

Volume

Metric units	U.S. customary units	U.S. customary units	Metric units
1 m^3 = 1000 L	= 1.30795 yd^3	1 yd^3	= 0.76456 m^3
	= 35.3147 ft^3	1 ft^3	= 0.02832 m^3
			= 28.3168 L
1 L = 1 dm^3	= 0.03532 ft^3		
1 cm^3 = 1 mL	= 0.06102 in^3	1 in^3	= 16.3871 cm^3

Mass

Metric units	U.S. customary units	U.S. customary units	Metric units
1 tonne = 1000 kg	= 0.98421 long ton	1 long ton = 2240 lb	= 1.01605 tonne
	= 1.10231 short ton	1 short ton = 2000 lb	= 0.90718 tonne
1 kg = 1000 g	= 2.20462 lb	1 lb	= 0.45359 kg

Unit Conversion Table [continued]

Metric units	U.S. customary units	U.S. customary units	Metric units
Density			
1 kg/m³ = 0.001 g/cm³	= 1.68556 lb/yd³	1 lb/yd³	= 0.59328 kg/m³
	= 0.06243 lb/ft³	1 lb/ft³	= 16.0185 kg/m³
Force			
1 MN = 1000 kN	= 100.361 long tons	1 long ton	= 9.96402 kN
1 kN = 1000 N	= 112.404 short tons	1 short ton	= 8.89644 kN
	= 0.22481 kip	1 kip = 1000 lb	= 4.44822 kN
	= 224.809 lb	1 lb	= 4.44822 N
Pressure, stress			
1 MPa = 1 MN/m²	= 0.14504 kip/in²	1 kip/in² = 1000 lb/in²	= 15.4443 MPa
1 kPa = 1 kN/m²	= 0.14504 lb/in²		= 107.252 kPa
1 Pa = 1 N/m²	= 0.02089 lb/ft²		= 6.89476 MPa
		1 lb/in²	= 6.89476 Pa
		1 lb/ft²	= 47.8803 kPa
Velocity, speed			
1 km/h = 0.278 m/s	= 0.62137 mi/h	1 mi/h = 1.466 ft/s	= 1.6093 km/h
	= 0.540 knot		= 0.4470 m/s
1 m/s	= 3.28084 ft/s	1 knot	= 1.853 km/h
1 cm/s	= 0.39370 in/s	1 ft/s	= 0.3048 m/s
		1 in/s	= 2.540 cm/s
Volume rate of flow			
1 m³/s = 1000 L/s	= 35.3147 ft³/s	1 ft³/s	= 0.02832 m³/s
1 L/s	= 2.11888 ft³/min	1 ft³/min	= 0.47195 L/s
Work, energy, heat			
1 MJ = 10^6 Nm	= 0.27778 kWh	1 kWh	= 3.6 MJ
		1 Btu	= 1.0551 kJ
		1 ft·lb = 0.3238 cal	= 1.3558 J

Metric units	U.S. customary units	U.S. customary units	Metric units
1 kJ = 1000 J 1 J = 0.2388 cal	= 0.94782 Btu = 0.73756 ft·lb		
Power, heat flow rate			
1 kW 1 W = 1 J/s	= 1.34102 hp = 3.41214 Btu/h	1 hp = 550 ft·lb/s 1 Btu/h	= 0.74570 kW = 0.29307 W
Calorific value			
1 kJ/kg = 1 J/g 1 kJ/m³	= 0.42992 Btu/lb = 0.02684 Btu/ft³	1 Btu/lb 1 Btu/ft³	= 2.326 kJ/kg = 37.2589 kJ/m³
Heat flux			
1 W/m²	= 0.316998 Btu/(ft²·h)	1 Btu/(ft²·h)	= 3.15459 W/m²
Thermal conductance			
1 W/(m²·°C)	= 0.17611 Btu/(ft²·h·°F)	1 Btu/(ft²·h·°F)	= 5.6783 W/(m²·°C)
Thermal resistance			
1 (m²·°C)/W	= 5.678 (ft²·h·°F)/Btu	1 (ft²·h·°F)/Btu	= 0.17611 (m²·°C)
Thermal conductivity			
1 W/(m·°C)	= 0.57779 Btu/(ft·h·°F) = 6.93347 Btu·in/(ft²·h·°F)	1 Btu/(ft·h·°F) 1 Btu·in/(ft²·h·°F)	= 1.73073 W/(m·°C) = 0.14423 W/(m·°C)
Heat capacity			
1 kJ/(kg·°C) 1 kJ/(m³·°C)	= 0.23885 Btu/(lb·°F) = 0.01491 Btu/(ft³·°F)	1 Btu/(lb·°F) 1 Btu/(ft³·°F)	= 4.18680 kJ/(kg·°C) = 67.0661 kJ/(m³·°C)
Thermal diffusivity			
1 m²/s = 10⁶ mm²/s 1 mm²/s	= 10.7639 ft²/s = 0.03875 ft²/h	1 ft²/s 1 ft²/h	= 0.0929 m²/s = 25.806 mm²/s

INDEX

ABOUT THE AUTHORS

ESA M. ERANTI, an arctic engineering specialist at Finn-Stroi Oi in Finland, has considerable experience in a variety of structural design and development projects. A native of Finland, he is a member of several Finnish boards and working groups managing and promoting cold region engineering research. Mr. Eranti received an undergraduate degree in civil engineering at Helsinki University of Technology and a master's degree in civil engineering from the State University of New York at Buffalo.

GEORGE C. LEE, Ph.D., is Professor and Dean, Faculty of Engineering and Applied Sciences, State University of New York at Buffalo. He is also the founder and associate director of the Calspan-UB Research Center in Buffalo, a structural engineering consultant, and the coauthor of *Structural Analysis and Design* (with R.L. Ketter and S.P. Prawel) and *Design of Single Story Rigid Frames* (with R.L. Ketter and T.L. Hsu). Dr. Lee received his B.S. degree in civil engineering from National Taiwan University and his master's and doctor's degrees from Lehigh University.